Valery Serov, Markus Harju

Complex Analysis and Special Functions

Also of Interest

Topics in Complex Analysis
Dan Romik, 2023
ISBN 978-3-11-079678-0, e-ISBN (PDF) 978-3-11-079681-0

Topics in Complex Analysis
Joel L. Schiff, 2022
ISBN 978-3-11-075769-9, e-ISBN (PDF) 978-3-11-075782-8
in: De Gruyter Studies in Mathematics
ISSN 0179-0986

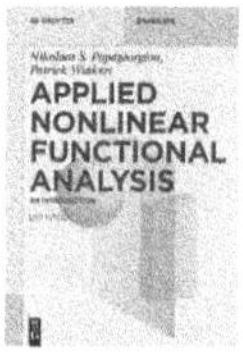

Applied Nonlinear Functional Analysis
An Introduction
Nikolaos S. Papageorgiou, Patrick Winkert, 2024
ISBN 978-3-11-128421-7, e-ISBN (PDF) 978-3-11-128695-2

Applications of Complex Variables
Asymptotics and Integral Transforms
Foluso Ladeinde, 2024
ISBN 978-3-11-135090-5, e-ISBN (PDF) 978-3-11-135117-9

Theta functions, elliptic functions and π
Heng Huat Chan, 2020
ISBN 978-3-11-054071-0, e-ISBN (PDF) 978-3-11-054191-5

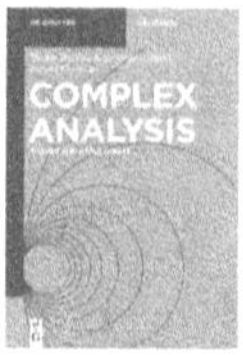

Complex Analysis
Theory and Applications
Teodor Bulboacă, Santosh B. Joshi, Pranay Goswami, 2019
ISBN 978-3-11-065782-1, e-ISBN (PDF) 978-3-11-065786-9

Valery Serov, Markus Harju

Complex Analysis and Special Functions

Cauchy Formula, Elliptic Functions and Laplace's Method

DE GRUYTER

Mathematics Subject Classification 2020
Primary: 30-01, 30D20, 33E05; Secondary: 42C10, 44A10

Authors
Dr. Sci. Valery Serov
Emeritus Professor of Applied Mathematics
University of Oulu
Finland
valeri.serov@oulu.fi

University Lecturer Markus Harju
Applied and Computational Mathematics
University of Oulu
Finland
markus.harju@oulu.fi

ISBN 978-3-11-163211-7
e-ISBN (PDF) 978-3-11-163227-8
e-ISBN (EPUB) 978-3-11-163232-2

Library of Congress Control Number: 2024947837

Bibliographic information published by the Deutsche Nationalbibliothek
The Deutsche Nationalbibliothek lists this publication in the Deutsche Nationalbibliografie; detailed bibliographic data are available on the Internet at http://dnb.dnb.de.

Cover image: Natalya Yudina (M) / iStock / Getty Images Plus
Typesetting: VTeX UAB, Lithuania

www.degruyter.com
Questions about General Product Safety Regulation:
productsafety@degruyterbrill.com

Preface

Present book can be considered as a standard university course in complex analysis and special functions (including orthogonal polynomials and basic material of special functions such as Euler's Gamma and Beta functions, Bessel's functions, Weierstrass and Jacobi elliptic functions and some others) for mathematics and physics students. However, some subjects of the book, such as the stationary phase method and Laplace's method, Weierstrass elliptic functions and their applications to nonlinear ordinary differential equations are usually very advanced and often are outside of such type of courses in Complex Analysis. This allows this book to be considered more than just a university textbook, as it has many possible applications in applied mathematics and physics.

This book is mainly based on the courses given by the authors at the University of Oulu in recent years, and given by the first author at Lomonosov Moscow State University in the end of 1990s and the beginning of 2000s.

The book consists of three parts divided with respect to the usual content of complex analysis, orthogonal polynomials, and special functions. The first part includes complex numbers, analytic functions and Cauchy theorem. The second part concerns to the maximum modulus of analytic functions, Phragmen-Lindelöf principle, Liouville's theorem, Taylor's and Laurent's expansions, entire functions, and evaluation of several types of integrals and number series using the residue theory. The third and biggest part includes conformal mappings, Laplace transform, and special functions. We remark that special functions is based on the classical method of Frobenius and includes (as an application) Bessel's functions, orthogonal polynomials, and also Laplace's method. Moreover, the theory of Weierstrass and Jacobi functions and their application to nonlinear ordinary differential equations are considered very carefully together with famous Weierstrass' formula.

This book contains about 500 exercises that are integral part of the text, and more than 150 examples. Each part ends in a selection of exercises that an instructor can use for the exams. They are not only an integral part of the book, but also indispensable for the understanding each part of the book. It might be mentioned here that many exercises were borrowed and reworked from the excellent monographs of Titchmarsh [1] and Whittaker and Watson [2]. Within the text, the reader will also find problems, which range from very easy to somewhat difficult. It can be expected that a careful reader will complete all these exercises.

This book is intended for undergraduate level students majoring in pure and applied mathematics (also in physics), but even graduate students can find here very useful information, which previously could only be detected in scientific monographs.

Despite the fact that this text is standard for universities, however, there are some things that distinguish it from well-known texts in this subject. One difference in this text is the discussion of extended complex plane and the concept of complex infinity including Taylor's and Laurent's expansions at infinity. Another key aspect is the evaluation of improper integrals for multivalued functions of certain special form, and calculation of

https://doi.org/10.1515/9783111632278-201

number series by residue theory. In regard to orthogonal polynomials (Legendre, Hermite, Chebyshev, trigonometric) and Bessel's functions, they are considered here to a sufficient degree of generality and their asymptotic behavior with respect to different parameters are proved. In addition, we prove theorems on expanding continuous functions that have piecewise continuous derivative associated with an orthogonal system of eigenfunctions that correspond to given polynomials. But the major difference in this text (compared with known books) is in consideration of the method of stationary phase for real integrals and of the Laplace's method (saddle point method) for the complex curve integrals. It can be also mentioned here that the Cardano's formulae are considered in their general form for polynomials of degree three.

The systematic and careful consideration of the Weierstrass and Jacobi elliptic functions, and their many applications, can be considered as one of the most important features of this book. We have partly used here the approach of Whittaker and Watson [2].

Last but not least, we want to say that when writing the Laplace's method we were greatly influenced by the excellent book by Sveshnikov and Tikhonov [3] and when writing the part of special functions (especially some orthogonal polynomials and their properties) we were inspired by very advanced book by Nikiforov and Uvarov [4].

September 2024

Valery Serov, Lahti
Markus Harju, Oulu

Contents

Part I

1 Complex numbers and their properties

Definition 1.1. The ordered pair (x, y) of real numbers x and y is called a *complex number* $z = (x, y)$ if the following properties are satisfied:

1. $z_1 = z_2$ if and only if $x_1 = x_2$ and $y_1 = y_2$. In particular, $z = (x, y) = 0$ if and only if $x = y = 0$.
2. $z_1 \pm z_2 = (x_1 \pm x_2, y_1 \pm y_2)$.
3. $z_1 \cdot z_2 = (x_1x_2 - y_1y_2, x_1y_2 + x_2y_1)$.

We denote $x = \operatorname{Re} z, y = \operatorname{Im} z$ and call them real and imaginary parts, respectively.

The complex number $z = (x, 0)$ is identified with real number x, and complex number $z = (0, y)$ is called *purely imaginary*.

Definition 1.2. The complex numbers $(0, 0)$, $(1, 0)$, and $(0, 1)$ are called *zero, unit,* and *imaginary unit* and are identified with 0, 1, and i, respectively.

It is easy to check that

$$\mathrm{i}^2 = (-1, 0), \quad \mathrm{i}(b, 0) = (0, b). \tag{1.1}$$

Indeed,

$$\mathrm{i}^2 = (0, 1) \cdot (0, 1) = (-1, 0)$$

and

$$\mathrm{i}(b, 0) = (0, 1) \cdot (b, 0) = (0, b)$$

by Definition 1.1.

Since

$$z = (x, y) = (x, 0) + (0, y)$$

using (1.1), we obtain that

$$z = (x, 0) + (0, 1) \cdot (y, 0) = x + \mathrm{i}y$$

such that

$$z_1 + z_2 = x_1 + \mathrm{i}y_1 + x_2 + \mathrm{i}y_2 = (x_1 + x_2) + \mathrm{i}(y_1 + y_2)$$

and

$$\begin{aligned} z_1 \cdot z_2 &= (x_1 + \mathrm{i}y_1)(x_2 + \mathrm{i}y_2) = x_1x_2 + \mathrm{i}y_1x_2 + \mathrm{i}x_1y_2 + \mathrm{i}^2y_1y_2 \\ &= (x_1x_2 - y_1y_2) + \mathrm{i}(y_1x_2 + x_1y_2), \end{aligned}$$

https://doi.org/10.1515/9783111632278-001

i. e., these operations (addition and multiplication) are performed as in the usual analysis.

We denote the set of all complex numbers by $\mathbb{C}$.

The division is defined as the operation, which is inverse to multiplication. Namely, if $z_2 \neq 0$ (i. e., $x_2 \neq 0$ or $y_2 \neq 0$, so $x_2^2 + y_2^2 > 0$) then

$$\frac{z_1}{z_2} = a + ib \quad \text{if and only if} \quad z_1 = (a + ib)z_2.$$

It means that

$$x_1 + iy_1 = (a + ib)(x_2 + iy_2)$$

or

$$\begin{cases} x_1 = ax_2 - by_2, \\ y_1 = bx_2 + ay_2. \end{cases}$$

Solving this for a and b gives

$$a = \frac{x_1x_2 + y_1y_2}{x_2^2 + y_2^2}, \quad b = \frac{y_1x_2 - x_1y_2}{x_2^2 + y_2^2}.$$

Hence,

$$\frac{z_1}{z_2} = \frac{x_1x_2 + y_1y_2}{x_2^2 + y_2^2} + i\frac{y_1x_2 - x_1y_2}{x_2^2 + y_2^2}. \tag{1.2}$$

Definition 1.3. For given complex number $z = x + iy$:
1. The number $\overline{z} := x - iy$ is called the *complex conjugate* to z.
2. The nonnegative (real) number $|z| := \sqrt{x^2 + y^2}$ is called the *modulus* of z.

The following properties can be checked straightforwardly:

$$\begin{gathered}
\overline{z_1 \pm z_2} = \overline{z_1} \pm \overline{z_2}, \\
\overline{z_1 \cdot z_2} = \overline{z_1} \cdot \overline{z_2}, \\
\overline{\left(\frac{z_1}{z_2}\right)} = \frac{\overline{z_1}}{\overline{z_2}}, \\
\operatorname{Re} z = \frac{z + \overline{z}}{2}, \quad \operatorname{Im} z = \frac{z - \overline{z}}{2i}, \\
|z| = 0 \quad \text{if and only if} \quad z = 0, \\
|z|^2 = z \cdot \overline{z}, \quad |z| = |\overline{z}|, \quad |z_1 \cdot z_2| = |z_1| \cdot |z_2|, \\
\left|\frac{z_1}{z_2}\right| = \frac{|z_1|}{|z_2|}, \quad \text{but}
\end{gathered}$$

$$
\begin{aligned}
\big||z_1| - |z_2|\big| \le |z_1 \pm z_2| \le |z_1| + |z_2|,\\
|\operatorname{Re} z| \le |z|, \quad |\operatorname{Im} z| \le |z|.
\end{aligned} \tag{1.3}
$$

Problem 1.4.

1. Prove that

$$\frac{z_1}{z_2} = \frac{z_1\overline{z_2}}{|z_2|^2}, \quad z_2 \neq 0.$$

   Compare with (1.2).
2. Prove that

$$|z_1 \pm z_2|^2 = |z_1|^2 + |z_2|^2 \pm 2|z_1| \cdot |z_2| \cos\alpha,$$

   where α is the angle between the two vectors $z_1 = (x_1, y_1)$ and $z_2 = (x_2, y_2)$ on the plane $\mathbb{R}^2$; see Figure 1.1
3. Prove the inequalities (1.3).

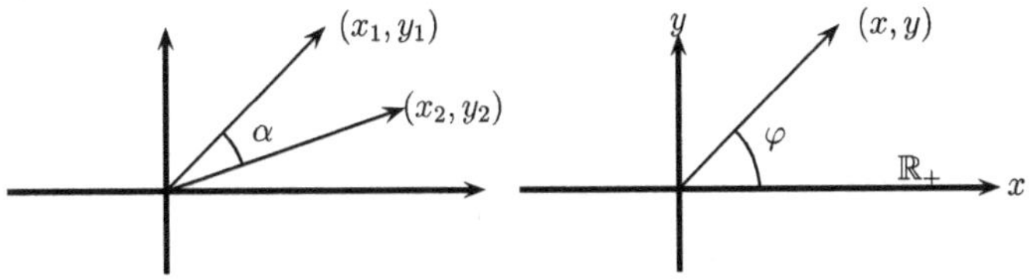

Figure 1.1: Angle between vectors and argument of a complex number.

Definition 1.5. The angle φ formed by the vector $z = (x, y)$, $z \neq 0$, and the positive real line $\mathbb{R}_+$ is said to be an *argument* of z and denoted by $\varphi = \operatorname{Arg} z$, $z \neq 0$; see Figure 1.1. The argument of $z = 0$ is not defined.

Remark. It is clear that $\operatorname{Arg} z$ is not defined uniquely. More precisely, it is defined up to $2\pi n$, $n = 0, \pm 1, \pm 2, \dots$, i. e.,

$$\operatorname{Arg} z = \varphi + 2\pi n,$$

where $\varphi \in (0, 2\pi]$ or $\varphi \in (-\pi, \pi]$. This value of φ is called the *main argument* and it is denoted as

$$\arg z = \varphi.$$

Let us assume in the future that

$$\arg z = \varphi \quad \text{with } \varphi \in (-\pi, \pi].$$

In this case, the Pythagorian theorem says that

$$\operatorname{Re} z = |z| \cos\varphi \quad \text{and} \quad \operatorname{Im} z = |z| \sin\varphi,$$

i. e.,

$$z = |z|(\cos\varphi + \mathrm{i}\sin\varphi), \quad z \neq 0. \tag{1.4}$$

Problem 1.6. Prove that:
1. $z_1 = z_2$ if and only if $|z_1| = |z_2|$ and $\varphi_1 = \varphi_2$.
2.

$$\begin{aligned} \varphi \in (0,\pi) &\quad \text{if and only if} \quad \operatorname{Im} z > 0, \\ \varphi \in (-\pi,0) &\quad \text{if and only if} \quad \operatorname{Im} z < 0, \\ \varphi = 0 &\quad \text{if and only if} \quad \operatorname{Im} z = 0, \operatorname{Re} z > 0, \\ \varphi = \pi &\quad \text{if and only if} \quad \operatorname{Im} z = 0, \operatorname{Re} z < 0. \end{aligned}$$

Problem 1.7. Prove the following statements:
1. $\arg \overline{z} = -\arg z$.
2.

$$\arg z = \begin{cases} \arctan \frac{\operatorname{Im} z}{\operatorname{Re} z}, & \operatorname{Re} z > 0, \\ \arctan \frac{\operatorname{Im} z}{\operatorname{Re} z} + \pi, & \operatorname{Re} z < 0, \operatorname{Im} z \geq 0, \\ \arctan \frac{\operatorname{Im} z}{\operatorname{Re} z} - \pi, & \operatorname{Re} z < 0, \operatorname{Im} z < 0, \\ \frac{\pi}{2}, & \operatorname{Re} z = 0, \operatorname{Im} z > 0, \\ -\frac{\pi}{2}, & \operatorname{Re} z = 0, \operatorname{Im} z < 0. \end{cases}$$

Problem 1.8. Prove the following properties:
1. $z_1 \cdot z_2 = |z_1| \cdot |z_2|(\cos(\varphi_1 + \varphi_2) + \mathrm{i}\sin(\varphi_1 + \varphi_2))$.
2. $z_1/z_2 = |z_1|/|z_2|(\cos(\varphi_1 - \varphi_2) + \mathrm{i}\sin(\varphi_1 - \varphi_2))$.
3. $z^n = |z|^n(\cos(n\varphi) + \mathrm{i}\sin(n\varphi))$ (*De Moivre formula*).

We will use the shorthand notation (which will be proved later)

$$\mathrm{e}^{\mathrm{i}\varphi} := \cos\varphi + \mathrm{i}\sin\varphi.$$

Then (1.4) can be written as

$$z = |z|\mathrm{e}^{\mathrm{i}\varphi}. \tag{1.5}$$

Definition 1.9. The form (1.5) is called the *trigonometric representation* of complex numbers.

The equality (1.5) is also called *Euler's formula*. Using (1.5), we may rewrite the above formulas in a shorter way:

$$\begin{aligned} z_1 \cdot z_2 &= |z_1| \cdot |z_2| e^{i(\varphi_1+\varphi_2)}, \\ z_1/z_2 &= |z_1|/|z_2| e^{i(\varphi_1-\varphi_2)}, \\ z^n &= |z|^n e^{in\varphi}. \end{aligned}$$

Definition 1.10. The complex number z_0 is said to be the *root of* nth *degree* of the complex number z if

$$z_0^n = z.$$

We denote this by $z_0 = \sqrt[n]{z}$. There are n solutions of the above equation and they are given by

$$(z_0)_k = |z|^{1/n} e^{i(\varphi/n+2\pi k/n)}, \quad k = 0, 1, \dots, n-1. \tag{1.6}$$

Problem 1.11. Prove (1.6) using the De Moivre formula.

Let us consider in the Euclidean space $\mathbb{R}^3$ the sphere S with center $(0, 0, 1/2)$ and radius $1/2$ in the coordinate system (ξ, η, ζ), i. e.,

$$\xi^2 + \eta^2 + (\zeta - 1/2)^2 = 1/4$$

or

$$\xi^2 + \eta^2 + \zeta^2 - \zeta = 0. \tag{1.7}$$

Let us draw a ray from the point $P = (0, 0, 1)$, which intersects the sphere S at the point $M = (\xi, \eta, \zeta)$ and complex plane $\mathbb{C}$ at the point $z = x + iy$.

The point M is called *stereographic projection* of the complex number z on the sphere S (Figure 1.2). Since the vectors $\overrightarrow{PM}$ and $\overrightarrow{Pz}$ are colinear, we have

$$\frac{\xi}{x} = \frac{\eta}{y} = \frac{1-\zeta}{1}.$$

Thus, using (1.7), we have

$$x = \frac{\xi}{1-\zeta}, \quad y = \frac{\eta}{1-\zeta}$$

so that

$$\xi = \frac{x}{1+|z|^2}, \quad \eta = \frac{y}{1+|z|^2}, \quad \zeta = \frac{|z|^2}{1+|z|^2}. \tag{1.8}$$

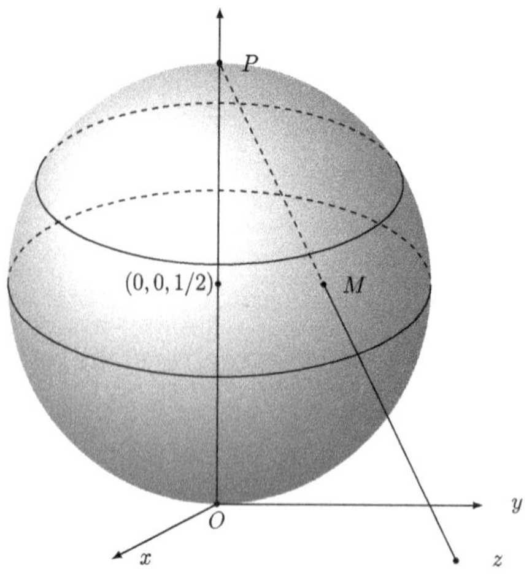

Figure 1.2: Stereographic projection.

Definition 1.12. The formulas (1.8) are called the formulas of the stereographic projection.

The formulas (1.8) allow us to introduce the notion of infinity, i. e., complex number $z = \infty$ as follows. Since there is one-to-one correspondence between $\mathbb{C}$ and $S \setminus P$, then we may supplement this correspondence by one more, namely

$$P(0,0,1) \longleftrightarrow \infty.$$

In this case,

$$S \longleftrightarrow \overline{\mathbb{C}} := \mathbb{C} \cup \{\infty\}$$

and, by (1.8),

$$\frac{1}{\infty} = 0, \quad \frac{1}{0} = \infty, \quad z \cdot \infty = \infty, \quad z \neq 0, \quad z + \infty = \infty, \quad \frac{z}{\infty} = 0, \quad z \neq \infty. \tag{1.9}$$

Remark. The set $\overline{\mathbb{C}}$ is called the *extended complex plane.*

Problem 1.13. Prove that the spherical distance between $z_1, z_2 \in \overline{\mathbb{C}}$ can be calculated as

$$\rho_S(z_1, z_2) = \frac{|z_1 - z_2|}{\sqrt{1 + |z_1|^2}\sqrt{1 + |z_2|^2}}.$$

The *neighborhood* of $z_0 \in \mathbb{C}$ is defined as

$$U_\delta(z_0) = \{z \in \mathbb{C} : |z - z_0| < \delta\}$$

and the neighborhood of $z_0 = \infty$ is defined as

$$U_R(\infty) = \{z \in \mathbb{C} : |z| > R\}.$$

Definition 1.14.

1. The complex number $z_0 \in \overline{\mathbb{C}}$ is called the *limiting point* of a set $M \subset \overline{\mathbb{C}}$ if for any $\delta > 0$ it is true that

$$(U_\delta(z_0) \setminus z_0) \cap M \neq \emptyset$$

   (or for any $R > 0$ it holds that $(U_R(\infty) \setminus \infty) \cap M \neq \emptyset$).
2. The set $M \subset \overline{\mathbb{C}}$ is called *closed* if it contains all its limiting points.
3. Denoting all limiting points of $M \subset \overline{\mathbb{C}}$ by M' we define the *closure* of M as

$$\overline{M} = M \cup M'.$$

4. The *boundary* ∂M of the set $M \subset \overline{\mathbb{C}}$ is defined as

$$\partial M = \overline{M} \cap (\overline{\overline{\mathbb{C}} \setminus M}).$$

5. The point $z_0 \in \overline{\mathbb{C}}$ is called *interior* of a set M if there exists $U_\delta(z_0)$ (or $U_R(\infty)$) such that $U_\delta(z_0) \subset M$ (or $U_R(\infty) \subset M$). If all points of M are interior, then M is called an *open set*.

Problem 1.15. Prove that $M \subset \overline{\mathbb{C}}$ is open if and only if $\overline{\mathbb{C}} \setminus M$ is closed.

Definition 1.16. The complex number $z_0 \in \mathbb{C}$ is said to be the *limit of sequence* $\{z_n\}_{n=1}^{\infty} \subset \mathbb{C}$, denoted by $z_0 = \lim_{n\to\infty} z_n$, if for any $\varepsilon > 0$ there is $n_0 = n_0(\varepsilon, z_0) \in \mathbb{N}$ such that

$$|z_n - z_0| < \varepsilon$$

for all $n \geq n_0$.

We say that $\infty = \lim_{n\to\infty} z_n$ if for any $R > 0$ there is $n_0 = n_0(R) \in \mathbb{N}$ such that $|z_n| > R$ for all $n \geq n_0$.

Proposition 1.17.

1. $z_0 = \lim_{n\to\infty} z_n$, $z_0 \neq \infty$ *if and only if*

$$\operatorname{Re} z_0 = \lim_{n\to\infty} \operatorname{Re} z_n \quad \textit{and} \quad \operatorname{Im} z_0 = \lim_{n\to\infty} \operatorname{Im} z_n.$$

2. $\infty = \lim_{n\to\infty} z_n$ *if and only if* $\lim_{n\to\infty} |z_n| = \infty$.

Proof. 1. If $z_0 = \lim_{n\to\infty} z_n$, then for any $\varepsilon > 0$ there exists $n_0(\varepsilon) \in \mathbb{N}$ such that

$$|z_n - z_0|^2 < \varepsilon^2, \quad n \geq n_0.$$

It means that

$$(\operatorname{Re} z_n - \operatorname{Re} z_0)^2 + (\operatorname{Im} z_n - \operatorname{Im} z_0)^2 < \varepsilon^2, \quad n \geq n_0.$$

It follows that

$$|\operatorname{Re} z_n - \operatorname{Re} z_0| < \varepsilon, \quad |\operatorname{Im} z_n - \operatorname{Im} z_0| < \varepsilon, \quad n \geq n_0$$

or

$$\operatorname{Re} z_0 = \lim_{n\to\infty} \operatorname{Re} z_n, \quad \operatorname{Im} z_0 = \lim_{n\to\infty} \operatorname{Im} z_n.$$

Conversely, if $a = \lim_{n\to\infty} \operatorname{Re} z_n$ and $b = \lim_{n\to\infty} \operatorname{Im} z_n$ then for any $\varepsilon > 0$ there exist $n_1(\varepsilon), n_2(\varepsilon) \in \mathbb{N}$ such that

$$|\operatorname{Re} z_n - a| < \varepsilon/2, \quad n \geq n_1,$$
$$|\operatorname{Im} z_n - b| < \varepsilon/2, \quad n \geq n_2.$$

Denoting $n_0 = \max(n_1, n_2)$, we obtain for all $n \geq n_0$ that

$$|z_n - (a + ib)| \leq |\operatorname{Re} z_n - a| + |\operatorname{Im} z_n - b| < \varepsilon/2 + \varepsilon/2 = \varepsilon.$$

2. It follows immediately from Definition 1.16. □

Remark. In part (2) of Proposition 1.17, we cannot say anything more. Indeed, let z_n be defined as follows:

$$z_n = \begin{cases} n + i/n, & n = 2k, \\ 1/n + in, & n = 2k + 1. \end{cases}$$

Then $|z_n| = \sqrt{n^2 + 1/n^2} \to \infty$ as $n \to \infty$ but $\operatorname{Re} z_n \not\to \infty$ and $\operatorname{Im} z_n \not\to \infty$.

The Bolzano–Weierstrass principle

If the sequence of complex numbers $\{z_n\}_{n=1}^{\infty}$ is bounded, i. e., there exists $M > 0$ such that

$$|z_n| \leq M, \quad n = 1, 2, \ldots$$

then there is a subsequence z_{k_n}, which converges to some point $z_0 \in \mathbb{C}$, i. e.,

$$\lim_{n\to\infty} z_{k_n} = z_0.$$

Indeed, since $|z_n| \leq M$ then $|\operatorname{Re} z_n| \leq M$ and $|\operatorname{Im} z_n| \leq M$. Using the Bolzano–Weierstrass principle to the real sequence $\operatorname{Re} z_n$, we find $\operatorname{Re} z_{k_n}$ such that there exists $a \in \mathbb{R}$ with

$$a = \lim_{n\to\infty} \operatorname{Re} z_{k_n}.$$

If we consider now $\operatorname{Im} z_{k_n}$, then it is also bounded, and hence there exists a subsequence, say $\operatorname{Im} z_{k_n}^{(1)}$, which has a limit

$$b = \lim_{n\to\infty} \operatorname{Im} z_{k_n}^{(1)}.$$

Thus,

$$\lim_{n\to\infty} (\operatorname{Re} z_{k_n}^{(1)} + \mathrm{i} \operatorname{Im} z_{k_n}^{(1)}) = \lim_{n\to\infty} \operatorname{Re} z_{k_n}^{(1)} + \mathrm{i} \lim_{n\to\infty} \operatorname{Im} z_{k_n}^{(1)} = a + ib.$$

Suppose the sequence of complex numbers $\{z_n\}_{n=1}^{\infty}$ is not bounded, i. e., for any $M > 0$ there exists $n_M \in \mathbb{N}$ such that $|z_{n_M}| > M$. Then there is a subsequence z_{k_n} such that

$$\lim_{n\to\infty} |z_{k_n}| = \infty.$$

The proof of this fact is the same as in real analysis.

There is one more useful property:

$$z_n \to \infty$$

(i. e., $|z_n| \to \infty$) if and only if

$$\lim_{n\to\infty} \frac{1}{z_n} = 0.$$

Cauchy criterion

The sequence of complex numbers $\{z_n\}_{n=1}^{\infty}$ converges if and only if it is a *Cauchy sequence*, i. e., for any $\varepsilon > 0$ there exists $n_0(\varepsilon)$ such that

$$|z_n - z_m| < \varepsilon, \quad n, m \geq n_0.$$

The proof follows from the Cauchy criterion of real analysis.

Arithmetic operations with convergent sequences

If

$$\lim_{n\to\infty} z_n = z_0, \quad \lim_{n\to\infty} w_n = w_0$$

then

$$\lim_{n\to\infty} (z_n \pm w_n) = z_0 \pm w_0,$$

$$\lim_{n\to\infty} z_n \cdot w_n = z_0 \cdot w_0,$$

$$\lim_{n\to\infty} \frac{z_n}{w_n} = \frac{z_0}{w_0}, \quad w_0 \neq 0.$$

If

$$\lim_{n\to\infty} z_n = \infty, \quad \lim_{n\to\infty} w_n = \infty$$

then

$$\lim_{n\to\infty} z_n \cdot w_n = \infty.$$

Problem 1.18.

1. Let $\lim_{n\to\infty} z_n = z_0$, $z_0 \neq 0$, $z_0 \neq \infty$ and $\lim_{n\to\infty} w_n = \infty$. Prove that

$$\lim_{n\to\infty} z_n \cdot w_n = \infty, \quad \lim_{n\to\infty} (z_n \pm w_n) = \infty, \quad \lim_{n\to\infty} z_n/w_n = 0.$$

2. Let $\lim_{n\to\infty} z_n = \infty$ and $\lim_{n\to\infty} w_n = \infty$. Prove that the limits

$$\lim_{n\to\infty} (z_n \pm w_n) \quad \text{and} \quad \lim_{n\to\infty} z_n/w_n$$

   may or may not exist.

Series

The series of the complex numbers

$$\sum_{k=1}^{\infty} z_k$$

is said to be *convergent* if the limit

$$\lim_{n\to\infty} \sum_{k=1}^{n} z_k$$

exists. Then this limit is denoted by

$$\sum_{k=1}^{\infty} z_k.$$

It is equivalent to the convergence of the real series

$$\sum_{k=1}^{\infty} \operatorname{Re} z_k \quad \text{and} \quad \sum_{k=1}^{\infty} \operatorname{Im} z_k$$

and in that case

$$\sum_{k=1}^{\infty} z_k = \sum_{k=1}^{\infty} \operatorname{Re} z_k + \mathrm{i} \sum_{k=1}^{\infty} \operatorname{Im} z_k.$$

The series $\sum_{k=1}^{\infty} z_k$ is said to be *absolutely convergent* if

$$\sum_{k=1}^{\infty} |z_k| < \infty$$

or

$$\sum_{k=1}^{\infty} |\operatorname{Re} z_k| < \infty \quad \text{and} \quad \sum_{k=1}^{\infty} |\operatorname{Im} z_k| < \infty.$$

The latter conditions follow from

$$|z| \le |\operatorname{Re} z| + |\operatorname{Im} z| \quad \text{and} \quad |\operatorname{Re} z|, |\operatorname{Im} z| \le |z|.$$

The absolute convergence implies convergence but not vice versa.

Example 1.19 (*Geometric series*). Since

$$\sum_{k=0}^{n} z^k = \frac{1 - z^{n+1}}{1 - z}, \qquad z \ne 1$$

then the limit

$$\lim_{n\to\infty} \sum_{k=0}^{n} z^k$$

exists if and only if $\lim_{n\to\infty} z^{n+1}$ exists and $z \ne 1$. But the latter limit exists if and only if $|z| < 1$ and in that case it equals 0. Thus, the series

$$\sum_{k=0}^{\infty} z^k$$

converges if and only if $|z| < 1$ and

$$\sum_{k=0}^{\infty} z^k = \frac{1}{1 - z}. \tag{1.10}$$

Example 1.20 (Exponential function). The *exponential function* $\mathrm{e}^z, z \in \mathbb{C}$ can be defined as the following series:

$$\mathrm{e}^z := \sum_{n=0}^{\infty} \frac{z^n}{n!}. \tag{1.11}$$

From real analysis, we know that

$$\sum_{n=0}^{\infty} \frac{|z|^n}{n!} = e^{|z|}.$$

Therefore, the series (1.11) is well-defined for all $z \in \mathbb{C}$. Even more is true. For $z = x \in \mathbb{R}$, we know that

$$e^x = \sum_{n=0}^{\infty} \frac{x^n}{n!}.$$

Using (1.11), we obtain for purely imaginary $z = iy$ that

$$\begin{aligned} e^{iy} &= \sum_{n=0}^{\infty} \frac{(iy)^n}{n!} = \sum_{k=0}^{\infty} \frac{(iy)^{2k}}{(2k)!} + \sum_{k=0}^{\infty} \frac{(iy)^{2k+1}}{(2k+1)!} \\ &= \sum_{k=0}^{\infty} \frac{(-1)^k y^{2k}}{(2k)!} + i \sum_{k=0}^{\infty} \frac{(-1)^k y^{2k+1}}{(2k+1)!} = \cos y + i \sin y. \end{aligned}$$

This proves formula (1.5).

Now we would like to show that actually the function (1.11) can be represented (or understood) as

$$e^z = e^x (\cos y + i \sin y),$$

where e^x, $\cos y$, and $\sin y$ are from real analysis. Indeed, by the binomial formula,

$$\begin{aligned} e^z = e^{x+iy} &= \sum_{n=0}^{\infty} \frac{(x+iy)^n}{n!} = \sum_{n=0}^{\infty} \frac{1}{n!} \sum_{k=0}^{n} \binom{n}{k} x^k (iy)^{n-k} \\ &= \sum_{n=0}^{\infty} \frac{1}{n!} \sum_{k=0}^{n} \frac{n!}{k!(n-k)!} x^k (iy)^{n-k} = \sum_{k=0}^{\infty} \sum_{n=k}^{\infty} \frac{x^k}{k!} \frac{(iy)^{n-k}}{(n-k)!} = \sum_{k=0}^{\infty} \sum_{m=0}^{\infty} \frac{x^k}{k!} \frac{(iy)^m}{m!} \\ &= \sum_{k=0}^{\infty} \frac{x^k}{k!} \sum_{m=0}^{\infty} \frac{(iy)^m}{m!} = e^x (\cos y + i \sin y), \end{aligned}$$

in particular,

$$e^{iz} = e^{-y} (\cos x + i \sin x)$$

and

$$\cos z = \frac{e^{iz} + e^{-iz}}{2}, \qquad \sin z = \frac{e^{iz} - e^{-iz}}{2i} \tag{1.12}$$

if $\cos z$ and $\sin z$ for $z \in \mathbb{C}$ are defined as

$$\cos z := \sum_{n=0}^{\infty} \frac{(-1)^n z^{2n}}{(2n)!}, \quad \sin z := \sum_{n=0}^{\infty} \frac{(-1)^n z^{2n+1}}{(2n+1)!}. \tag{1.13}$$

These formulas are justified in Example 3.16 and remark after it.

Problem 1.21. Show that:
1. $e^{z_1}e^{z_2} = e^{z_1+z_2}$.
2. $e^{z+i2\pi k} = e^z$, $k \in \mathbb{Z}$.
3. $e^{-z} = 1/e^z$ or $e^z = 1/e^{-z}$.
4. $(e^z)^n = e^{nz}$, $n \in \mathbb{Z}$.
5. $|e^z| = e^x \le e^{|z|}$, $z = x + iy$.
6. $|e^z| = \begin{cases} e^{|z|}, & \text{if and only if } \operatorname{Re} z \ge 0, \operatorname{Im} z = 0, \\ e^{-|z|}, & \text{if and only if } \operatorname{Re} z \le 0, \operatorname{Im} z = 0. \end{cases}$
7. $|\cos z| \le \frac{e^y+e^{-y}}{2}$, $|\sin z| \le \frac{e^y+e^{-y}}{2}$, $z = x + iy$.

Products

An infinite product of complex numbers

$$\prod_{k=1}^{\infty} p_k = p_1 p_2 \cdots p_k \cdots$$

is said to be convergent if there exists a finite limit

$$\lim_{n\to\infty} \prod_{k=1}^{n} p_k =: P;$$

otherwise, it diverges.

Remark. If one of the terms p_k is zero, then trivially

$$\prod_{k=1}^{\infty} p_k = 0.$$

That is why we will assume that $p_k \ne 0$ for all $k \in \mathbb{N}$.

If the product

$$\prod_{k=1}^{\infty} p_k = P$$

converges, i. e., $P \ne \infty$ and $P \ne 0$, then $\lim_{k\to\infty} p_k = 1$. Indeed, since

$$\prod_{k=1}^{n+1} p_k = p_{n+1} \prod_{k=1}^{n} p_k,$$

we have

$$p_{n+1} = \frac{\prod_{k=1}^{n+1} p_k}{\prod_{k=1}^{n} p_k}.$$

Therefore,

$$\lim_{n\to\infty} p_{n+1} = \frac{\lim_{n\to\infty} \prod_{k=1}^{n+1} p_k}{\lim_{n\to\infty} \prod_{k=1}^{n} p_k} = \frac{P}{P} = 1.$$

This is the *necessary condition of convergence of product.*

There is one simple case.

Example 1.22. If all $p_k \geq 1$, then the product $\prod_{k=1}^{\infty} p_k$ and the series $\sum_{k=1}^{\infty}(p_k-1)$ converge or diverge simultaneously. Indeed, for such values of p_k one can show using induction that the following double inequality holds:

$$\sum_{k=1}^{n}(p_k - 1) \leq \prod_{k=1}^{n} p_k \leq e^{\sum_{k=1}^{n}(p_k-1)}.$$

This proves the result due to the monotonicity of the corresponding terms.

The product $\prod_{k=1}^{\infty} p_k$ is said to be *convergent absolutely* if the product $\prod_{k=1}^{\infty} |p_k|$ converges.

Proposition 1.23. *The product $\prod_{k=1}^{\infty} p_k$ converges absolutely if and only if the series $\sum_{k=1}^{\infty} \log|p_k|$ converges, where $\log(\cdot)$ is the usual real logarithm.*

Proof. We use the following inequality for usual real logarithm:

$$\frac{p-1}{2p-1} < \log p < \frac{p-1}{2-p}, \quad 1 < p < 2.$$

Let now $P := \prod_{k=1}^{\infty} |p_k|$. Then

$$\lim_{n\to\infty} g_n = 1 \quad \text{for } g_n := \frac{\prod_{k=1}^{n} |p_k|}{P},$$

where without loss of generality we may assume that $g_n > 1$ for all $n \in \mathbb{N}$. Further,

$$\log g_n = \log \frac{\prod_{k=1}^{n} |p_k|}{P} = \sum_{k=1}^{n} \log|p_k| - \log P.$$

Applying the inequality for the logarithm from above, we obtain

$$\frac{g_n - 1}{2g_n - 1} < \sum_{k=1}^{n} \log|p_k| - \log P < \frac{g_n - 1}{2 - g_n}.$$

This inequality immediately implies that

$$\lim_{n\to\infty} \sum_{k=1}^{n} \log |p_k| = \log P.$$

Conversely, if $\sum_{k=1}^{\infty} \log |p_k|$ converges, then $\lim_{n\to\infty} g_n = 1$, i. e., $\prod_{k=1}^{\infty} |p_k|$ converges. □

Example 1.24. We show that the product $\prod_{k=1}^{\infty} \frac{\sin(z/k)}{z/k}$ converges absolutely for any fixed complex $z \neq 0$. Based on Proposition 1.23, we consider the series

$$\sum_{k=1}^{\infty} \log\left|\frac{\sin(z/k)}{z/k}\right|.$$

Since z is fixed then using Example 1.20, we have for k large enough that

$$\log\left|\frac{\sin(z/k)}{z/k}\right| = \log\left|1 - \frac{\lambda_z}{k^2}\right|, \quad |\lambda_z| \leq C,$$

where a constant C is independent on $k \geq 1$ and depends only on z, which is fixed. Using now the asymptotic behavior of usual real logarithm $\log(1 + u)$ for small u, we obtain that the convergence of the series $\sum_{k=1}^{\infty} \log |\frac{\sin(z/k)}{z/k}|$ is equivalent to the convergence of the number series $\sum_{k=1}^{\infty} \frac{1}{k^2}$. Hence, we obtained what is needed.

Problem 1.25. Show that the product

$$\prod_{k=2}^{\infty}\left(1 - \frac{k^k}{(z(k-1))^k}\right)$$

converges absolutely for $|z| > 1$.

Problem 1.26. Using the fact that (see Example 15.14) for any $z \in \mathbb{C}$,

$$\frac{\sin z}{z} = \prod_{k=1}^{\infty}\left(1 - \frac{z^2}{k^2\pi^2}\right)$$

show that for any $m \in \mathbb{N}$

$$\lim_{n\to\infty} \prod_{k=-n,k\neq 0}^{mn}\left(1 + \frac{z}{k\pi}\right) = m^{z/\pi}\frac{\sin z}{z}.$$

Example 1.27. We show that the product $\prod_{k=1}^{\infty}(1 - \frac{z}{k})e^{z/k}$ converges absolutely for any $z \in \mathbb{C}$. Based on Proposition 1.23, we consider the series

$$\begin{aligned}\sum_{k=1}^{\infty} \log\left|\left(1 - \frac{z}{k}\right)e^{z/k}\right| &= \sum_{k=1}^{\infty}\left(\log\left|1 - \frac{z}{k}\right| + \log|e^{z/k}|\right)\\ &= \sum_{k=1}^{\infty}\left(\frac{1}{2}\log\left(1 - \frac{2x}{k} + \frac{x^2+y^2}{k^2}\right) + \frac{x}{k}\right), \quad z = x + iy.\end{aligned}$$

Using now the asymptotic behavior of usual real logarithm $\log(1 + u)$ for small u, we obtain that the convergence of the latter series is equivalent to the convergence of the series

$$\sum_{k=1}^{\infty}\left(\frac{1}{2}\left(-\frac{2x}{k}+O\left(\frac{1}{k^2}\right)\right)+\frac{x}{k}\right)=\sum_{k=1}^{\infty}O\left(\frac{1}{k^2}\right),$$

which is absolutely convergent. Thus, we obtained what is needed. Here $f = O(g)$ means that

$$|f(z)| \le C|g(z)|$$

with some constant $C > 0$.

Problem 1.28. Show that

$$\left(1-\frac{z}{\pi}\right)\left(1-\frac{z}{2\pi}\right)\left(1+\frac{z}{\pi}\right)\left(1-\frac{z}{3\pi}\right)\left(1-\frac{z}{4\pi}\right)\left(1+\frac{z}{2\pi}\right)\cdots=\frac{\sin z}{z}\cdot e^{-z/\pi\log 2}.$$

Problem 1.29. Show that the product $\prod_{k=1}^{\infty}|1+\frac{i}{k}|$ converges but the product $\prod_{k=1}^{\infty}(1+\frac{i}{k})$ does not converge. Explain this phenomena.

2 Functions of complex variable

The complex-valued function of one real variable is the mapping

$$f : (a, b) \to \mathbb{C} \quad \text{or} \quad f : [a, b] \to \mathbb{C}$$

such that

$$z = f(t) = f_1(t) + \mathrm{i}f_2(t),$$

where $t \in (a, b)$ or $t \in [a, b]$. Here, the open interval (a, b) might be infinite but the closed interval $[a, b]$ is considered only for finite a and b.

The notions of limit, continuity, differentiability, and integrability are defined coordinatewise, i. e., for two real-valued functions $f_1(t)$ and $f_2(t)$ of one real variable t.

Definition 2.1.
1. The continuous mapping $f : [a, b] \to \mathbb{C}$, $z = f(t)$ is called the *Jordan curve* if $z(t_1) \neq z(t_2)$ for $t_1 \neq t_2$. If in addition $z(a) = z(b)$, then this curve is called *closed.*
2. The Jordan curve is called *piecewise smooth* if there are points

$$a = t_0 < t_1 < \cdots < t_n = b$$

such that $z = f(t)$ is continuously differentiable on the intervals $[t_{j-1}, t_j]$ for $j = 1, 2, \ldots, n$ and $f'(t) \neq 0$.
3. If $n = 1$ above, then the Jordan curve is called *smooth.*

We will use the following statement proved by Jordan (we accept it like axiom, without proof).

Any closed Jordan curve γ divides $\overline{\mathbb{C}}$ into two domains (regions): internal (not containing $z = \infty$) and external (containing $z = \infty$). They are denoted as $\operatorname{int}\gamma$ and $\operatorname{ext}\gamma$, respectively, so that

$$\overline{\mathbb{C}} = \operatorname{int}\gamma \cup \gamma \cup \operatorname{ext}\gamma.$$

Definition 2.2.
1. A set $D \subset \mathbb{C}$ is called *connected* if for any points $z_1, z_2 \in D$ there is a Jordan curve connecting these points and lying in D.
2. A set $D \subset \mathbb{C}$ is called a *domain* if it is connected and open.

We consider a complex-valued function w of one complex variable z as follows. Let us have two copies of the complex plane, one in z and one in w. Let D be a domain in z and G a domain in w. Then a function $w = f(z)$ is the mapping

$$f : D \to G$$

https://doi.org/10.1515/9783111632278-002

such that

$$w = u + \mathrm{i}v = f(z) = f_1(x, y) + \mathrm{i}f_2(x, y),$$

see Figure 2.1. This is equivalent to the definition of two real-valued functions u and v of two real variables x and y such that $w = f(z)$ if and only if

$$u(x, y) = \operatorname{Re} w \quad \text{and} \quad v(x, y) = \operatorname{Im} w.$$

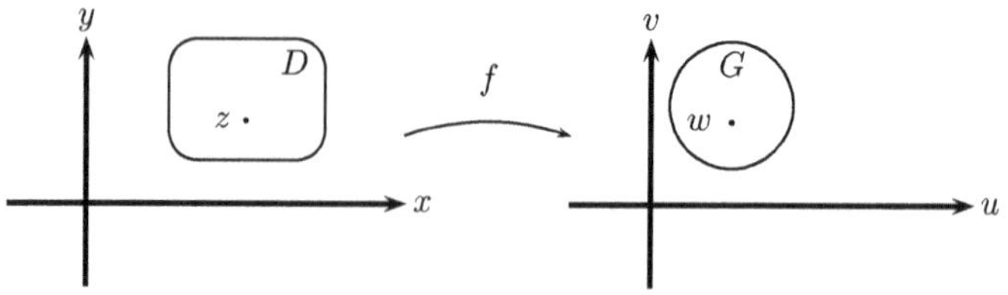

Figure 2.1: Mapping $f : D \to G$ between two copies of the complex plane.

In particular, we have that ($z_0 \neq \infty$)

$$b = \lim_{z \to z_0} f(z), \quad b \neq \infty \tag{2.1}$$

if and only if

$$\operatorname{Re} b = \lim_{(x,y) \to (x_0, y_0)} \operatorname{Re} f(z) \quad \text{and} \quad \operatorname{Im} b = \lim_{(x,y) \to (x_0, y_0)} \operatorname{Im} f(z).$$

Also,

$$\lim_{z \to z_0} f(z) = \infty$$

if and only if

$$\lim_{z \to z_0} |f(z)| = +\infty,$$

i. e., for any $R > 0$ there exists $\delta(R) > 0$ such that $|f(z)| > R$ whenever $|z - z_0| < \delta$.

Here, (2.1) means that for any $\varepsilon > 0$ there is $\delta = \delta(\varepsilon, z_0) > 0$ such that

$$|f(z) - b|_{\mathbb{C}} < \varepsilon$$

whenever $|z - z_0| < \delta$, i. e., $|(x, y) - (x_0, y_0)|_{\mathbb{R}^2} < \delta$. Therefore, the arithmetic operations for complex-valued functions of one complex variable are satisfied. Namely, let us assume that

$$\lim_{z \to z_0} f(z) = a \quad \text{and} \quad \lim_{z \to z_0} g(z) = b.$$

Then:

1.
$$\lim_{z\to z_0}(f(z)\pm g(z)) = a\pm b,$$

i. e.,

$$\lim_{(x,y)\to(x_0,y_0)}(\operatorname{Re} f\pm\operatorname{Re} g) = \operatorname{Re} a\pm\operatorname{Re} b$$

and

$$\lim_{(x,y)\to(x_0,y_0)}(\operatorname{Im} f\pm\operatorname{Im} g) = \operatorname{Im} a\pm\operatorname{Im} b.$$

2.
$$\lim_{z\to z_0} f(z)\cdot g(z) = a\cdot b,$$

i. e.,

$$\lim_{(x,y)\to(x_0,y_0)}\operatorname{Re}(f\cdot g) = \operatorname{Re}(a\cdot b)$$

and

$$\lim_{(x,y)\to(x_0,y_0)}\operatorname{Im}(f\cdot g) = \operatorname{Im}(a\cdot b).$$

3.
$$\lim_{z\to z_0} f(z)/g(z) = a/b, \quad \text{if}\, b\neq 0,$$

i. e.,

$$\lim_{(x,y)\to(x_0,y_0)}\operatorname{Re}(f/g) = \operatorname{Re}(a/b)$$

and

$$\lim_{(x,y)\to(x_0,y_0)}\operatorname{Im}(f/g) = \operatorname{Im}(a/b).$$

Definition 2.3. A function $w = f(z)$ is called *univalent* if $f : D \to G$ onto (is surjective) and if for any $z_1, z_2 \in D$, $z_1 \neq z_2$,

$$w_1 = f(z_1) \neq w_2 = f(z_2) \quad \text{(injectivity)}.$$

In this case, there is an *inverse function* f^{-1}, which maps as

$$f^{-1} : G \to D$$

onto (surjectively) such that $f^{-1}(w) = z$ if $w = f(z)$, i. e.,

$$z = f^{-1}(f(z)), \quad w = f(f^{-1}(w)), \quad z \in D, w \in G.$$

This inverse function f^{-1} is also univalent (bijective).

Summarizing, we have

$$z = f^{-1}(f(z)) \quad \text{for all } z \in D$$

and

$$w = f(f^{-1}(w)) \quad \text{for all } w \in G.$$

Definition 2.4. A function $w = f(z)$ is:

1. *Continuous* at $z = z_0 \neq \infty$ if $f(z)$ is well-defined in a neighborhood $U_\delta(z_0)$ and if for any $\varepsilon > 0$ there exists $\delta(\varepsilon, z_0) > 0$ such that
$$|f(z) - f(z_0)| < \varepsilon$$
whenever $|z - z_0| < \delta$.
2. Continuous at $z = \infty$ if $f(z)$ is well-defined for $|z| > A$ and there exists $b \in \mathbb{C}$ such that for any $\varepsilon > 0$ there is $R(\varepsilon, b) > 0$ such that
$$|f(z) - b| < \varepsilon$$
for any $|z| > R$. In that case, $f(\infty) = b$.
3. Continuous on the set $A \subset \overline{\mathbb{C}}$ if it is continuous at any point $z_0 \in A$.
4. *Uniformly continuous* on the set $A \subset \overline{\mathbb{C}}$ if for any $\varepsilon > 0$ there exists $\delta(\varepsilon) > 0$ such that
$$|f(z_1) - f(z_2)| < \varepsilon$$
whenever $|z_1 - z_2| < \delta$ and $z_1, z_2 \in A$.

Remark. Since ($z_0 \neq \infty$)

$$|z - z_0|_{\mathbb{C}} < \delta$$

if and only if

$$|(x, y) - (x_0, y_0)|_{\mathbb{R}^2} < \delta$$

and

$$|f(z) - f(z_0)| < \varepsilon$$

if and only if

$$|u(x,y) - u(x_0,y_0)| < \varepsilon, \quad \text{and} \quad |v(x,y) - v(x_0,y_0)| < \varepsilon,$$

then the continuity of $f(z)$ is equivalent to the continuity of $\operatorname{Re} f(z)$ and $\operatorname{Im} f(z)$ as the functions of two variables (x, y).

Problem 2.5. Show that $e^z \neq 0$ for any $z \in \mathbb{C}$ and the limit $\lim_{z\to\infty} e^z$ does not exist (finite or infinite).

Problem 2.6. Investigate the continuity of the functions

$$z^2/|z|^2, \quad (z \operatorname{Re} z)/|z|, \quad (\operatorname{Im} z)/z, \quad e^{-1/z^2}$$

at 0.

Example 2.7. A *linear-fractional* (*bilinear*) function is defined for $z \in \mathbb{C}$ as

$$w = \frac{az + b}{cz + d}, \quad ad - bc \neq 0, c \neq 0. \tag{2.2}$$

It is well-defined for $z \neq -d/c$. Since

$$w_1 - w_2 = \frac{az_1 + b}{cz_1 + d} - \frac{az_2 + b}{cz_2 + d} = \frac{(ad - bc)(z_1 - z_2)}{(cz_1 + d)(cz_2 + d)},$$

then this function is univalent in the domain $\mathbb{C} \setminus \{-d/c\}$. The inverse function $z = z(w)$ is also bilinear and defined by

$$z = \frac{dw - b}{a - cw}$$

and it is well-defined (and univalent) in the domain $\mathbb{C} \setminus \{a/c\}$. If we define

$$w(-d/c) = \infty \quad \text{and} \quad w(\infty) = a/c$$

then the bilinear function maps $\overline{\mathbb{C}}$ onto $\overline{\mathbb{C}}$ bijectively. The same is true for the inverse function.

Let us show that the bilinear function (2.2) is continuous everywhere in $\overline{\mathbb{C}} \setminus \{-d/c\}$. Indeed, if first $z_0 \neq -d/c$, $z_0 \neq \infty$, then

$$|w(z) - w(z_0)| = \left|\frac{(ad - bc)(z - z_0)}{(cz + d)(cz_0 + d)}\right| = \frac{|ad - bc||z - z_0|}{|cz_0 + d + c(z - z_0)||cz_0 + d|}.$$

Since $|cz_0 + d| > 0$, then we may choose $|z - z_0| < \delta$ and $|z - z_0| \le \frac{|cz_0+d|}{2|c|}$. In this case,

$$|cz + d| = \big|cz_0 + d + c(z - z_0)\big| \ge |cz_0 + d| - |c||z - z_0| \ge \frac{|cz_0 + d|}{2}$$

and

$$\big|w(z) - w(z_0)\big| < \frac{|ad - bc|\delta}{\frac{|cz_0+d|^2}{2}} \le \varepsilon.$$

If for arbitrary $\varepsilon > 0$, we will define

$$\delta = \min\left\{\frac{|cz_0 + d|}{2|c|}, \frac{\varepsilon|cz_0 + d|^2}{2|ad - bc|}\right\},$$

then the condition $|z - z_0| < \delta$ implies $|w(z) - w(z_0)| < \varepsilon$, i. e., the bilinear function is continuous at any such point z_0.

In the case $z_0 = \infty$, we may choose $|z| > 2|d/c|$ and obtain

$$\begin{aligned}\big|w(z) - w(\infty)\big| &= \left|\frac{az + b}{cz + d} - \frac{a}{c}\right| = \frac{|ad - bc|}{|c||cz + d|} = \frac{|ad - bc|}{|c|^2|z + d/c|}\\ &\le \frac{|ad - bc|}{|c|^2(|z| - |d/c|)} \le \frac{2|ad - bc|}{|c|^2|z|} < \varepsilon.\end{aligned}$$

Hence, if for arbitrary $\varepsilon > 0$ we will choose

$$R = \frac{2|ad - bc|}{|c|^2\varepsilon},$$

then the condition $|z| > R$ implies $|w(z) - w(\infty)| < \varepsilon$, i. e., the bilinear function is continuous also at ∞.

Remark. For $c = 0$, the bilinear function reduces to the linear function

$$w(z) = \frac{a}{d}z + \frac{b}{d}, \quad d \ne 0.$$

It is easy to check that this is continuous on $\mathbb{C}$ (but not at ∞) and univalent on $\mathbb{C}$.

Example 2.8. The squared function is defined for $z \in \overline{\mathbb{C}}$ as

$$w = w(z) = z^2, \quad w(\infty) = \infty. \tag{2.3}$$

Since

$$w_1 - w_2 = z_1^2 - z_2^2 = (z_1 - z_2)(z_1 + z_2),$$

then $w_1 = w_2$ if and only if $z_1 = z_2$ or $z_1 = -z_2$. Thus, the squared function (2.3) is not univalent on $\overline{\mathbb{C}}$.

But if we consider two subdomains,

$$D_+ = \{z \in \overline{\mathbb{C}} : \operatorname{Im} z > 0\}$$

and

$$D_- = \{z \in \overline{\mathbb{C}} : \operatorname{Im} z < 0\},$$

then in each of these two subdomains the squared function (2.3) is univalent. It is very easy to check that in both domains $z_1 \neq -z_2$. Indeed, $z_1 = -z_2$ if and only if $\operatorname{Re} z_1 = -\operatorname{Re} z_2$ and $\operatorname{Im} z_1 = -\operatorname{Im} z_2$, i. e., these equalities are impossible in D_+ or in D_-.

In order to define the inverse of $w = z^2$ in $D_\pm$, we proceed as follows:

$$w_1 + iw_2 = z^2 = x^2 - y^2 + 2ixy$$

if and only if

$$w_1 = x^2 - y^2, \quad x = \frac{w_2}{2y}.$$

So,

$$w_1 = \frac{w_2^2}{4y^2} - y^2$$

or

$$4y^4 + 4y^2 w_1 - w_2^2 = 0.$$

Hence,

$$y^2 = \frac{-w_1 + \sqrt{w_1^2 + w_2^2}}{2}.$$

It yields

$$y = \sqrt{\frac{\sqrt{w_1^2 + w_2^2} - w_1}{2}} \quad \text{in } D_+$$

and

$$y = -\sqrt{\frac{\sqrt{w_1^2 + w_2^2} - w_1}{2}} \quad \text{in } D_-.$$

Consequently,

$$x = \frac{w_2}{\sqrt{2}\sqrt{\sqrt{w_1^2 + w_2^2} - w_1}} \quad \text{in } D_+$$

and

$$x = -\frac{w_2}{\sqrt{2}\sqrt{\sqrt{w_1^2 + w_2^2} - w_1}} \quad \text{in } D_-.$$

Remark. As we can see, in $D_\pm$, $w_2 = 0$ if and only if $x = 0$, i. e., $\operatorname{Im} w = 0$ if and only if $\operatorname{Re} z = 0$ and in this case $\operatorname{Re} w = -(\operatorname{Im} z)^2$, i. e., $w_1 = -y^2 < 0$.

So, finally we have

$$z_+ = \frac{w_2}{\sqrt{2}\sqrt{\sqrt{w_1^2 + w_2^2} - w_1}} + \mathrm{i}\frac{\sqrt{\sqrt{w_1^2 + w_2^2} - w_1}}{\sqrt{2}},$$

$$z_- = -\frac{w_2}{\sqrt{2}\sqrt{\sqrt{w_1^2 + w_2^2} - w_1}} - \mathrm{i}\frac{\sqrt{\sqrt{w_1^2 + w_2^2} - w_1}}{\sqrt{2}}.$$

We may simplify these formulas to obtain

$$z_+ = \sqrt{\frac{w_1 + |w|}{2}} + \mathrm{i}\frac{w_2}{\sqrt{2(w_1 + |w|)}}, \quad z_- = -z_+. \tag{2.4}$$

In these formulas, z_+ is called $(\sqrt{w})_+$ with $\operatorname{Im} z_+ > 0$ and z_- is called $(\sqrt{w})_-$ with $\operatorname{Im} z_+ < 0$ so that we have two *branches* for inverse function.

For the case $x = 0$, we obtain easily from the remark above that

$$z_+ = \mathrm{i}\sqrt{-w_1} \quad \text{and} \quad z_- = -\mathrm{i}\sqrt{-w_1}. \tag{2.5}$$

For the case $\operatorname{Im} z = 0$, we have a real-valued (and nonnegative) function of one real variable x, i. e.,

$$w_1 = x^2.$$

Its inverse also has two branches

$$x_+ = \sqrt{w_1}, \quad x_- = -\sqrt{w_1}, \quad w_1 \geq 0. \tag{2.6}$$

The formulas (2.4)–(2.6) can be written shortly (compare with (1.6)) as

$$z_{\pm} = \sqrt{|w|}e^{i \arg w/2} \quad \text{and} \quad z_{\mp} = \sqrt{|w|}e^{i(\arg w/2+\pi)} = -\sqrt{|w|}e^{i \arg w/2}, \tag{2.7}$$

where $\arg w \in (-\pi, \pi]$. Here, $\pm$ depend on $\arg w$. More precisely, if $\arg w \in (0, \pi)$, then $z_+ \in D_+$ and $z_- \in D_-$, but if $\arg w \in (-\pi, 0)$, then $z_+ \in D_-$ and $z_- \in D_+$.

Problem 2.9. Show that (2.4)–(2.6) and (2.7) are equivalent.

The squared function (2.3) is continuous at any point $z_0 \in \mathbb{C}$ since

$$|w(z) - w(z_0)| = |z^2 - z_0^2| = |z - z_0||z + z_0| < \delta|z + z_0| < \delta(\delta + 2|z_0|) = \varepsilon.$$

So, if for arbitrary $\varepsilon > 0$, we choose

$$\delta = -|z_0| + \sqrt{|z_0|^2 + \varepsilon} > 0,$$

then the condition $|z - z_0| < \delta$ implies $|w(z) - w(z_0)| < \varepsilon$. So, $w = z^2$ is continuous at $z_0 \neq \infty$. At $z_0 = \infty$, this function is not continuous since $w(\infty) = \infty$.

Cardano's formulae

We consider the *depressed cubic equation:*

$$z^3 + pz + q = 0, \quad p, q \in \mathbb{C}, \tag{2.8}$$

and we solve it as follows. Assuming that $z = u + v$, we obtain (for particular solution) the system

$$\begin{cases} u^3 \cdot v^3 = -\frac{p^3}{27}, \\ u^3 + v^3 = -q. \end{cases}$$

The solutions of this system are given by

$$u^3 = -\frac{q}{2} \pm \sqrt{\Delta}, \quad v^3 = -\frac{q}{2} \mp \sqrt{\Delta},$$

where $\Delta = \frac{q^2}{4} + \frac{p^3}{27}$ and is called the *discriminant* of equation (2.8). It must be mentioned here that the value for the square roots is chosen with respect to Definition 1.10 (see also Example 2.8). One root is then equal to

$$z_1 = \left(-\frac{q}{2} + \sqrt{\Delta}\right)^{\frac{1}{3}} + \left(-\frac{q}{2} - \sqrt{\Delta}\right)^{\frac{1}{3}}. \tag{2.9}$$

When the first root of (2.8) is defined, then this equation can be rewritten as

$$(z - z_1)(z^2 + z_1 z + p + z_1^2) = 0.$$

Therefore, two other roots of (2.8) are given by

$$z_{2,3} = -\frac{z_1}{2} \pm \sqrt{-\frac{3z_1^2}{4} - p}. \tag{2.10}$$

Problem 2.10. Show that the formulae (2.9) and (2.10) are the same as

$$z_j = \epsilon_j \left(-\frac{q}{2} + \sqrt{\Delta}\right)^{\frac{1}{3}} + \epsilon_j^2 \left(-\frac{q}{2} - \sqrt{\Delta}\right)^{\frac{1}{3}}, \quad j = 1, 2, 3, \tag{2.11}$$

where ϵ_j are three different roots of equation $z^3 = 1$ with $\epsilon_1 = 1$. These formulae are known as the *Cardano's formulae* for the depressed cubic equation.

Remark. The general cubic equation (and its roots)

$$az^3 + bz^2 + cz + d = 0, \quad a \neq 0, a, b, c, d \in \mathbb{C}$$

can be reduced to the depressed cubic equation (2.8) (and its roots) if one uses a new variable $\zeta = z + \frac{b}{3a}$.

Remark. Based on Cardano's formulae, the general equation of degree 4 can be also solved as follows. First, the equation of degree 4 reads as

$$z^4 + a_1 z^3 + a_2 z^2 + a_3 z + a_4 = 0,$$

and using a new variable $\zeta = z + \frac{a_1}{4}$, it can be reduced to the following equation of degree 4:

$$\zeta^4 + a\zeta^2 + b\zeta + c = 0,$$

where $a = a_2 - \frac{3a_1^2}{8}$, $b = \frac{a_1^3}{8} - \frac{a_1 a_2}{2} + a_3$, and $c = -\frac{3a_1^4}{256} + \frac{a_1^2 a_2}{16} - \frac{a_1 a_3}{4} + a_4$. Second, using the representation

$$\zeta^4 + a\zeta^2 + b\zeta + c = (\zeta^2 + \alpha\zeta + \beta_1)(\zeta^2 - \alpha\zeta + \beta_2),$$

and considering the cubic equation

$$w^3 + 2aw^2 + (a^2 - 4c)w - b^2 = 0,$$

and its roots (by Cardano's formulae) $\alpha_1, \alpha_2, \alpha_3$, one can easily show that the roots $\zeta_1, \zeta_2, \zeta_3, \zeta_4$ of the equation of degree 4 from above are given by

$$\zeta_1 = \frac{1}{2}(\sqrt{\alpha_1} + \sqrt{\alpha_2} + \sqrt{\alpha_3}), \quad \zeta_2 = \frac{1}{2}(\sqrt{\alpha_1} - \sqrt{\alpha_2} - \sqrt{\alpha_3}),$$

$$\zeta_3 = \frac{1}{2}(-\sqrt{\alpha_1} + \sqrt{\alpha_2} - \sqrt{\alpha_3}), \quad \zeta_4 = \frac{1}{2}(-\sqrt{\alpha_1} - \sqrt{\alpha_2} + \sqrt{\alpha_3}).$$

Problem 2.11. Investigate the function $w = z^3$ by the same manner as in Example 2.8. Hint: Use Cardano's formulae for the depressed cubic equation.

Problem 2.12. Assume that p, q from equation (2.8) are real. Show that:
1. If the discriminant $\Delta < 0$, then equation (2.8) has three different real roots.
2. If the discriminant $\Delta > 0$, then equation (2.8) has one real root and two complex conjugate roots.
3. If the discriminant $\Delta = 0$, then equation (2.8) has three real roots and at least one is double.
4. If the discriminant $\Delta < 0$, then three different real roots of equation (2.8) can be expressed as

$$z_j = 2\sqrt{-\frac{p}{3}}\cos\left(\frac{1}{3}\arccos\left(\frac{3q}{2p}\sqrt{-\frac{3}{p}}\right) - \frac{2\pi j}{3}\right), \quad j = 0, 1, 2,$$

where cos and arccos are well-known real-valued functions. These formulae are called the *Viete formulae* for the depressed cubic equation.

Example 2.13. The *Zhukovski function* is defined for any $z \neq 0$ and $z \neq \infty$ as

$$w(z) = \frac{1}{2}\left(z + \frac{1}{z}\right) \tag{2.12}$$

or $z^2 - 2zw + 1 = 0$. We define

$$w(\infty) = w(0) = \infty.$$

Since

$$w(z_1) - w(z_2) = \frac{1}{2}(z_1 - z_2)\left(1 - \frac{1}{z_1 z_2}\right),$$

then $w(z_1) \neq w(z_2)$ if and only if $z_1 \neq z_2$ and $z_1 z_2 \neq 1$. Thus, the Zhukovski function (2.12) is univalent if and only if $z_1 z_2 \neq 1$, e. g., if either $|z| < 1$ or $|z| > 1$, i. e., in the domains

$$D_1 = \{z \in \mathbb{C} : |z| < 1\}, \quad D_2 = \{z \in \mathbb{C} : |z| > 1\}.$$

On the unit circle $|z| = 1$, there are always two different points z_1 and z_2 such that $z_1 z_2 = 1$. Indeed, if $z_1 = e^{i\varphi_1}$, $\varphi_1 \in (-\pi, \pi)$, then if we consider $z_2 = e^{-i\varphi_1}$, and we have $z_1 z_2 = 1$, but $z_1 \neq z_2$. In this consideration, the case $z_1 = e^{i\pi} = -1$ is excluded.

For any $z = re^{i\varphi}$, we have that

$$w(z) = \frac{1}{2}\left(re^{i\varphi} + \frac{1}{r}e^{-i\varphi}\right) = \frac{1}{2}\left(r + \frac{1}{r}\right)\cos\varphi + \frac{i}{2}\left(r - \frac{1}{r}\right)\sin\varphi.$$

It implies

$$|w|^2 = \frac{1}{4}\left(r + \frac{1}{r}\right)^2 + \frac{\cos\varphi - 1}{2},$$

and hence,

$$|w|^2 \le \frac{1}{4}\left(r + \frac{1}{r}\right)^2, \quad |w|^2 \ge \frac{1}{4}\left(r + \frac{1}{r}\right)^2 - 1.$$

Using (2.12), we obtain that the inverse function is given by

$$z = w - \sqrt{w^2 - 1} \quad \text{in } D_1$$

and by

$$z = w + \sqrt{w^2 - 1} \quad \text{in } D_2$$

depending on the choice of $\sqrt{w^2 - 1}$.

The Zhukovski function is continuous at any point $z_0 \neq 0, \infty$. Indeed, for such z_0 we have

$$\begin{aligned}
|w(z) - w(z_0)| &= \frac{1}{2}|z - z_0|\left|1 - \frac{1}{zz_0}\right| = \frac{1}{2}|z - z_0|\left|1 - \frac{1}{((z - z_0) + z_0)z_0}\right| \\
&\le \frac{|z - z_0|}{2}\left(1 + \frac{1}{|(z - z_0) + z_0||z_0|}\right) \\
&\le \frac{|z - z_0|}{2}\left(1 + \frac{1}{|z_0|(|z_0| - |z - z_0|)}\right) \\
&\le \frac{|z - z_0|}{2}\left(1 + \frac{1}{|z_0||z_0|/2}\right)
\end{aligned}$$

if $|z - z_0| \le |z_0|/2$. Thus, for any $\varepsilon > 0$ and $|z - z_0| < \min(\delta, |z_0|/2)$, we have

$$|w(z) - w(z_0)| < \frac{\delta}{2}\left(1 + \frac{2}{|z_0|^2}\right) = \varepsilon.$$

So, choosing

$$\delta = \min\left(\frac{2\varepsilon}{1 + 2/|z_0|^2}, \frac{|z_0|}{2}\right),$$

the condition $|z - z_0| < \delta$ implies $|w(z) - w(z_0)| < \varepsilon$. At $z = 0$ or $z = \infty$, the Zhukovski function is not continuous since $w(0) = w(\infty) = \infty$.

Problem 2.14. Show that the Zhukovski function maps real numbers into real numbers and purely imaginary numbers to purely imaginary numbers.

Problem 2.15. Show that the Zhukovski function maps the unit circle $|z| = 1$ into $\cos(\arg z)$.

Example 2.16. Consider the exponential function $w(z) = e^z$ (see Example 1.20). Since

$$w(z_1) - w(z_2) = e^{z_2}(e^{z_1 - z_2} - 1)$$

and since (see Problem 1.21(2)) $e^{i2\pi k} = 1$, $k \in \mathbb{Z}$, and e^z is never equal to zero, then the exponential function is univalent if and only if $z \in D_k = \{z : 2\pi k < \operatorname{Im} z < 2\pi(k+1)\}$, $k \in \mathbb{Z}$.

For $z = x + iy$ and $w = w_1 + iw_2$, we obtain that

$$w_1 = e^x \cos y, \quad w_2 = e^x \sin y$$

and this implies that

$$\begin{cases} w_1^2 + w_2^2 = e^{2x}, \\ \frac{w_2}{w_1} = \tan y. \end{cases}$$

It is equivalent to

$$\begin{cases} x = \log |w|, \\ y = \arctan \frac{w_2}{w_1} + \pi k, \quad k \in \mathbb{Z}, \end{cases}$$

where log and arctan are real-valued functions of a real variable. These formulae give us the inverse function $z(w)$.

The exponential function is continuous at any point $z_0 \neq \infty$. Indeed, we have

$$\begin{aligned} |w(z) - w(z_0)| &\le |e^x - e^{x_0}| + e^{x_0}(|\cos y - \cos y_0| + |\sin y - \sin y_0|) \\ &\le e^{x_0}(e^{|x - x_0|} - 1) + 2e^{x_0}|y - y_0| \le 4e^{x_0}|z - z_0|. \end{aligned}$$

Thus, for any $\epsilon > 0$ and $|z - z_0| < \delta$ we have

$$|w(z) - w(z_0)| \le 4\delta e^{x_0} = \epsilon, \quad \delta = \frac{\epsilon}{4e^{x_0}}.$$

This establishes the continuity of the exponential function. At $z = \infty$, the exponential function cannot be defined.

Remark. It can be mentioned here that the exponential function maps real numbers into positive real numbers and pure imaginary numbers to the unit circle $|w| = 1$.

Example 2.17. Consider the function $w(z) = y^2 + ix^2$, $z = x + iy$. Since

$$w(z_1) - w(z_2) = (y_1 - y_2)(y_1 + y_2) + i(x_1 - x_2)(x_1 + x_2),$$

then $w(z_1) \neq w(z_2)$ if and only if $z_1 \neq z_2$, or $z_1 \neq -z_2$, or $z_1 \neq \overline{z_2}$, or $z_1 \neq -\overline{z_2}$. Thus, this function is univalent if and only if z belongs to one of the four quadrants on the complex plane $\mathbb{C}$. Since for this function $\operatorname{Re} w \geq 0$ and $\operatorname{Im} w \geq 0$, then the inverse function is given for each of these four quadrants by

$$z = \begin{cases} \sqrt{w_2} + \mathrm{i}\sqrt{w_1}, & \text{for } D_1, \\ -\sqrt{w_2} + \mathrm{i}\sqrt{w_1}, & \text{for } D_2, \\ -\sqrt{w_2} - \mathrm{i}\sqrt{w_1}, & \text{for } D_3, \\ \sqrt{w_2} - \mathrm{i}\sqrt{w_1}, & \text{for } D_4, \end{cases}$$

where $w = w_1 + \mathrm{i}w_2$ and where $D_j, j = 1,2,3,4$ denote the corresponding quadrant on the complex plane $\mathbb{C}$ with respect to z.

This "quasiquadratic" function is continuous at any point $z_0 \neq \infty$. Indeed, for such z_0 and for any $\epsilon > 0$, we have ($|z - z_0| < \delta$),

$$\begin{aligned} |w(z) - w(z_0)| &= |y^2 - y_0^2 + \mathrm{i}(x^2 - x_0^2)| \\ &\leq |y - y_0|(|y - y_0| + 2|y_0|) + |x - x_0|(|x - x_0| + 2|x_0|) \\ &< 2\delta^2 + 4\delta|z_0| = \epsilon. \end{aligned}$$

So, choosing

$$\delta = -|z_0| + \sqrt{|z_0|^2 + \frac{\epsilon}{2}} > 0,$$

we obtain continuity of $w(z)$ at z_0. At $z = \infty$, this function is not continuous since $w(\infty) = \infty$.

Remark. The latter "quasiquadratic" function can be written in terms of variable z as

$$w(z) = \frac{1 + \mathrm{i}}{2}|z|^2 - \frac{1 - \mathrm{i}}{4}(z^2 + \overline{z}^2).$$

As a consequence of the notion of limit, we may formulate and prove (as in real analysis) the following general statements.

Proposition 2.18. *Assume that f and g are continuous at some point z_0 (or on a set A). Then:*

1. $f \pm g$,
2. $f \cdot g$,
3. $\frac{f}{g}$, *if $g(z_0) \neq 0$ (or $g(z) \neq 0$ for all $z \in A$),*
4. $|f|$

are continuous at z_0 (or on the set A).

Proposition 2.19. *Let $w = f(z)$ be continuous on a set A and $g(w)$ continuous on the set $f(A)$. Then the* composite function

$$\eta = g(f(z)) = (g \circ f)(z)$$

is continuous on the set A.

Corollary 2.20. *If $w = f(z)$ is univalent and continuous on a domain D, then the inverse function $z = f^{-1}(w)$ is continuous on the domain $G = f(D)$.*

Proof. Since for any $z \in D$, we have

$$z = f^{-1}(f(z))$$

and f is continuous on D, then $f^{-1}(w)$ is continuous on $G = f(D)$ because z is continuous. □

Weierstrass theorems

1. If $D \subset \mathbb{C}$ is compact (i. e., closed and bounded) and f is continuous on D, then f is bounded and uniformly continuous on D.
2. The previous statement holds also for compact $D \subset \overline{\mathbb{C}}$ (see stereographic projection).
3. If $D \subset \overline{\mathbb{C}}$ is compact and f is continuous on D, then $|f|$ achieves maximum and minimum on D.

3 Analytic functions (differentiability)

Definition 3.1. Let $w = f(z)$ be well-defined on a domain $D \subset \mathbb{C}$ and $z_0 \in D$. If the limit

$$\lim_{D \ni z \to z_0} \frac{f(z) - f(z_0)}{z - z_0}$$

exists, then this limit is called the *derivative* of $f(z)$ at the point z_0 and it is denoted as $f'(z_0)$. In this case, f is called *differentiable* at z_0 with

$$\lim_{D \ni z \to z_0} \frac{f(z) - f(z_0)}{z - z_0} = f'(z_0). \tag{3.1}$$

We say that $f'(\infty)$ exists if f is continuous at $z = \infty$ and there is $g'(0)$ for $g(z) = f(1/z)$. This is equivalent to

$$g'(0) = \lim_{\zeta \to \infty} \zeta[f(\zeta) - f(\infty)] =: f'(\infty).$$

This definition is equivalent to the existence of the limit

$$\lim_{\substack{x \to x_0 \\ y \to y_0}} \frac{u(x,y) - u(x_0,y_0) + \mathrm{i}(v(x,y) - v(x_0,y_0))}{(x - x_0) + \mathrm{i}(y - y_0)}.$$

In particular, if $x = x_0$ and $y \to y_0, y \neq y_0$ the latter limit equals

$$\begin{aligned} &\lim_{y \to y_0} \frac{u(x_0,y) - u(x_0,y_0) + \mathrm{i}(v(x_0,y) - v(x_0,y_0))}{\mathrm{i}(y - y_0)} \\ &\quad = \frac{1}{\mathrm{i}} \frac{\partial u}{\partial y}(x_0,y_0) + \frac{\partial v}{\partial y}(x_0,y_0) = \frac{\partial v}{\partial y}(x_0,y_0) - \mathrm{i}\frac{\partial u}{\partial y}(x_0,y_0). \end{aligned} \tag{3.2}$$

In the case $y = y_0$ and $x \to x_0, x \neq x_0$, the limit equals

$$\begin{aligned} &\lim_{x \to x_0} \frac{u(x,y_0) - u(x_0,y_0) + \mathrm{i}(v(x,y_0) - v(x_0,y_0))}{x - x_0} \\ &\quad = \frac{\partial u}{\partial x}(x_0,y_0) + \mathrm{i}\frac{\partial v}{\partial x}(x_0,y_0). \end{aligned} \tag{3.3}$$

Since the limit (3.1) is unique, we obtain from (3.2) and (3.3) that we must necessarily have

$$\frac{\partial u}{\partial x}(x_0,y_0) = \frac{\partial v}{\partial y}(x_0,y_0) \quad \text{and} \quad \frac{\partial u}{\partial y}(x_0,y_0) = -\frac{\partial v}{\partial x}(x_0,y_0). \tag{3.4}$$

The equalities (3.4) are called the *Cauchy–Riemann conditions*. We have proved that they are necessary for existence of $f'(z)$. Actually they are in some sense also sufficient. More

https://doi.org/10.1515/9783111632278-003

precisely, let $u(x,y)$ and $v(x,y)$ be differentiable at the point (x_0,y_0). If the conditions (3.4) are satisfied, then $f'(z_0)$ exists. Indeed, we have

$$\begin{aligned} &u(x,y)-u(x_0,y_0)\\ &\quad=\frac{\partial u}{\partial x}(x_0,y_0)(x-x_0)+\frac{\partial u}{\partial y}(x_0,y_0)(y-y_0)+o\big(\sqrt{(x-x_0)^2+(y-y_0)^2}\big) \end{aligned}$$

and

$$\begin{aligned} &v(x,y)-v(x_0,y_0)\\ &\quad=\frac{\partial v}{\partial x}(x_0,y_0)(x-x_0)+\frac{\partial v}{\partial y}(x_0,y_0)(y-y_0)+o\big(\sqrt{(x-x_0)^2+(y-y_0)^2}\big), \end{aligned}$$

where $o(\cdot)$ means that $o(s)/s\to 0$ as $s\to 0$. Therefore, we have using (3.4),

$$\begin{aligned} u(x,y)&-u(x_0,y_0)+\mathrm{i}(v(x,y)-v(x_0,y_0))\\ &=\frac{\partial u}{\partial x}(x_0,y_0)(x-x_0)+\frac{\partial u}{\partial y}(x_0,y_0)(y-y_0)\\ &\quad+\mathrm{i}\Big(\frac{\partial v}{\partial x}(x_0,y_0)(x-x_0)+\frac{\partial v}{\partial y}(x_0,y_0)(y-y_0)\Big)\\ &\quad+o\big(\sqrt{(x-x_0)^2+(y-y_0)^2}\big)\\ &=\Big[\frac{\partial u}{\partial x}(x_0,y_0)+\mathrm{i}\frac{\partial v}{\partial x}(x_0,y_0)\Big][(x-x_0)+\mathrm{i}(y-y_0)]\\ &\quad+o\big(\sqrt{(x-x_0)^2+(y-y_0)^2}\big) \end{aligned}$$

or

$$\frac{f(z)-f(z_0)}{z-z_0}=\frac{\partial f}{\partial x}(x_0,y_0)+\frac{o(|z-z_0|)}{z-z_0}.$$

This representation implies that the limit

$$\lim_{z\to z_0}\frac{f(z)-f(z_0)}{z-z_0}=\frac{\partial f}{\partial x}(x_0,y_0)=f'(z_0) \tag{3.5}$$

exists. In a similar manner, we obtain

$$\lim_{z\to z_0}\frac{f(z)-f(z_0)}{z-z_0}=-\mathrm{i}\frac{\partial f}{\partial y}(x_0,y_0)=f'(z_0). \tag{3.6}$$

Thus, we have proved the following fundamental result.

Theorem 3.2. *The function $w=f(z)$ is differentiable at the point z_0 if and only if $\operatorname{Re} f(z)$ and $\operatorname{Im} f(z)$ are differentiable at the point (x_0,y_0) as real-valued functions of two real variables x and y and the Cauchy–Riemann conditions* (3.4) *are satisfied.*

Remark. Formulas (3.5) and (3.6) imply that

$$\begin{aligned} f'(z_0) &= \frac{1}{2}\left(\frac{\partial f}{\partial x} - \mathrm{i}\frac{\partial f}{\partial y}\right) =: \frac{\partial f}{\partial z}, \\ 0 &= \frac{1}{2}\left(\frac{\partial f}{\partial x} + \mathrm{i}\frac{\partial f}{\partial y}\right) =: \frac{\partial f}{\partial \overline{z}}. \end{aligned} \tag{3.7}$$

Hence, the Cauchy–Riemann conditions are equivalent to

$$\frac{\partial f}{\partial z}(z_0) = f'(z_0) \quad \text{and} \quad \frac{\partial f}{\partial \overline{z}}(z_0) = 0. \tag{3.8}$$

Example 3.3. Consider the function

$$f(z) = \overline{z}.$$

Then $u(x,y) = x$ and $v(x,y) = -y$. The partial derivatives in this case are

$$\frac{\partial u}{\partial x} = 1, \quad \frac{\partial u}{\partial y} = 0, \quad \frac{\partial v}{\partial x} = 0, \quad \frac{\partial v}{\partial y} = -1$$

so that

$$1 = \frac{\partial u}{\partial x} \neq \frac{\partial v}{\partial y} = -1, \quad 0 = \frac{\partial u}{\partial y} = -\frac{\partial v}{\partial x} = 0.$$

Thus, Cauchy–Riemann conditions are not satisfied and, therefore, $f(z) = \overline{z}$ has no derivative.

Example 3.4. Let us consider the function (see the remark after Example 2.17)

$$f(z) = y^2 + \mathrm{i}x^2, \quad z = x + \mathrm{i}y.$$

Then

$$\frac{\partial u}{\partial x} = 0, \quad \frac{\partial u}{\partial y} = 2y, \quad \frac{\partial v}{\partial x} = 2x, \quad \frac{\partial v}{\partial y} = 0.$$

Hence, the Cauchy–Riemann conditions are

$$0 = \frac{\partial u}{\partial x} = \frac{\partial v}{\partial y} = 0 \quad \text{and} \quad 2y = \frac{\partial u}{\partial y} = -\frac{\partial v}{\partial x} = -2x,$$

i. e., $x = -y$. Thus, (3.4) are satisfied at any point $z_0 = x_0 - \mathrm{i}x_0$, $x_0 \in \mathbb{R}$, and the limit (at any such point)

$$\lim_{z \to z_0} \frac{f(z) - f(z_0)}{z - z_0}$$

exists and equals to $2ix_0$, i. e., (see also (3.7) and (3.8)),

$$f'(z_0) = 2\mathrm{i}\,\mathrm{Re}\,z_0, \quad z_0 = x_0 - \mathrm{i}x_0.$$

Compared with the latter function, the function $f(z) = z\bar{z} = |z|^2$ is differentiable only at the point $z_0 = 0$ and $f'(0) = 0$. This can be checked straightforwardly using (3.7) or (3.8).

Problem 3.5. Show that function $f(z)$ is not differentiable at $z = 0$ but the Cauchy–Riemann conditions are satisfied at this point, when:

1. $f(z) = \sqrt{|xy|}$,
2. $f(z) = \frac{xy^2 z}{x^2+y^4}$,

where $z = x + \mathrm{i}y$.

Problem 3.6. Let

$$f(z) = R(x,y)\mathrm{e}^{\mathrm{i}\theta(x,y)},$$

where R and θ are real-valued. Prove that Cauchy–Riemann conditions can be written in this case as

$$\frac{\partial R}{\partial x} = R\frac{\partial \theta}{\partial y} \quad \text{and} \quad \frac{\partial R}{\partial y} = -R\frac{\partial \theta}{\partial x}. \tag{3.9}$$

Problem 3.7. Let

$$w = \frac{az+b}{cz+d}, \quad ad \neq bc, c \neq 0$$

be a bilinear function. Show that

$$w'(z) = -\frac{bc-ad}{(cz+d)^2}$$

for any $z \neq -d/c$.

Problem 3.8. Let

$$w = \frac{az+b}{cz+d}, \quad ad \neq bc, c \neq 0.$$

Show that $w'(\infty) = (bc - ad)/c^2$.

Problem 3.9. Let

$$w = \mathrm{e}^z = \mathrm{e}^x(\cos y + \mathrm{i}\sin y).$$

Show that

$$(e^z)' = e^z$$

at any point $z \neq \infty$.

Problem 3.10. Let us consider the Zhukovski function

$$w = \frac{1}{2}\left(z + \frac{1}{z}\right).$$

Show that

$$w'(z) = \frac{1}{2}\left(1 - \frac{1}{z^2}\right), \quad z \neq 0, z \neq \infty.$$

Show also that $w'(\infty)$ does not exist but

$$\lim_{z\to 0} w'(z) = \infty, \quad \lim_{z\to\infty} w'(z) = 1/2.$$

Proposition 3.11. *If $w = f(z)$ is differentiable at $z = z_0$, then $f(z)$ is also continuous at z_0 but not vice versa.*

Proof. Since the limit

$$\lim_{z\to z_0} \frac{f(z) - f(z_0)}{z - z_0} = f'(z_0)$$

exists, then

$$f(z) - f(z_0) = f'(z_0)(z - z_0) + o(z - z_0).$$

This implies that

$$\lim_{z\to z_0} f(z) = f(z_0).$$

The function $f(z) = \bar{z}$ provides an example of a function, which is continuous but not differentiable. □

Problem 3.12. Let $f(z) = e^{-\frac{1}{z^4}}$, $f(0) = 0$. Show that $f(z)$ is differentiable everywhere (including $z = \infty$) except $z = 0$, but the Cauchy–Riemann conditions are satisfied everywhere in $\overline{\mathbb{C}}$.

Proposition 3.13. *Let $\eta(z) = g(f(z))$ be the composition of functions $w = f(z)$ and $\eta = g(w)$. If $f(z)$ is differentiable at $z = z_0$ and $g(w)$ is differentiable at $w_0 = f(z_0)$, then $\eta(z)$ is differentiable at $z = z_0$ and*

$$\eta'(z_0) = g'(w_0)f'(z_0) = g'(f(z_0))f'(z_0). \tag{3.10}$$

Proof. By definition, we have

$$\frac{\eta(z)-\eta(z_0)}{z-z_0} = \frac{g(f(z))-g(f(z_0))}{z-z_0} = \frac{g(w)-g(w_0)}{w-w_0}\cdot\frac{f(z)-f(z_0)}{z-z_0},$$

where $w=f(z)$ and $w_0=f(z_0)$. If $z\to z_0$, then $w\to w_0$ by Proposition 3.11. Then due to conditions of this proposition, we have

$$\lim_{z\to z_0}\frac{\eta(z)-\eta(z_0)}{z-z_0} = \lim_{w\to w_0}\frac{g(w)-g(w_0)}{w-w_0}\cdot\lim_{z\to z_0}\frac{f(z)-f(z_0)}{z-z_0} = g'(w_0)f'(z_0)$$

or $\eta'(z_0)=g'(w_0)f'(z_0)$. □

Corollary 3.14. *Let $w=f(z)$ be univalent on a domain D. Then f is differentiable on D if and only if the inverse function $z=f^{-1}(w)$ is differentiable on $G=f(D)$ and*

$$f'(z)=\frac{1}{(f^{-1})'(w)},\quad w=f(z). \tag{3.11}$$

In particular, both derivatives are not equal to zero.

Proof. The claim follows from the representations

$$z=f^{-1}(f(z)),\quad z\in D\quad\text{and}\quad w=f(f^{-1}(w)),\quad w\in G,$$

and Proposition 3.13. Indeed,

$$1=(z)'=(f^{-1})'(w)f'(z),$$

where $w=f(z)$ and both derivatives are not equal to zero necessarily. □

Example 3.15. Consider the Zhukovski function

$$w=\frac{1}{2}\left(z+\frac{1}{z}\right).$$

Then (3.11) leads to

$$(f^{-1})'(w)=\frac{2}{1-1/z^2}=\frac{2}{1-1/(2wz-1)}=\frac{2wz-1}{wz-1}=1+\frac{w}{w-1/z},$$

where $z=w\pm\sqrt{w^2-1}$. So,

$$(f^{-1})'(w)=1\pm\frac{w}{\sqrt{w^2-1}}$$

depending on the domains D_1 and D_2; see Example 2.13. In the domains D_1 and D_2, we have $w\neq\pm1$ and, therefore, the latter formula is well-defined.

Example 3.16. Let us introduce some new functions (compare with (1.12) and (1.13)):

$$\begin{aligned} \sin z &:= \frac{e^{iz} - e^{-iz}}{2i}, \quad \cos z := \frac{e^{iz} + e^{-iz}}{2}, \\ \sinh z &:= \frac{e^{z} - e^{-z}}{2}, \quad \cosh z := \frac{e^{z} + e^{-z}}{2}. \end{aligned} \tag{3.12}$$

These functions are compositions of e^z and e^{iz}. That is why we have

$$\begin{aligned} (\sin z)' &= \frac{(e^{iz})' - (e^{-iz})'}{2i} = \frac{ie^{iz} + ie^{-iz}}{2i} = \frac{e^{iz} + e^{-iz}}{2} = \cos z, \\ (\cos z)' &= \frac{(e^{iz})' + (e^{-iz})'}{2} = \frac{ie^{iz} - ie^{-iz}}{2} = -\frac{e^{iz} - e^{-iz}}{2i} = -\sin z, \\ (\sinh z)' &= \frac{(e^{z})' - (e^{-z})'}{2} = \frac{e^{z} + e^{-z}}{2} = \cosh z, \\ (\cosh z)' &= \frac{(e^{z})' + (e^{-z})'}{2} = \frac{e^{z} - e^{-z}}{2} = \sinh z. \end{aligned}$$

There are also some useful equalities:

$$\cos^2 z + \sin^2 z = \frac{e^{2iz} + 2 + e^{-2iz}}{4} - \frac{e^{2iz} - 2 + e^{-2iz}}{4} = 1$$

and

$$\cosh^2 z - \sinh^2 z = \frac{(e^{z} + e^{-z})^2}{4} - \frac{(e^{z} - e^{-z})^2}{4} = 1.$$

Also, we obtain the equalities

$$\begin{aligned} \cos(-z) = \cos z, \quad \sin(-z) &= -\sin(z), \\ e^{iz} &= \cos z + i \sin z, \\ e^{-iz} &= \cos z - i \sin z. \end{aligned} \tag{3.13}$$

Remark. Since

$$e^z = \sum_{n=0}^{\infty} \frac{z^n}{n!}, \quad z \in \mathbb{C},$$

then

$$e^{iz} = \sum_{n=0}^{\infty} \frac{(iz)^n}{n!}, \quad e^{-iz} = \sum_{n=0}^{\infty} \frac{(-iz)^n}{n!}.$$

So, using (3.12), we obtain

$$\begin{aligned}\cos z &= \frac{1}{2}\sum_{n=0}^{\infty}\frac{\mathrm{i}^n + (-\mathrm{i})^n}{n!}z^n \\ &= \frac{1}{2}\left(\sum_{k=0}^{\infty}\frac{\mathrm{i}^{2k} + (-\mathrm{i})^{2k}}{(2k)!}z^{2k} + \sum_{k=0}^{\infty}\frac{\mathrm{i}^{2k+1} + (-\mathrm{i})^{2k+1}}{(2k+1)!}z^{2k+1}\right) \\ &= \sum_{k=0}^{\infty}\frac{(-1)^k}{(2k)!}z^{2k}\end{aligned}$$

because

$$\mathrm{i}^{2k} + (-\mathrm{i})^{2k} = (-1)^k + (-1)^k = 2(-1)^k$$

and

$$\mathrm{i}^{2k+1} + (-\mathrm{i})^{2k+1} = \mathrm{i}(-1)^k - \mathrm{i}(-1)^k = 0.$$

So,

$$\cos z = \sum_{k=0}^{\infty}\frac{(-1)^k}{(2k)!}z^{2k}, \quad z \in \mathbb{C}. \tag{3.14}$$

In a similar fashion, we obtain

$$\sin z = \sum_{k=0}^{\infty}\frac{(-1)^k}{(2k+1)!}z^{2k+1}, \quad z \in \mathbb{C}. \tag{3.15}$$

Problem 3.17. Show that

$$\cosh z = \sum_{k=0}^{\infty}\frac{z^{2k}}{(2k)!}, \quad \sinh z = \sum_{k=0}^{\infty}\frac{z^{2k+1}}{(2k+1)!}, \quad z \in \mathbb{C}. \tag{3.16}$$

Problem 3.18. Show that:
1. $\cos z = \cosh(\mathrm{i}z)$ and $\sin z = -\mathrm{i}\sinh(\mathrm{i}z)$,
2. $|\mathrm{e}^z| = \mathrm{e}^x$ and $\overline{(\mathrm{e}^z)} = \mathrm{e}^{\bar{z}}$ for $z = x + \mathrm{i}y$,
3. $|\cos z| = \sqrt{\cosh^2 y - \sin^2 x}$,
4. $|\sin z| = \sqrt{\sinh^2 y + \sin^2 x}$,
5. $|\cos z|^2 + |\sin z|^2 = \cosh^2 y + \sinh^2 y = 1 + 2\sinh^2 y$,
6. $|\sin z| \le \cosh y$, $|\cos z| \le \cosh y$.

Problem 3.19. Calculate the derivative of the function $f(z) = \mathrm{e}^{z^2}$ using (3.10).

Problem 3.20. Calculate the derivative of the inverse function for $w = z^n$ using (3.11).

Definition 3.21. A function $f(z)$ is said to be:

1. *Analytic* in a domain $D \subset \mathbb{C}$ if for each $z \in D$ the derivative $f'(z)$ exists and is continuous in D. The set of all analytic functions in D will be denoted by $H(D)$.
2. Analytic at the point $z_0 \in D \subset \mathbb{C}$ if $f(z)$ is analytic in some neighborhood $U_\delta(z_0) \subset D$ of z_0.
3. Analytic at $z = \infty$ if $g(z) = f(1/z)$ is analytic at the point $z = 0$.

From this definition and the definition of the derivative, it follows that:
1. If $f_1, f_2 \in H(D)$, then
$$f_1 \pm f_2, f_1 \cdot f_2, \frac{f_1}{f_2} \in H(D),$$
too. In the last case, we assume $f_2 \neq 0$.
2. If $f \in H(D)$ and $g \in H(G)$, where $G = f(D)$, then $g \circ f \in H(D)$.

Example 3.22. The function
$$P_n(z) := a_0 + a_1 z + \cdots + a_n z^n,$$
where $a_0, a_1, \ldots, a_n \in \mathbb{C}$, $a_n \neq 0$ is called the *polynomial* of degree n. It is clear that $P_n(z) \in H(\mathbb{C})$ but it is not analytic at $z = \infty$ if $n \geq 1$.

If $P_n(z_0) = 0$, then z_0 is called the root of this polynomial and in that case $P_n(z) = (z - z_0)P_{n-1}(z)$, where P_{n-1} is a polynomial of degree $n - 1$.

Problem 3.23. Let
$$P_n(z) := a_0 + a_1 z + \cdots + a_n z^n$$
be a polynomial with the coefficients $a_0, a_1, \ldots, a_n$ that satisfy the conditions
$$a_0 \geq a_1 \geq \cdots \geq a_n \geq 0.$$
Prove that if $P_n(z)$ is not identically equal to 0 then all its roots z_0 satisfy the inequality $|z_0| > 1$.

Example 3.24. The function
$$R(z) = \frac{P_n(z)}{Q_m(z)}, \quad Q_m(z) \neq 0,$$
where P_n and Q_m are polynomials of degree n and m, respectively, is called the *rational function*. It follows that $R(z)$ is analytic everywhere in
$$\mathbb{C} \setminus \{z_0^{(1)}, \ldots, z_0^{(k)}\},$$
where

$$P_n(z_0^{(j)}) \neq 0 \quad \text{and} \quad Q_m(z_0^{(j)}) = 0.$$

Example 3.25. The tangent function is defined by

$$\tan z := \frac{\sin z}{\cos z}, \qquad \cos z \neq 0.$$

The zeros of $\cos z$ satisfy $e^{iz} + e^{-iz} = 0$. So, $e^{2iz} = -1$ or

$$e^{2ix} = -e^{2y}.$$

Comparing real and imaginary parts, we see that

$$\cos 2x = -e^{2y}, \quad \sin 2x = 0,$$

or

$$2x = \pi k, \quad k \in \mathbb{Z}, \quad \cos(\pi k) = -e^{2y}.$$

So,

$$x = \pi k/2, \quad (-1)^k = -e^{2y}, \quad k \in \mathbb{Z},$$

or

$$x = \pi k/2, \quad 1 = e^{2y}, \quad k = \pm 1, \pm 3, \ldots.$$

Thus,

$$y = 0, \quad x = \frac{\pi}{2}(2m+1), \quad m \in \mathbb{Z}.$$

We denote

$$z_m = -\frac{\pi}{2} + m\pi + i0, \quad m \in \mathbb{Z}.$$

Since $\sin z_m = \pm 1 \neq 0$, then $\tan z$ is analytic everywhere in $\mathbb{C}$ except at z_m. In this domain,

$$(\tan z)' = \frac{(\sin z)' \cos z - \sin z (\cos z)'}{\cos^2 z} = \frac{\cos^2 z + \sin^2}{\cos^2 z} = \frac{1}{\cos^2 z}.$$

Problem 3.26. Show that $\sin z = 0$ if and only if $z = \pi k + i0$, $k \in \mathbb{Z}$.

Example 3.27. The function

$$\tanh z := \frac{\sinh z}{\cosh z}, \qquad \cosh z \neq 0$$

is called a hyperbolic tangent function. The zeros of $\cosh z$ satisfy

$$e^{2x} = -e^{-i2y}$$

or $x = 0, y = \pi/2 + \pi m, m \in \mathbb{Z}$. Hence, $\tanh z$ is analytic everywhere in $\mathbb{C} \setminus \{z_m\}_{m=-\infty}^{\infty}$, where

$$z_m = 0 + i(\pi/2 + \pi m), \quad m \in \mathbb{Z},$$

and

$$(\tanh z)' = \frac{1}{\cosh^2 z}.$$

Problem 3.28. Show that $\sinh z = 0$ if and only if $z = 0 + i\pi k, k \in \mathbb{Z}$.

Example 3.29. Let us consider again (see Chapter 2) the exponential function

$$w = e^z$$

and let us try to find its inverse. Since

$$w = |w|e^{i \arg w}, \quad w \neq 0, \arg w \in (-\pi, \pi],$$

and $e^z = e^x e^{iy}$, then

$$|w| = e^x \quad \text{and} \quad \arg w = y + 2\pi k, \quad k \in \mathbb{Z}.$$

So, e^z is never equal to zero, and thus

$$x = \log|w| \quad \text{and} \quad y = \arg w + 2\pi k, \quad k \in \mathbb{Z}.$$

That is why

$$z = \log|w| + i \arg w + i2\pi k, \quad k \in \mathbb{Z}.$$

We see that the inverse of the function $w = e^z$ is not single-valued, namely we have infinitely many branches

$$z_k = \log|w| + i \arg w + i2\pi k, \quad k \in \mathbb{Z}.$$

The multivalued function is

$$z = \operatorname{Log} w := \log|w| + i \arg w + i2\pi k, \quad k \in \mathbb{Z}.$$

Its main branch is

$$z = \log w := \log|w| + \mathrm{i}\arg w, \quad \arg w \in (-\pi, \pi].$$

The *logarithmic function* $w = \operatorname{Log} z$, $z \neq 0$ is analytic everywhere in $\mathbb{C} \setminus \overline{\mathbb{R}_-}$ since $\arg z$ has a jump over negative real axis. Moreover,

$$(\operatorname{Log} z)' = \frac{1}{(\mathrm{e}^w)'} = \frac{1}{\mathrm{e}^w} = \frac{1}{z}.$$

Therefore, it is also continuous in $\mathbb{C} \setminus \overline{\mathbb{R}_-}$ (compare with Corollary 3.14).

Remark. Since

$$\operatorname{Log} z = \log z + \mathrm{i}2\pi k, \quad k \in \mathbb{Z},$$

then the derivative of $\operatorname{Log} z$ is the same ($(\operatorname{Log} z)' = 1/z$) for all branches of the multivalued logarithmic function.

Example 3.30. The function

$$z^{m/n} := \mathrm{e}^{\frac{m}{n}\operatorname{Log} z}, \quad z \neq 0$$

is called the *rational power function*. Since

$$\operatorname{Log} z = \log|z| + \mathrm{i}\arg z + \mathrm{i}2\pi k, \quad k \in \mathbb{Z},$$

then

$$z^{m/n} = \mathrm{e}^{\frac{m}{n}(\log|z| + \mathrm{i}\arg z + \mathrm{i}2\pi k)} = \mathrm{e}^{\frac{m}{n}\log|z|}\mathrm{e}^{\mathrm{i}\frac{m}{n}\arg z + \mathrm{i}\frac{2\pi km}{n}}.$$

The expression

$$\mathrm{e}^{\mathrm{i}\frac{2\pi km}{n}}$$

has different values only for $k = 0, 1, \ldots, n-1$ (we have assumed that m/n is an irreducible fraction). That is why we have n different branches of

$$z^{m/n} = |z|^{\frac{m}{n}}\mathrm{e}^{\mathrm{i}(\frac{m}{n}\arg z + \frac{2\pi km}{n})}, \quad k = 0, 1, \ldots, n-1.$$

Its derivative is

$$(z^{m/n})' = (\mathrm{e}^{\frac{m}{n}\operatorname{Log} z})' = \mathrm{e}^{\frac{m}{n}\operatorname{Log} z}\frac{m}{n}(\operatorname{Log} z)' = \frac{m}{n}z^{m/n-1}.$$

Example 3.31. The function

$$z^{\alpha} := \mathrm{e}^{\alpha \operatorname{Log} z}, \quad z \neq 0, \alpha \in \mathbb{R} \setminus \mathbb{Q}$$

is called the *irrational power function.* It is actually equal to

$$z^{\alpha} = e^{\alpha(\log|z| + i\arg z + i2\pi k)} = |z|^{\alpha} e^{i\alpha\arg z + i\alpha 2\pi k}, \quad k \in \mathbb{Z},$$

and we have infinitely many branches since α is not a rational number. Its derivative is

$$(z^{\alpha})' = (e^{\alpha \operatorname{Log} z})' = \alpha e^{\alpha \operatorname{Log} z}(\operatorname{Log} z)' = \alpha z^{\alpha-1}.$$

The definition of irrational power can be easily generalized for any complex power $\alpha = \alpha_1 + i\alpha_2$. Namely the function

$$z^{\alpha} := e^{\alpha \operatorname{Log} z}, \quad z \neq 0, \alpha \in \mathbb{C}$$

is called the *general power function.* As before, it is equal to

$$\begin{aligned} z^{\alpha} &= e^{\alpha \operatorname{Log} z} = e^{\alpha(\log|z| + i\arg z + i2\pi k)} = e^{(\alpha_1 + i\alpha_2)(\log|z| + i\arg z + i2\pi k)} \\ &= e^{\alpha_1 \log|z| - \alpha_2(\arg z + 2\pi k)} e^{i(\alpha_2 \log|z| + \alpha_1(\arg z + 2\pi k))} \end{aligned}$$

and we have infinitely many branches. The derivative is again $(z^{\alpha})' = \alpha z^{\alpha-1}$.

Example 3.32. Let us find the inverse of $w = \sin z$. From

$$w = \frac{e^{iz} - e^{-iz}}{2i},$$

we obtain

$$2iw = e^{iz} - \frac{1}{e^{iz}}$$

or $(e^{iz})^2 - 2iwe^{iz} - 1 = 0$. It implies

$$e^{iz} = iw + \sqrt{1 - w^2}.$$

So,

$$iz = \operatorname{Log}(iw + \sqrt{1 - w^2})$$

or

$$z = -i \operatorname{Log}(iw + \sqrt{1 - w^2}),$$

where Log denotes the multivalued function. The inverse of $\sin z$ is hence

$$z = z(w) = -i \operatorname{Log}(iw + \sqrt{1 - w^2}) =: \arcsin w,$$

and it has infinitely many branches. Its derivative is

$$\frac{\mathrm{d}}{\mathrm{d}w}\arcsin w = \frac{\mathrm{d}}{\mathrm{d}w}\frac{1}{\mathrm{i}}\operatorname{Log}(\mathrm{i}w + \sqrt{1-w^2}) = \frac{1}{\mathrm{i}}\frac{1}{\mathrm{i}w+\sqrt{w^2-1}}\left(\mathrm{i} - \frac{w}{\sqrt{1-w^2}}\right)$$
$$= \frac{1}{\mathrm{i}}\frac{1}{\mathrm{i}w+\sqrt{w^2-1}}\frac{\mathrm{i}\sqrt{1-w^2}-w}{\sqrt{1-w^2}} = \frac{1}{\sqrt{1-w^2}}, \quad w \neq \pm 1.$$

Problem 3.33. Show that:
1. $\sin(z_1+z_2) = \sin z_1 \cos z_2 + \cos z_1 \sin z_2$,
2. $\cos(z_1+z_2) = \cos z_1 \cos z_2 - \sin z_1 \sin z_2$,
3. $\sinh(z_1+z_2) = \sinh z_1 \cosh z_2 + \cosh z_1 \sinh z_2$,
4. $\cosh(z_1+z_2) = \cosh z_1 \cosh z_2 + \sinh z_1 \sinh z_2$.

Example 3.34. Consider the following function:

$$f(z) = \frac{\sin z}{\cosh y}, \quad z = x + \mathrm{i}y.$$

Based on Problems 3.18 and 3.33, we obtain that

$$f(z) = \sin x + \mathrm{i}\cos x \tanh y =: u(x,y) + \mathrm{i}v(x,y).$$

This representation shows that $f(z)$, as a function of two variables x and y, is infinitely many times differentiable since $u(x,y)$ and $v(x,y)$ are. However, this function is not analytic on the complex plane because Cauchy–Riemann conditions are not satisfied. Indeed,

$$\frac{\partial u(x,y)}{\partial x} = \cos x, \quad \frac{\partial v(x,y)}{\partial y} = \frac{\cos x}{\cosh^2 y},$$
$$\frac{\partial u(x,y)}{\partial y} = 0, \quad \frac{\partial v(x,y)}{\partial x} = -\sin x \tanh y.$$

The same is true about the function

$$f(z) = \frac{\cos z}{\sinh y}, \quad z = x + \mathrm{i}y, y \neq 0.$$

Problem 3.35. Show that, as sets,

$$\operatorname{Log}(z_1 \cdot z_2) = \operatorname{Log} z_1 + \operatorname{Log} z_2$$

for any $z_1 \neq 0$ and $z_2 \neq 0$.

Problem 3.36. Let function $f(z)$ be defined as follows:

$$f(z) := \sum_{n=0}^{\infty} \frac{\mathrm{e}^{\mathrm{i}2\pi nz}}{(a+n)^s}, \quad a > 0, s \in \mathbb{C},$$

where $(a+n)^s$ is understood as $e^{s\log(a+n)}$ with a real logarithm. Prove that $f(z)$ is analytic in $\{z : \operatorname{Im} z > 0\}$.

We will finish this chapter by the following a very useful rule, which is called *L'Hôpital's rule.*

Proposition 3.37. *Suppose f and g are analytic at z_0. If $f(z_0) = g(z_0) = 0$ but $g'(z_0) \neq 0$, then*

$$\lim_{z\to z_0} \frac{f(z)}{g(z)} = \frac{f'(z_0)}{g'(z_0)}.$$

Proof. Because $g'(z_0) \neq 0$, then g is not identically equal to zero and there is a neighborhood $U_\delta(z_0)$ in which $g'(z) \neq 0$. Therefore, the quotient

$$\frac{f(z)}{g(z)} = \frac{f(z) - f(z_0)}{g(z) - g(z_0)}$$

is defined for all $z \in U_\delta(z_0)$ and

$$\lim_{z\to z_0} \frac{f(z)}{g(z)} = \lim_{z\to z_0} \frac{f(z) - f(z_0)}{g(z) - g(z_0)} = \lim_{z\to z_0} \frac{\frac{f(z)-f(z_0)}{z-z_0}}{\frac{g(z)-g(z_0)}{z-z_0}} = \frac{f'(z_0)}{g'(z_0)}.$$ □

Problem 3.38. Use L'Hôpital's rule to evaluate the limits

$$\lim_{z\to 0} \frac{\log^2(1+z)}{z^2} \quad \text{and} \quad \lim_{z\to 0} \frac{1-\cos z}{\sin^2 z}.$$

Example 3.39. Let us show that

$$\lim_{z\to 0} z^\epsilon \operatorname{Log} z = 0, \quad \epsilon > 0,$$

where we consider any fixed branches of multivalued functions z^ϵ and $\operatorname{Log} z$ (see Examples 3.31 and 3.29, respectively), but it cannot be calculated using L'Hôpital's rule. Indeed, since we have ($r = |z|$),

$$z^\epsilon \operatorname{Log} z = e^{\epsilon(\log r + i \operatorname{Arg} z)}(\log r + i \operatorname{Arg} z),$$

where $r > 0$, $\operatorname{Arg} z \in (-\pi + 2\pi k, \pi + 2\pi k]$, $k \in \mathbb{Z}$, then using the fact that for the real logarithm we have

$$\lim_{r\to +0} r^\epsilon \log r = 0, \quad \epsilon > 0,$$

and the fact that $\operatorname{Arg} z$ is bounded, we obtain the needed limit for complex $z \to 0$. If we will try to use the L'Hôpital's rule, we would proceed as

$$\lim_{z\to 0} z^{\epsilon}\log z = \lim_{z\to 0}\frac{z^{\epsilon}}{\frac{1}{\log z}} = -\epsilon \lim_{z\to 0}\frac{z^{\epsilon}}{\frac{1}{\log^2 z}} = \frac{\epsilon^2}{2}\lim_{z\to 0}\frac{z^{\epsilon}}{\frac{1}{\log^3 z}} = \cdots$$

and (as we can see) this procedure does not give us the needed result. It means that L'Hôpital's rule is only sufficient but not necessary for the existence of the limit.

By the same manner, we can obtain that

$$\lim_{z\to\infty}\frac{\operatorname{Log} z}{z^{\epsilon}} = 0, \quad \epsilon > 0.$$

Problem 3.40. Let function $F(z), z \in \mathbb{C}$ be formally defined by the integral

$$F(z) := \int_0^{\infty} \frac{1}{(e^x - z)x^{\alpha}}\,dx, \quad -1 \le \alpha < 0.$$

Prove that:
1. $F(z)$ is analytic in an open unit disk $\{z : |z| < 1\}$.
2. $F(z)$ is differentiable for all z, $|z| = 1, z \neq 1$.
3. $F(z)$ is not differentiable for $z = 1$; however, the integral is convergent, i. e., $F(1)$ is well-defined.

4 Integration of functions of complex variable (curve integration)

Let γ be a smooth Jordan curve, i. e.,

$$\gamma : z = z(t), \quad t \in [a, b].$$

Assuming that $f(z)$ is a continuous function, we may define two types of *curve integrals* along γ as

$$\begin{aligned}
\int_\gamma f(z)\mathrm{d}z &:= \int_a^b f(z(t))z'(t)\mathrm{d}t \\
&= \int_a^b (u(z(t)) + \mathrm{i}v(z(t)))(x'(t) + \mathrm{i}y'(t))\mathrm{d}t \\
&= \int_a^b [u(x(t), y(t))x'(t) - v(x(t), y(t))y'(t)]\mathrm{d}t \\
&\quad + \mathrm{i}\int_a^b [v(x(t), y(t))x'(t) + u(x(t), y(t))y'(t)]\mathrm{d}t \\
&= \int_\gamma (u(x, y)\mathrm{d}x - v(x, y)\mathrm{d}y) + \mathrm{i}\int_\gamma (v(x, y)\mathrm{d}x + u(x, y)\mathrm{d}y)
\end{aligned} \tag{4.1}$$

and

$$\begin{aligned}
\int_\gamma f(z)|\mathrm{d}z| &:= \int_a^b f(z(t))\sqrt{(x'(t))^2 + (y'(t))^2}\mathrm{d}t = \int_a^b f(z(t))|z'(t)|\mathrm{d}t \\
&= \int_a^b u(x(t), y(t))\sqrt{(x'(t))^2 + (y'(t))^2}\mathrm{d}t \\
&\quad + \mathrm{i}\int_a^b v(x(t), y(t))\sqrt{(x'(t))^2 + (y'(t))^2}\mathrm{d}t.
\end{aligned} \tag{4.2}$$

The first integral (4.1) is called a *curve integral of the second kind*, and the second integral (4.2) is called the *curve integral of the first kind*.

Example 4.1. Let $f(z) = z$.

https://doi.org/10.1515/9783111632278-004

1. Let $\gamma : z(t) = t + \mathrm{i}t^2$ for $t \in [0,1]$. Then

$$\int_\gamma f(z)\mathrm{d}z = \int_\gamma z\mathrm{d}z = \int_0^1 (t + \mathrm{i}t^2)(1 + 2\mathrm{i}t)\mathrm{d}t = \int_0^1 (t + 3\mathrm{i}t^2 - 2t^3)\mathrm{d}t$$
$$= \left(\frac{t^2}{2} + 3\mathrm{i}\frac{t^3}{3} - 2\frac{t^4}{4}\right)\Bigg|_0^1 = \frac{1}{2} + \mathrm{i} - \frac{1}{2} = \mathrm{i}.$$

2. Let $\gamma : z(t) = \alpha + \mathrm{i}t$ for $t \in [0,1]$. Then

$$\int_\gamma f(z)\mathrm{d}z = \int_\gamma z\mathrm{d}z = \int_0^1 (\alpha + \mathrm{i}t)\mathrm{i}\mathrm{d}t = \mathrm{i}\alpha - \frac{1}{2}.$$

3. Let $\gamma : z(t) = t^2 + \mathrm{i}\beta$ for $t \in [0,1]$. Then

$$\int_\gamma f(z)\mathrm{d}z = \int_\gamma z\mathrm{d}z = \int_0^1 (t^2 + \mathrm{i}\beta)2t\mathrm{d}t = \left(2\frac{t^4}{4} + \mathrm{i}\beta t^2\right)\Bigg|_0^1 = \frac{1}{2} + \mathrm{i}\beta.$$

Remark. It can be easily checked that in all integrations in Example 4.1 the final result depends only on the value of the function $z^2/2$ at the ends of the curve γ. Namely the result is

$$\frac{(z(1))^2}{2} - \frac{(z(0))^2}{2}.$$

We will make this more precise later.

Example 4.2. Let $f(z) = z$.

1. Let $\gamma : z(t) = \alpha + \mathrm{i}t$ for $t \in [0,1]$. Then

$$\int_\gamma f(z)|\mathrm{d}z| = \int_\gamma z|\mathrm{d}z| = \int_0^1 (\alpha + \mathrm{i}t)\mathrm{d}t = \alpha + \frac{\mathrm{i}}{2}.$$

2. Let $\gamma : z(t) = t^2 + \mathrm{i}\beta$ for $t \in [0,1]$. Then

$$\int_\gamma f(z)|\mathrm{d}z| = \int_\gamma z|\mathrm{d}z| = \int_0^1 (t^2 + \mathrm{i}\beta)2t\mathrm{d}t = \left(2\frac{t^4}{4} + \mathrm{i}\beta t^2\right)\Bigg|_0^1 = \frac{1}{2} + \mathrm{i}\beta.$$

Example 4.3. Let $\gamma : z(t) = a + re^{\mathrm{i}t}$, $t \in (-\pi, \pi]$. Then

1.

$$
\begin{aligned}
\int_\gamma (z-a)^n \mathrm{d}z &= \int_{-\pi}^{\pi} (r\cos t + r\mathrm{i}\sin t)^n r(-\sin t + \mathrm{i}\cos t)\mathrm{d}t \\
&= r^{n+1} \int_{-\pi}^{\pi} (\cos(nt) + \mathrm{i}\sin(nt))(-\sin t + \mathrm{i}\cos t)\mathrm{d}t \\
&= r^{n+1} \int_{-\pi}^{\pi} [-\cos(nt)\sin t - \sin(nt)\cos t]\mathrm{d}t \\
&\quad + \mathrm{i}r^{n+1} \int_{-\pi}^{\pi} [\cos(nt)\cos t - \sin(nt)\sin t]\mathrm{d}t \\
&= -r^{n+1} \int_{-\pi}^{\pi} \sin(n+1)t\mathrm{d}t + \mathrm{i}r^{n+1} \int_{-\pi}^{\pi} \cos(n+1)t\mathrm{d}t \\
&= r^{n+1}\left(\frac{\cos(n+1)t}{n+1} + \mathrm{i}\frac{\sin(n+1)t}{n+1}\right)\Bigg|_{-\pi}^{\pi} \\
&= \begin{cases} 0, & n \neq -1, \\ 2\pi\mathrm{i}, & n = -1. \end{cases}
\end{aligned}
$$

2.

$$
\begin{aligned}
\int_\gamma (z-a)^n |\mathrm{d}z| &= \int_{-\pi}^{\pi} r^n \mathrm{e}^{\mathrm{i}nt} r\mathrm{d}t \\
&= r^{n+1}\left(\int_{-\pi}^{\pi} \cos(nt)\mathrm{d}t + \mathrm{i}\int_{-\pi}^{\pi} \sin(nt)\mathrm{d}t\right) \\
&= \begin{cases} 0, & n \neq 0, \\ 2\pi r, & n = 0. \end{cases}
\end{aligned}
$$

Problem 4.4. Let $f(z) = \bar{z}$. Evaluate

$$\int_\gamma \bar{z}\mathrm{d}z,$$

where:

1. $\gamma : z(t) = t + \mathrm{i}t^2, t \in [0,1]$,
2. $\gamma : z(t) = \alpha + \mathrm{i}t, t \in [0,1]$,
3. $\gamma : z(t) = t^2 + \mathrm{i}\beta, t \in [0,1]$,
4. $\gamma = \gamma_1 \cup \gamma_2$, where $\gamma_1 : z(t) = t + \mathrm{i}t^2$ and $\gamma_2 : z(t) = (1-t) + \mathrm{i}(1-t)$ for $t \in [0,1]$.

If γ is a piecewise smooth Jordan curve (see Definition 2.1, part (2)), then the integrals along this curve are defined as

$$\begin{aligned}\int_\gamma f(z)\mathrm{d}z &:= \sum_{j=0}^{n-1}\int_{t_j}^{t_{j+1}} f(z(t))z'(t)\mathrm{d}t,\\ \int_\gamma f(z)|\mathrm{d}z| &:= \sum_{j=0}^{n-1}\int_{t_j}^{t_{j+1}} f(z(t))|z'(t)|\mathrm{d}t.\end{aligned} \tag{4.3}$$

Using the properties of the Riemann integral, we obtain that:

1. $$\int_\gamma (c_1 f_1(z) + c_2 f_2(z))\mathrm{d}z = c_1\int_\gamma f_1(z)\mathrm{d}z + c_2\int_\gamma f_2(z)\mathrm{d}z.$$

2. $$\int_{\gamma_1\cup\gamma_2} f(z)\mathrm{d}z = \int_{\gamma_1} f(z)\mathrm{d}z + \int_{\gamma_2} f(z)\mathrm{d}z.$$

3. $$\int_\gamma (c_1 f_1(z) + c_2 f_2(z))|\mathrm{d}z| = c_1\int_\gamma f_1(z)|\mathrm{d}z| + c_2\int_\gamma f_2(z)|\mathrm{d}z|.$$

4. $$\int_{\gamma_1\cup\gamma_2} f(z)|\mathrm{d}z| = \int_{\gamma_1} f(z)|\mathrm{d}z| + \int_{\gamma_2} f(z)|\mathrm{d}z|.$$

If $\gamma : z(t), t \in [a,b]$ is a piecewise smooth Jordan curve, we can run the curve *backwards* as follows. Let us consider the curve

$$\gamma_1 : \tilde{z} = \tilde{z}(s) = z(a+b-s), \quad s \in [a,b].$$

The curve γ_1 is denoted by $-\gamma$, i. e., $\gamma_1 = -\gamma$ and

$$\begin{aligned}\int_{\gamma_1} f(\tilde{z})\mathrm{d}\tilde{z} &= \int_a^b f(\tilde{z}(s))\tilde{z}'(s)\mathrm{d}s = -\int_a^b f(z(a+b-s))z'(a+b-s)\mathrm{d}s\\ &= \int_b^a f(z(t))z'(t)\mathrm{d}t = -\int_a^b f(z(t))z'(t)\mathrm{d}t = -\int_\gamma f(z)\mathrm{d}z,\end{aligned}$$

i. e.,

$$\int_{-\gamma} f(z)\mathrm{d}z = -\int_{\gamma} f(z)\mathrm{d}z.$$

Definition 4.5. A function $f(z)$ is said to have a *primitive* $F(z)$ on $D \subset \mathbb{C}$ if $F(z)$ is differentiable on D and $F'(z) = f(z)$ everywhere on D.

Theorem 4.6. *If a continuous function $f(z)$ has a primitive $F(z)$ on $D \subset \mathbb{C}$, then for any smooth Jordan curve $\gamma : z(t), t \in [a, b]$ in D, it holds that*

$$\int_{\gamma} f(z)\mathrm{d}z = F(z(b)) - F(z(a)). \tag{4.4}$$

Thus, this integral does not depend on γ, but on the endpoints of γ. In particular, if γ is closed and f has a primitive, then

$$\int_{\gamma} f(z)\mathrm{d}z = 0. \tag{4.5}$$

Proof. Let $\gamma : z(t), t \in [a, b]$ be a smooth Jordan curve. Then for any continuous function $f(z)$, the composition $f(z(t))$ and the product $f(z(t))z'(t)$ are continuous and

$$\int_{\gamma} f(z)\mathrm{d}z = \int_a^b f(z(t))z'(t)\mathrm{d}t.$$

But $f(z(t))z'(t) = (F(z(t)))'$, where F is a primitive of f. Hence,

$$\int_a^b f(z(t))z'(t)\mathrm{d}t = \int_a^b (F(z(t)))'\mathrm{d}t = F(z(t))|_a^b = F(z(b)) - F(z(a)).$$

This proves the theorem. □

Corollary 4.7. *If $\gamma : z(t), t \in [a, b]$ is a piecewise smooth Jordan curve, then*

$$\int_{\gamma} f(z)\mathrm{d}z = F(z(b)) - F(z(a)),$$

too, where F is a primitive of f in the domain D.

Proof. By (4.3), we have

$$\int_{\gamma} f(z)\mathrm{d}z = \sum_{j=0}^{n-1} \int_{t_j}^{t_{j+1}} f(z(t))z'(t)\mathrm{d}t = \sum_{j=0}^{n-1} (F(z(t_{j+1})) - F(z(t_j)))$$
$$= F(z(t_n)) - F(z(t_0)) = F(z(b)) - F(z(a))$$

and this proves the claim. □

Theorem 4.8. *Let $\gamma : z(t), t \in [a, b]$ be a piecewise smooth Jordan curve and let f be a continuous function. Then*

$$\left|\int_{\gamma} f(z)\mathrm{d}z\right| \le \int_{\gamma} |f(z)||\mathrm{d}z| \le \max_{z\in\gamma} |f(z)| L(\gamma), \tag{4.6}$$

where $L(\gamma) = \int_{\gamma} |\mathrm{d}z|$ denotes the length *of γ.*

Proof. We have

$$\int_{\gamma} f(z)\mathrm{d}z = \int_{a}^{b} f(z(t))z'(t)\mathrm{d}t.$$

Since this Riemann integral can be understood as limit of integral sums, then we obtain

$$\left|\int_{a}^{b} f(z(t))z'(t)\mathrm{d}t\right| = \left|\lim_{\Delta t\to 0} \sum_{j=1}^{n} f(z(t_j^*))z'(t_j^*)\Delta t_j\right|$$
$$\le \lim_{\Delta t\to 0} \sum_{j=1}^{n} |f(z(t_j^*))||z'(t_j^*)|\Delta t_j$$
$$= \int_{a}^{b} |f(z(t))||z'(t)|\mathrm{d}t = \int_{\gamma} |f(z)||\mathrm{d}z|$$
$$\le \max_{z\in\gamma} |f(z)| \int_{\gamma} 1|\mathrm{d}z| = \max_{z\in\gamma} |f(z)| L(\gamma).$$ □

Theorem 4.9 (Change of variable). *Let $g(z)$ be analytic in the domain $D \subset \mathbb{C}$ and $f(D) \subset D'$. Suppose that $\gamma : z(t), t \in [a, b]$ is a piecewise smooth Jordan curve in D and $\gamma' = g(\gamma)$, $\tilde{z}(t) = g(z(t))$, $t \in [a, b]$ is the transformed curve in D'. Then for all continuous functions f on D, we have*

$$\int_{\gamma} f(g(z))g'(z)\mathrm{d}z = \int_{\gamma'} f(w)\mathrm{d}w. \tag{4.7}$$

Proof. We know that

$$\int_{\gamma} f(g(z))g'(z)\mathrm{d}z = \int_{a}^{b} f(g(z(t)))g'(z(t))z'(t)\mathrm{d}t = \int_{a}^{b} f(g(z(t)))(g(z(t)))'\mathrm{d}t$$
$$= \int_{a}^{b} f(\tilde{z}(t))\tilde{z}'(t)\mathrm{d}t = \int_{\gamma'} f(w)\mathrm{d}w. \qquad \square$$

Example 4.10. Let $\gamma : z(t) = t + \mathrm{i}t^2$, $t \in [0,1]$. Using $g(z) = z^2$, we get

$$\int_{\gamma} \sin(z^2)z\mathrm{d}z = \frac{1}{2}\int_{\gamma'} \sin w\mathrm{d}w = -\frac{1}{2}\cos w\Big|_{w(0)}^{w(1)}$$
$$= -\frac{1}{2}\cos w\Big|_{0}^{2\mathrm{i}} = \frac{1}{2}(1 - \cos(2\mathrm{i})) = \frac{1}{2}(1 - \cosh 2).$$

Here, we have used the notation $w(t) = g(z(t))$.

Problem 4.11. Let $\gamma : z(t) = a + \mathrm{i}t^2$, $t \in [0,1]$. Evaluate

$$\int_{\gamma} \mathrm{e}^{\sin z}\cos z\mathrm{d}z.$$

Problem 4.12. Let $\gamma : z(t) = 1 + \mathrm{i}t$, $t \in [0,1]$. Evaluate

$$\int_{\gamma} \frac{\log z}{z}\mathrm{d}z.$$

Problem 4.13. Let function $f(x)$ be continuous on the interval $[a,b]$. Show that the functions

$$F(z) := \int_{a}^{b} f(t)\sin(zt)\mathrm{d}t, \quad G(z) := \int_{a}^{b} f(t)\cos(zt)\mathrm{d}t$$

are analytic for all $z \in \mathbb{C}$.

Problem 4.14. Let function $f(x)$ be continuous on the interval $[a,b]$. Show that the function

$$H(z) := \int_{a}^{b} \frac{f(t)}{t - z}\mathrm{d}t$$

is analytic for all $z \in \mathbb{C} \setminus [a,b]$ and find $H'(z)$ for all such z.

5 Cauchy theorem and Cauchy integral formulae

Definition 5.1. A bounded domain $D \subset \mathbb{C}$ is called *simply connected* if for any closed Jordan curve $\gamma \subset D$ the internal domain ($\operatorname{int}\gamma$) belongs to D. Otherwise, D is called *multiply connected*. The number of connected components of the boundary is said to be the *connected order* of D.

Theorem 5.2 (Cauchy theorem). *Let D be a bounded simply connected domain with the boundary ∂D, which is a piecewise smooth closed Jordan curve γ. Then for any function $f \in H(D)$, which is continuous in $\overline{D}$, we have*

$$\int_{\gamma} f(z)\mathrm{d}z = 0.$$

Proof. Since $f \in C(\overline{D})$, then $\int_{\gamma} f(z)\mathrm{d}z$ is well-defined and it is equal to

$$\int_{\gamma} f(z)\mathrm{d}z = \int_{\gamma} (u + \mathrm{i}v)(\mathrm{d}x + \mathrm{i}\mathrm{d}y) = \int_{\gamma} u\mathrm{d}x - v\mathrm{d}y + \mathrm{i}\int_{\gamma} v\mathrm{d}x + u\mathrm{d}y.$$

Using now Green's theorem (or Stokes's theorem), we obtain that the integrals on the right-hand side are equal to

$$\iint_{D} \left(-\frac{\partial v}{\partial x} - \frac{\partial u}{\partial y}\right)\mathrm{d}x\mathrm{d}y + \mathrm{i}\iint_{D} \left(\frac{\partial u}{\partial x} - \frac{\partial v}{\partial y}\right)\mathrm{d}x\mathrm{d}y = 0$$

because of Cauchy–Riemann conditions. Thus, the theorem is proved. □

Remark. If the domain D is simply connected, then the Cauchy theorem holds not only for the boundary ∂D, but also for any closed piecewise smooth Jordan curve γ such that $\gamma \subset D$.

Corollary 5.3. *Let D be a bounded $(n+1)$-connected domain such that $\partial D = \bigcup_{j=0}^{n} \gamma_j$, where γ_j are closed piecewise smooth Jordan curves, $\operatorname{int}\gamma_j \cap \operatorname{int}\gamma_k = \emptyset$, $k \neq j$, and $\gamma_1, \ldots, \gamma_n \subset \operatorname{int}\gamma_0$. If $f \in H(D) \cap C(\overline{D})$, then*

$$\int_{\partial D} f(z)\mathrm{d}z = \int_{\gamma_0} f(z)\mathrm{d}z - \sum_{j=1}^{n} \int_{\gamma_j} f(z)\mathrm{d}z = 0.$$

Proof. By the conditions of this corollary, domain D has the form depicted in Figure 5.1.

Let us join γ_j, $j = 1, 2, \ldots, n$ with γ_0 by the smooth Jordan curves Γ_j such that any Γ_j, $j = 1, 2, \ldots, n$ is passed twice in opposite directions. In this case, we obtain simply

https://doi.org/10.1515/9783111632278-005

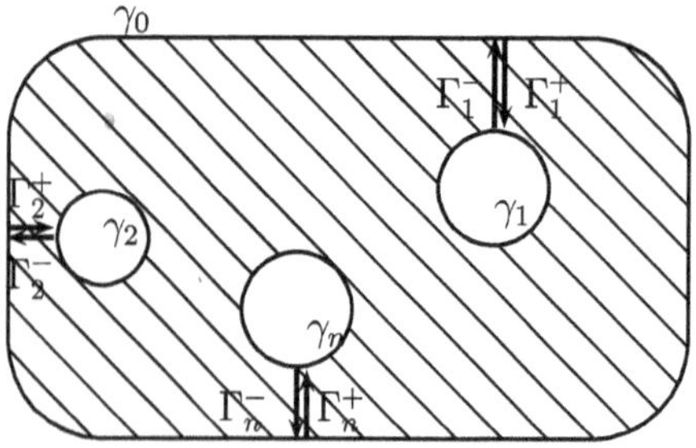

Figure 5.1: Domain *D*.

connected domain D_1 with the boundary

$$\partial D_1 = \left(\bigcup_{j=0}^{n} \gamma_j\right) \cup \left(\bigcup_{j=1}^{n} \Gamma_j^{\pm}\right).$$

Thus, applying the Cauchy theorem to domain D_1, we obtain

$$\begin{aligned} 0 = \int_{\partial D_1} f(z)\mathrm{d}z &= \int_{\partial D} f(z)\mathrm{d}z + \sum_{j=1}^{n} \int_{\Gamma_j^+} f(z)\mathrm{d}z + \sum_{j=1}^{n} \int_{\Gamma_j^-} f(z)\mathrm{d}z \\ &= \int_{\gamma} f(z)\mathrm{d}z - \sum_{j=1}^{n} \int_{\gamma_j} f(z)\mathrm{d}z. \end{aligned}$$

Here, we have used the fact that

$$\int_{\Gamma_j^+} f(z)\mathrm{d}z + \int_{\Gamma_j^-} f(z)\mathrm{d}z = 0$$

and that the positive direction of integration is the direction in which the internal domain is on the left. □

If domain *D* is multiply connected, then the Cauchy theorem does not hold for arbitrary closed piecewise smooth Jordan curve. In this case, it is necessary to integrate over the whole boundary of *D*. Indeed, let

$$D = \{z : 1 < |z| < 3\}$$

and $\gamma = \{z : |z| = 2\}$. Then $\gamma \subset D$ but

$$\int_{\gamma} \frac{1}{z}\mathrm{d}z = 2\pi\mathrm{i}.$$

However, the integral over the whole boundary ∂D is equal to zero. Indeed,

$$\int_{\partial D} \frac{1}{z}\mathrm{d}z = \int_{|z|=3} \frac{1}{z}\mathrm{d}z - \int_{|z|=1} \frac{1}{z}\mathrm{d}z = 2\pi\mathrm{i} - 2\pi\mathrm{i} = 0.$$

Corollary 5.4. *Let D be a domain, which satisfies either the conditions of Theorem 5.2 or Corollary 5.3. If $f \in H(D) \cap C(\overline{D})$ except the points $z_1, \dots, z_m \in D$ with*

$$\lim_{z \to z_k} (z - z_k) f(z) = 0, \quad k = 1, 2, \dots, m,$$

then

$$\int_{\partial D} f(z)\mathrm{d}z = 0.$$

Proof. For simplicity and without loss of generality, we assume that $m = 1$. Then for any $\varepsilon > 0$, there is $\delta(z_1, \varepsilon) > 0$ such that for all z with $0 < |z - z_1| < \delta$ it follows that

$$|z - z_1||f(z)| < \varepsilon.$$

Let $D_1 := D \setminus \{z : |z - z_1| \le \delta\}$ assuming that $\delta > 0$ is so small that $\{z : |z - z_1| \le \delta\} \subset D$. Then for the domain D_1, the Cauchy theorem holds and, therefore,

$$0 = \int_{\partial D_1} f(z)\mathrm{d}z = \int_{\partial D} f(z)\mathrm{d}z - \int_{|z-z_1|=\delta} f(z)\mathrm{d}z.$$

But

$$\begin{aligned}\left| \int_{|z-z_1|=\delta} f(z)\mathrm{d}z \right| &\le \int_{|z-z_1|=\delta} |f(z)||\mathrm{d}z| \\ &= \int_{|z-z_1|=\delta} |z - z_1||f(z)| \frac{|\mathrm{d}z|}{|z - z_1|} < \varepsilon \frac{1}{\delta} \int_{|z-z_1|=\delta} |\mathrm{d}z| = 2\pi\varepsilon.\end{aligned}$$

Since $\varepsilon > 0$ is arbitrary, then we may let $\varepsilon \to 0$ and obtain

$$\begin{aligned} 0 &= \lim_{\varepsilon \to 0} \int_{\partial D_1} f(z)\mathrm{d}z = \lim_{\varepsilon \to 0} \Big(\int_{\partial D} f(z)\mathrm{d}z - \int_{|z-z_1|=\delta} f(z)\mathrm{d}z \Big) \\ &= \int_{\partial D} f(z)\mathrm{d}z - \lim_{\varepsilon \to 0} \int_{|z-z_1|=\delta} f(z)\mathrm{d}z = \int_{\partial D} f(z)\mathrm{d}z. \end{aligned}$$ □

Example 5.5. Let us evaluate $\int_{|z|=1} \log z\mathrm{d}z$. Using the parametrization $z = \mathrm{e}^{\mathrm{i}\theta}$, $\theta \in [-\pi, \pi]$, and integration by parts, we obtain

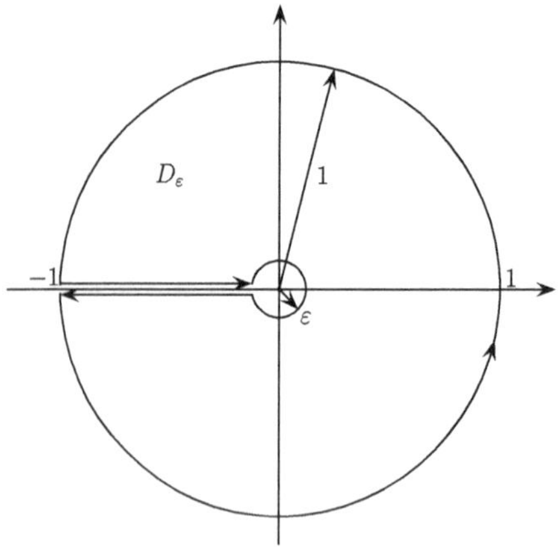

Figure 5.2: Domain D_ε.

$$\begin{aligned}\int_{|z|=1} \log z \mathrm{d}z &= \int_{-\pi}^{\pi} (\log|z| + \mathrm{i}\theta)\mathrm{i}e^{\mathrm{i}\theta}\mathrm{d}\theta = \mathrm{i}^2 \int_{-\pi}^{\pi} \theta e^{\mathrm{i}\theta}\mathrm{d}\theta \\ &= -\int_{-\pi}^{\pi} \theta\cos\theta \mathrm{d}\theta - \mathrm{i}\int_{-\pi}^{\pi} \theta \sin\theta \mathrm{d}\theta = -2\mathrm{i}\int_{0}^{\pi} \theta\sin\theta \mathrm{d}\theta \\ &= 2\mathrm{i}\left(\theta\cos\theta\Big|_0^\pi - \int_0^\pi \cos\theta \mathrm{d}\theta\right) = -2\pi\mathrm{i}.\end{aligned}$$

It shows that the Cauchy theorem does not hold in this case. But we know that $\log z$ is analytic and has a removable singularity at $z = 0$ (see Definition 10.9). This phenomenon can be explained as follows: $\log z$ has a jump $2\pi\mathrm{i}$ over the negative real line, i. e., it is not continuous in the unit disk and, therefore, it is not analytic. Even more is true, it is not univalent there. In order to eliminate this problem, we proceed as follows. Let us consider the domain D_ε shown in Figure 5.2 for $\varepsilon > 0$ small enough.

In this domain D_ε, the function $\log z$ is not only analytic but also univalent. Applying the Cauchy theorem (see Theorem 5.2), we obtain

$$\begin{aligned}0 &= \int_{\partial D_\varepsilon} \log z \mathrm{d}z \\ &= \int_{-\pi}^{\pi} (\log 1 + \mathrm{i}\theta)\mathrm{i}e^{\mathrm{i}\theta}\mathrm{d}\theta + \int_{-1}^{-\varepsilon} (\log|x| + \mathrm{i}\pi)\mathrm{d}x \\ &\quad - \int_{-\pi}^{\pi} (\log\varepsilon + \mathrm{i}\theta)\mathrm{i}\varepsilon e^{\mathrm{i}\theta}\mathrm{d}\theta + \int_{-\varepsilon}^{-1} (\log|x| - \mathrm{i}\pi)\mathrm{d}x\end{aligned}$$

$$= -2\pi\mathrm{i} + \int_{-1}^{-\varepsilon} \log|x|\mathrm{d}x + \mathrm{i}\pi(1-\varepsilon) + 2\pi\mathrm{i}\varepsilon$$
$$+ \int_{-\varepsilon}^{-1} \log|x|\mathrm{d}x - \mathrm{i}\pi(-1+\varepsilon) = 0$$

for any $\varepsilon > 0$. Taking $\varepsilon \to +0$, we obtain that

$$\int_{\partial D} \log z\mathrm{d}z := \lim_{\varepsilon\to+0} \int_{\partial D_\varepsilon} \log z\mathrm{d}z = 0,$$

where ∂D is the unit circle with a cut along the negative real line.

Example 5.6. Let γ be a piecewise smooth closed Jordan curve and $z_0 \in \operatorname{int}\gamma$. Then

$$\int_\gamma \frac{1}{z-z_0}\mathrm{d}z = 2\pi\mathrm{i}.$$

Indeed, if we consider the domain (see Figure 5.3)

$$D_1 := \operatorname{int}\gamma \setminus \{z : |z-z_0| \le \delta\},$$

then by Corollary 5.3, we have

$$0 = \int_\gamma \frac{\mathrm{d}z}{z-z_0} - \int_{|z-z_0|=\delta} \frac{\mathrm{d}z}{z-z_0}.$$

But

$$\int_{|z-z_0|=\delta} \frac{\mathrm{d}z}{z-z_0} = \int_{-\pi}^{\pi} \frac{\mathrm{i}\delta e^{\mathrm{i}\theta}\mathrm{d}\theta}{\delta e^{\mathrm{i}\theta}} = 2\pi\mathrm{i}.$$

This example can be generalized to the multiply connected domain D also, i. e., if $z_0 \in D$ then

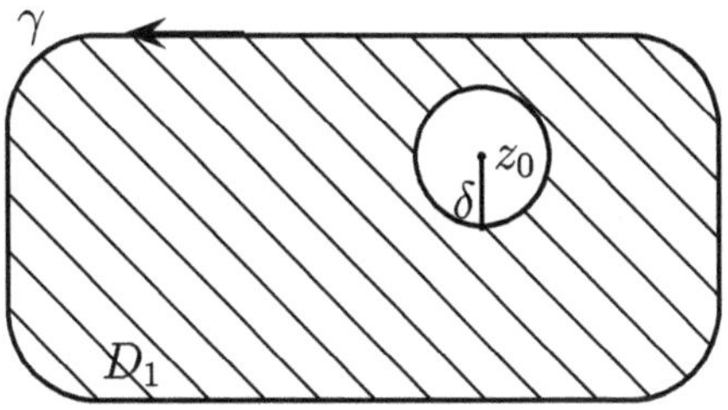

Figure 5.3: Domain D_1.

$$\int_{\partial D} \frac{1}{z - z_0} \mathrm{d}z = 2\pi \mathrm{i}.$$

There is a similar but more delicate example of application of the Cauchy theorem.

Example 5.7. Let γ be a piecewise smooth closed Jordan curve such that $0 \in \operatorname{int}\gamma$ and $\overline{\operatorname{int}\gamma} \subset \{z : |z| < 2\pi\}$. Then

$$\int_{\gamma} \frac{1}{\mathrm{e}^z - 1} \mathrm{d}z = 2\pi \mathrm{i}.$$

Indeed, if we consider for small enough $\epsilon > 0$, the domain

$$D := \operatorname{int}\gamma \setminus \{z : |z| \leq \epsilon\}$$

then the function $\frac{1}{\mathrm{e}^z - 1}$ will be analytic in D since all its singular points $z_k = \mathrm{i}2\pi k$, $k \in \mathbb{Z}$ are located outside of D. Applying now the Cauchy theorem (see Corollary 5.3), we obtain

$$0 = \int_{\gamma} \frac{\mathrm{d}z}{\mathrm{e}^z - 1} - \int_{|z|=\epsilon} \frac{\mathrm{d}z}{\mathrm{e}^z - 1}.$$

But

$$\int_{|z|=\epsilon} \frac{\mathrm{d}z}{\mathrm{e}^z - 1} = \int_{-\pi}^{\pi} \frac{\mathrm{i}\epsilon \mathrm{e}^{\mathrm{i}\theta}}{\mathrm{e}^{\epsilon \mathrm{e}^{\mathrm{i}\theta}} - 1} \mathrm{d}\theta.$$

Using the asymptotic of the real exponential function for small argument and Euler's formula, we have

$$\mathrm{e}^{\epsilon \mathrm{e}^{\mathrm{i}\theta}} - 1 = (1 + \epsilon \cos\theta + O(\epsilon^2))(1 + \mathrm{i}\epsilon \sin\theta + O(\epsilon^2)) - 1 = \epsilon \mathrm{e}^{\mathrm{i}\theta} + O(\epsilon^2)$$

uniformly in $\theta \in [-\pi, \pi]$. Thus, the latter integral will be equal to

$$\int_{-\pi}^{\pi} \frac{\mathrm{i}\epsilon \mathrm{e}^{\mathrm{i}\theta}}{\mathrm{e}^{\epsilon \mathrm{e}^{\mathrm{i}\theta}} - 1} \mathrm{d}\theta = \int_{-\pi}^{\pi} \frac{\mathrm{i}\epsilon \mathrm{e}^{\mathrm{i}\theta}}{\epsilon \mathrm{e}^{\mathrm{i}\theta} + O(\epsilon^2)} \mathrm{d}\theta \to 2\pi \mathrm{i}, \quad \epsilon \to 0.$$

Since the integral over the curve γ from above is independent on ϵ, then we obtain the desired result.

Theorem 5.8 (Cauchy integral formula). *Let $D \subset \mathbb{C}$ be a bounded domain with the boundary ∂D, which satisfies all conditions of Corollary 5.3. Then for any function $f \in H(D) \cap C(\overline{D})$ and any $z_0 \in \mathbb{C}$, we have*

$$\frac{1}{2\pi i}\int_{\partial D}\frac{f(z)}{z-z_0}dz=\begin{cases}0, & z_0\notin\overline{D},\\ f(z_0), & z_0\in D,\\ \frac{1}{2}f(z_0), & z_0\in\partial D.\end{cases}$$

Proof. If $z_0 \notin \overline{D}$, then the function

$$h(z) := \frac{f(z)}{z-z_0}$$

is analytic in D and continuous in $\overline{D}$. Then Corollary 5.3 leads to

$$0=\int_{\partial D} h(z)dz=\int_{\partial D}\frac{f(z)}{z-z_0}dz.$$

If $z_0 \in D$, then we consider the function

$$h(z) := \frac{f(z)-f(z_0)}{z-z_0}.$$

It is clear that $h \in H(D\backslash z_0)\cap C(\overline{D}\backslash z_0)$ and $\lim_{z\to z_0}(z-z_0)h(z) = 0$. Thus, using Corollary 5.4, we obtain

$$0=\frac{1}{2\pi i}\int_{\partial D} h(z)dz=\frac{1}{2\pi i}\int_{\partial D}\frac{f(z)}{z-z_0}dz-\frac{1}{2\pi i}\int_{\partial D}\frac{f(z_0)}{z-z_0}dz$$

or

$$\frac{1}{2\pi i}\int_{\partial D}\frac{f(z)}{z-z_0}dz=\frac{1}{2\pi i}f(z_0)\int_{\partial D}\frac{1}{z-z_0}dz.$$

But Example 5.6 implies that

$$\frac{1}{2\pi i}\int_{\partial D}\frac{f(z)}{z-z_0}dz=f(z_0).$$

If $z_0 \in \partial D$, then the integral on the left-hand side must be understood as the *principal value integral*,

$$\text{p.v.}\int_{\partial D}\frac{f(z)}{z-z_0}dz:=\lim_{\varepsilon\to+0}\int_{\partial D\backslash\{z:|z-z_0|<\varepsilon\}}\frac{f(z)}{z-z_0}dz$$

if this limit exists. For $z_0 \in \partial D$, we consider the domain (see Figure 5.4)

$$D_\varepsilon = D\setminus(D\cap\{z:|z-z_0|<\varepsilon\}).$$

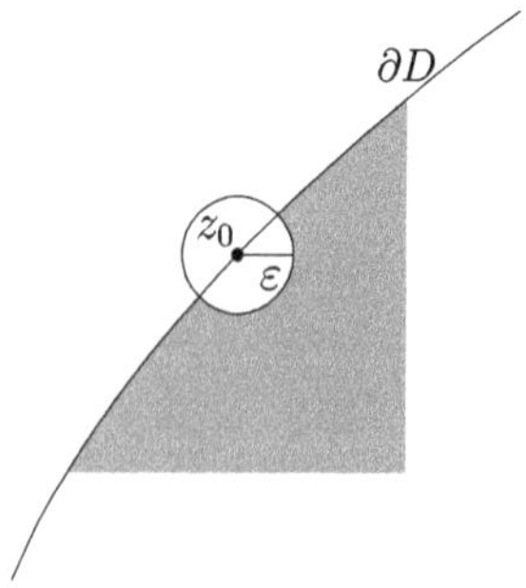

Figure 5.4: Domain D_ε.

It is clear that the function

$$h(z) = \frac{f(z) - f(z_0)}{z - z_0}$$

is analytic in D_ε and continuous up to the boundary of D_ε. Thus, using again Corollary 5.4 we obtain

$$\begin{aligned} 0 &= \frac{1}{2\pi \mathrm{i}} \int_{\partial D_\varepsilon} h(z)\mathrm{d}z \\ &= \frac{1}{2\pi \mathrm{i}} \int_{\partial D_\varepsilon} \frac{f(z)}{z - z_0}\mathrm{d}z + \frac{1}{2\pi \mathrm{i}} \int_{\partial D \setminus \{z:|z-z_0|<\epsilon\}} \frac{f(z)}{z - z_0}\mathrm{d}z \\ &\quad + \frac{1}{2\pi \mathrm{i}} \int_{D \cap \{z:|z-z_0|=\epsilon\}} \frac{f(z)}{z - z_0}\mathrm{d}z \to \text{p.v.} \int_{\partial D} \frac{f(z)}{z - z_0}\mathrm{d}z - \frac{f(z_0)}{2} \end{aligned}$$

as $\varepsilon \to +0$ since only half of the circle is presented (in the negative direction of integration). □

Example 5.9. Let us evaluate the integral

$$\int_\gamma \frac{\mathrm{e}^{z^2}}{z(z^2 + 4)}\mathrm{d}z,$$

where $\gamma = \{z : |z| = 3\}$. We parametrize this smooth closed Jordan curve as $\gamma : z(t) = 3\mathrm{e}^{\mathrm{i}t}$, $t \in [-\pi, \pi]$. Next,

$$\frac{1}{z(z^2 + 4)} = \frac{1}{z(z - 2\mathrm{i})(z + 2\mathrm{i})} = \frac{1}{4} \cdot \frac{1}{z} - \frac{1}{8} \cdot \frac{1}{z - 2\mathrm{i}} - \frac{1}{8} \cdot \frac{1}{z + 2\mathrm{i}}.$$

Hence, applying the Cauchy integral formula,

$$\int_\gamma \frac{e^{z^2}}{z(z^2+4)}dz = \frac{1}{4}\int_\gamma \frac{e^{z^2}}{z}dz - \frac{1}{8}\int_\gamma \frac{e^{z^2}}{z-2i}dz - \frac{1}{8}\int_\gamma \frac{e^{z^2}}{z+2i}dz$$

$$= 2\pi i\frac{1}{4}e^0 - 2\pi i\frac{1}{8}e^{(-2i)^2} - 2\pi i\frac{1}{8}e^{(2i)^2} = 2\pi i\frac{1-e^{-4}}{4} = \pi i\frac{1-e^{-4}}{2}.$$

Example 5.10. Let us evaluate the integral

$$\int_0^\pi e^{a\cos t}\cos(a\sin t)dt.$$

Since the integrand is even and sine is odd, we have

$$\int_0^\pi e^{a\cos t}\cos(a\sin t)dt = \frac{1}{2}\int_{-\pi}^{\pi} e^{a\cos t}\cos(a\sin t)dt$$

$$= \frac{1}{2}\int_{-\pi}^{\pi} e^{a\cos t}(\cos(a\sin t) + i\sin(a\sin t))dt$$

$$= \frac{1}{2}\int_{-\pi}^{\pi} e^{a\cos t}e^{ia\sin t}dt.$$

For $z(t) = e^{it}$, $t \in [-\pi, \pi]$, we have $dt = \frac{dz}{iz}$. Then the latter integral can be interpreted as the curve integral over the closed Jordan curve $\gamma : z(t) = e^{it}$, $t \in [-\pi, \pi]$. That is why it is equal to

$$\frac{1}{2}\int_{-\pi}^{\pi} e^{a(e^{it}+e^{-it})/2}e^{ia(e^{it}-e^{-it})/2i}dt = \frac{1}{2}\int_\gamma e^{\frac{a}{2}(z+1/z)}e^{\frac{a}{2}(z-1/z)}\frac{dz}{iz}$$

$$= \frac{1}{2i}\int_\gamma \frac{e^{az}}{z}dz = \frac{1}{2i}2\pi i e^0 = \pi$$

by the Cauchy integral formula.

Example 5.11. Let us evaluate the integral

$$\int_\gamma \frac{2z}{z^2+2}dz,$$

where $\gamma = \{z : |z - i| = 1\}$. First, we have

$$\frac{2z}{z^2+2} = \frac{1}{z - i\sqrt{2}} + \frac{1}{z + i\sqrt{2}}$$

and, therefore,

$$\int_\gamma \frac{2z}{z^2+2}\mathrm{d}z = \int_\gamma \frac{1}{z-\mathrm{i}\sqrt{2}}\mathrm{d}z + \int_\gamma \frac{1}{z+\mathrm{i}\sqrt{2}}\mathrm{d}z = 2\pi\mathrm{i}$$

since $\mathrm{i}\sqrt{2} \in \operatorname{int}\gamma$ but $-\mathrm{i}\sqrt{2} \notin \operatorname{int}\gamma$.

Let us consider now a piecewise smooth Jordan curve (not necessarily closed) γ and continuous function $f(z)$ on this curve. If $z \notin \gamma$, then the function

$$F(z) := \frac{1}{2\pi\mathrm{i}}\int_\gamma \frac{f(\zeta)}{\zeta - z}\mathrm{d}\zeta \tag{5.1}$$

is well-defined on $\mathbb{C} \setminus \gamma$. This function $F(z)$ is called a *Cauchy type integral.*

Theorem 5.12. *The Cauchy type integral* (5.1) *is an analytic function in* $\mathbb{C}\setminus\gamma$. *It has derivatives of any order* $n \in \mathbb{N}$ *and the formula*

$$F^{(n)}(z) = \frac{n!}{2\pi\mathrm{i}}\int_\gamma \frac{f(\zeta)}{(\zeta - z)^{n+1}}\mathrm{d}\zeta \tag{5.2}$$

holds.

Proof. Let $z \notin \gamma$ and $z + \Delta z \notin \gamma$, too. Then

$$\frac{F(z+\Delta z) - F(z)}{\Delta z} = \frac{1}{2\pi\mathrm{i}}\int_\gamma \frac{f(\zeta)}{(\zeta - z)(\zeta - z - \Delta z)}\mathrm{d}\zeta.$$

Since $z \notin \gamma$ and $z + \Delta z \notin \gamma$, then there is $\delta > 0$ and $d > 0$ such that $z + \Delta z \in U_\delta(z)$, $U_\delta(z) \cap \gamma = \emptyset$, and $|\zeta - z| \geq d > 0$, $|\zeta - z - \Delta z| \geq d > 0$ for any $\zeta \in \gamma$. Actually $d = \operatorname{dist}(\gamma, |\zeta - z| = \delta)$, see Figure 5.5.

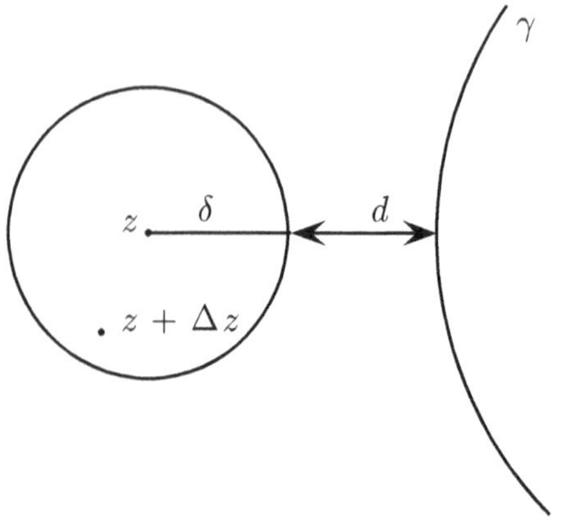

Figure 5.5: Distance from γ to circle around z.

In that case, we have

$$\begin{aligned}\left|\frac{F(z+\Delta z)-F(z)}{\Delta z}-\frac{1}{2\pi i}\int_\gamma \frac{f(\zeta)}{(\zeta-z)^2}d\zeta\right| &= \frac{1}{2\pi}\left|\int_\gamma \frac{\Delta z f(\zeta)}{(\zeta-z)^2(\zeta-z-\Delta z)}d\zeta\right| \\ &\le \frac{1}{2\pi}|\Delta z|\int_\gamma \frac{|f(\zeta)||d\zeta|}{|\zeta-z|^2|\zeta-z-\Delta z|} \\ &\le \frac{1}{2\pi}|\Delta z|M\frac{1}{d^3}\int_\gamma |d\zeta| = \frac{|\Delta z|ML}{2\pi d^3},\end{aligned}$$

where L is the length of γ and $M = \max_\gamma |f(\zeta)| < \infty$. Letting $\Delta z \to 0$, this estimate shows that

$$\lim_{\Delta z\to 0}\frac{F(z+\Delta z)-F(z)}{\Delta z} = \frac{1}{2\pi i}\int_\gamma \frac{f(\zeta)}{(\zeta-z)^2}d\zeta$$

or

$$F'(z) = \frac{1}{2\pi i}\int_\gamma \frac{f(\zeta)}{(\zeta-z)^2}d\zeta.$$

After this, (5.2) can be proved by induction. Indeed,

$$\begin{aligned}&\frac{F^{(n-1)}(z+\Delta z)-F^{(n-1)}(z)}{\Delta z}-\frac{n!}{2\pi i}\int_\gamma \frac{f(\zeta)d\zeta}{(\zeta-z)^{n+1}} \\ &= \frac{1}{2\pi i}\int_\gamma f(\zeta)\left[\frac{(n-1)!}{\Delta z}\left(\frac{1}{(\zeta-z-\Delta z)^n}-\frac{1}{(\zeta-z)^n}\right)-\frac{n!}{(\zeta-z)^{n+1}}\right]d\zeta \\ &= \frac{(n-1)!}{2\pi i}\int_\gamma f(\zeta)\left[\frac{(\zeta-z)^n-(\zeta-z-\Delta z)^n}{\Delta z(\zeta-z-\Delta z)^n(\zeta-z)^n}-\frac{n}{(\zeta-z)^{n+1}}\right]d\zeta \\ &= \frac{(n-1)!}{2\pi i}\int_\gamma f(\zeta)\left[\frac{\sum_{k=1}^n(-1)^{k+1}\binom{n}{k}(\Delta z)^k(\zeta-z)^{n-k}}{\Delta z(\zeta-z-\Delta z)^n(\zeta-z)^n}-\frac{n}{(\zeta-z)^{n+1}}\right]d\zeta \\ &= \frac{(n-1)!}{2\pi i}\int_\gamma f(\zeta) \\ &\quad\times\left[\frac{\sum_{j=0}^{n-1}(-1)^j\binom{n}{j+1}(\Delta z)^j(\zeta-z)^{n-j}-n\sum_{j=0}^n(-1)^j\binom{n}{j}(\Delta z)^j(\zeta-z)^{n-j}}{(\zeta-z-\Delta z)^n(\zeta-z)^{n+1}}\right]d\zeta \\ &= \frac{(n-1)!}{2\pi i}\int_\gamma f(\zeta) \\ &\quad\times\left[\frac{\sum_{j=1}^{n-1}(-1)^j(\Delta z)^j(\zeta-z)^{n-j}(\binom{n}{j+1}-n\binom{n}{j})+n(-1)^{n+1}(\Delta z)^n}{(\zeta-z-\Delta z)^n(\zeta-z)^{n+1}}\right]d\zeta.\end{aligned}$$

This representation implies that

$$\left|\frac{F^{(n-1)}(z+\Delta z)-F^{(n-1)}(z)}{\Delta z}-\frac{n!}{2\pi i}\int_{\gamma}\frac{f(\zeta)d\zeta}{(\zeta-z)^{n+1}}\right| \le \frac{(n-1)!}{2\pi}O(|\Delta z|)\int_{\gamma}\frac{|f(\zeta)||d\zeta|}{|(\zeta-z-\Delta z)^{n}||(\zeta-z)^{n+1}|} \le \frac{(n-1)!}{2\pi}O(|\Delta z|)\frac{ML}{d^{2n+1}}.$$

This estimate completes the proof of (5.2) by induction. □

Corollary 5.13 (Cauchy integral formula). *Let $D \subset \mathbb{C}$ be a domain (not necessarily simply connected) and $f \in H(D)$. Then f is infinitely many times differentiable in D and*

$$f^{(n)}(z)=\frac{n!}{2\pi i}\int_{\gamma}\frac{f(\zeta)}{(\zeta-z)^{n+1}}d\zeta \tag{5.3}$$

for any $n = 1, 2, \dots$, where γ is an arbitrary piecewise smooth closed Jordan curve such that $\operatorname{int}\gamma \subset D$ *and* $z \in \operatorname{int}\gamma$.

Proof. Let $z \in D$. Let also γ be an arbitrary piecewise smooth closed Jordan curve such that $\operatorname{int}\gamma \subset D$ and $z \in \operatorname{int}\gamma$. Then by the Cauchy integral formula (see Theorem 5.8), we have

$$f(z)=\frac{1}{2\pi i}\int_{\gamma}\frac{f(\zeta)d\zeta}{\zeta-z}.$$

But the right-hand side is a Cauchy type integral since f is continuous on γ. Applying Theorem 5.12, we obtain that for any $n = 1, 2, \dots$ we have

$$f^{(n)}(z)=\frac{n!}{2\pi i}\int_{\gamma}\frac{f(\zeta)d\zeta}{(\zeta-z)^{n+1}}$$

and f is infinitely many times differentiable in D. □

Remark. Formula (5.3) holds also for the boundary ∂D of a domain that satisfies all conditions of Corollary 5.3 if we assume that $f \in H(D) \cap C(\overline{D})$. Moreover, as the simplest case, formula (5.3) holds and is very applicable for $\gamma = \{\zeta : |\zeta - z| = \delta\}$ with $\delta > 0$ small enough, i. e.,

$$f^{(n)}(z)=\frac{n!}{2\pi i}\int_{|\zeta-z|=\delta}\frac{f(\zeta)d\zeta}{(\zeta-z)^{n+1}}=\frac{n!}{2\pi i}\int_{-\pi}^{\pi}\frac{f(z+\delta e^{i\theta})\delta i e^{i\theta}}{\delta^{n+1}e^{i\theta(n+1)}}d\theta = \frac{n!}{2\pi}\delta^{-n}\int_{-\pi}^{\pi}f(z+\delta e^{i\theta})e^{-i\theta n}d\theta.$$

Problem 5.14. Evaluate the derivative of $F(z)$ from (5.1) at $z = \infty$. Show first that $F(z)$ is continuous at $z = \infty$ and $F(\infty) = 0$. Show then that

$$F'(\infty) = -\frac{1}{2\pi \mathrm{i}} \int_{\gamma} f(\zeta)\mathrm{d}\zeta.$$

There is some generalization of the formulas (5.1) and (5.2).

Problem 5.15. Let function $f(z,\zeta)$ (the variable ζ might be considered as a complex parameter) be continuous for all z in the domain $D \subset \mathbb{C}$ and for all $\zeta \in \gamma$, where γ is a Jordan curve on the complex plane $\mathbb{C}$. Prove that if $f(z,\zeta)$ is analytic for each fixed $\zeta \in \gamma$, then the function

$$F(z) := \int_{\gamma} f(z,\zeta)\mathrm{d}\zeta$$

is analytic with respect to z on D and

$$F'(z) = \int_{\gamma} f'_z(z,\zeta)\mathrm{d}\zeta.$$

Similar formulas are valid for all derivatives of $F(z)$.

Example 5.16. Let us evaluate the integral

$$\int_{\gamma} \frac{\sin z}{(z-\pi/6)^3}\mathrm{d}z,$$

where $\gamma = \{z : |z| = 1\}$. Since $|\pi/6| < 1$, then applying (5.3) we obtain

$$\int_{\gamma} \frac{\sin z\mathrm{d}z}{(z-\pi/6)^3} = \frac{2\pi \mathrm{i}}{2!}(\sin z)''\big|_{z=\pi/6} = \pi\mathrm{i}(-\sin z)\big|_{z=\pi/6} = -\pi\mathrm{i}\sin(\pi/6) = -\frac{\pi\mathrm{i}}{2}.$$

Example 5.17. Let us evaluate the integral

$$\int_{\gamma} \frac{\mathrm{d}z}{(z-a)^4(z-b)},$$

where $\gamma = \{z : |z| = r\}$ and $|a| < r < |b|$. Since $z \neq b$ for all $|z| \le r$, then this integral is equal to

$$\int_{\gamma} \frac{\frac{1}{z-b}\mathrm{d}z}{(z-a)^4} = \frac{2\pi\mathrm{i}}{3!}\left(\frac{1}{z-b}\right)'''\bigg|_{z=a} = \frac{2\pi\mathrm{i}}{6}\left(-\frac{6}{(z-b)^4}\right)\bigg|_{z=a} = -\frac{2\pi\mathrm{i}}{(a-b)^4}.$$

Example 5.18. Let f be analytic in a simply connected domain D and $z_1, z_2 \in D$, $z_1 \neq z_2$. Then for any piecewise smooth closed Jordan curve γ such that $z_1, z_2 \in \operatorname{int}\gamma$, we have

$$\frac{f(z_2) - f(z_1)}{z_2 - z_1} = \frac{1}{2\pi i} \int_\gamma \frac{f(\zeta)d\zeta}{(\zeta - z_1)(\zeta - z_2)}.$$

Indeed, since

$$f(z_j) = \frac{1}{2\pi i} \int_\gamma \frac{f(\zeta)d\zeta}{\zeta - z_j}, \quad j = 1, 2,$$

then

$$\begin{aligned} f(z_2) - f(z_1) &= \frac{1}{2\pi i} \int_\gamma f(\zeta)\left(\frac{1}{\zeta - z_2} - \frac{1}{\zeta - z_1}\right)d\zeta \\ &= \frac{z_2 - z_1}{2\pi i} \int_\gamma \frac{f(\zeta)d\zeta}{(\zeta - z_1)(\zeta - z_2)}. \end{aligned}$$

Problem 5.19. Let $z_1 \neq z_2$ be two arbitrary complex numbers. Evaluate the integral

$$\int_\gamma \frac{z^2}{(z - z_1)(z - z_2)} dz$$

over the circle $\gamma : |z| = r$ for three different cases: $|z_1|, |z_2| < r$, $|z_1| < r$, $|z_2| > r$, and $|z_1|, |z_2| > r$.

Problem 5.20. Show that if $\operatorname{Re} z > 0$, then

$$\int_0^\infty (e^{-t} - e^{-tz})\frac{dt}{t} = \log z.$$

Hint: Integrate over the quadrilateral with vertices $\delta > 0$, $\rho > 1$ on the real line and δz, ρz in the right half-plane, and use the fact that the function $\frac{e^{-t} - e^{-tz}}{t}$ is analytic in $\{z : \operatorname{Re} z > 0\}$.

Problem 5.21. Using the Cauchy theorem, prove that

$$\int_0^\infty e^{-zt^2} dt = \frac{1}{2}\sqrt{\frac{\pi}{z}}$$

for $\operatorname{Re} z > 0$. In particular, for $x > 0$ we have

$$\int_0^\infty e^{-xt^2} \cos(yt^2)dt = \frac{\sqrt{\pi}}{2(x^2 + y^2)^{1/4}} \cos\left(\frac{1}{2} \arctan \frac{y}{x}\right)$$

and

$$\int_0^\infty e^{-xt^2} \sin(yt^2)dt = \frac{\sqrt{\pi}}{2(x^2+y^2)^{1/4}} \sin\left(\frac{1}{2}\arctan\frac{y}{x}\right).$$

Hint: Use the trigonometric formula $\tan\frac{\alpha}{2} = \frac{-1\pm\sqrt{1+\tan^2(\alpha)}}{\tan(\alpha)}$.

Example 5.22. Let us show that

$$\int_{|z|=1} e^z \overline{dz} = -2\pi i$$

or

$$\int_{|z|=1} \overline{e^z} dz = 2\pi i.$$

We have

$$\begin{aligned}\int_{|z|=1} e^{\overline{z}} dz &= \int_{|z|=1} e^{1/z} dz \\ &= -\int_{|\zeta|=1} e^{\zeta} d\left(\frac{1}{\zeta}\right) = \int_{|\zeta|=1} \frac{e^{\zeta}}{\zeta^2} d\zeta = 2\pi i (e^{\zeta})'|_{\zeta=0} = 2\pi i,\end{aligned}$$

where we have also changed the direction of integration when changing variables.

Problem 5.23. Using the same ideas as in Example 5.22, show that

$$\int_{|z|=1} \overline{\left(\frac{1}{z} + \frac{\sin z}{z^2}\right)} dz = -\frac{\pi i}{3}$$

or

$$\int_{|z|=1} \left(\frac{1}{z} + \frac{\sin z}{z^2}\right) \overline{dz} = \frac{\pi i}{3}.$$

Remark. The latter integrals lead to the following considerations. Let $f(z) = u(x,y) + iv(x,y)$ be an arbitrary function and let γ be an arbitrary curve (not necessary closed). Then, assuming that $f(z)$ is integrable along this curve, we obtain that

$$\int_\gamma \overline{f(z)} dz + \int_\gamma f(z) \overline{dz} = 2 \int_\gamma u(x,y)dx + v(x,y)dy.$$

Concerning the function $f(z) = \frac{1}{z} + \frac{\sin z}{z^2}$ from Problem 5.23, we conclude that

$$u(x,y) = \frac{x}{x^2+y^2} + \frac{2xy\cos x\sinh y + (x^2-y^2)\sin x\cosh y}{(x^2+y^2)^2}$$

and

$$v(x,y) = -\frac{y}{x^2+y^2} + \frac{(x^2-y^2)\cos x\sinh y - 2xy\sin x\cosh y}{(x^2+y^2)^2}$$

and, therefore,

$$\int_{|z|=1} u(x,y)\mathrm{d}x + v(x,y)\mathrm{d}y = 0$$

for these concrete $u(x,y)$ and $v(x,y)$. It might be mentioned also here that due to the Cauchy–Riemann conditions for the function $f(z) = \frac{1}{z} + \frac{\sin z}{z^2}$ we have that

$$\frac{\partial u(x,y)}{\partial x} = \frac{\partial v(x,y)}{\partial y}, \quad \frac{\partial u(x,y)}{\partial y} = -\frac{\partial v(x,y)}{\partial x},$$

and thus the differential form

$$u(x,y)\mathrm{d}x + v(x,y)\mathrm{d}y$$

is not an exact form (meaning that this differential form is not equal to differential of some function $F(x,y)$).

Exercises

1. Evaluate
   a) i^k, b) i^{-k}
   for $k = 0, 1, 2, \ldots$.
2. Find $\operatorname{Re} z$ and $\operatorname{Im} z$, when
   a) $z = (2+3\mathrm{i})(-3+2\mathrm{i})$, b) $z = \frac{4+2\mathrm{i}}{3-4\mathrm{i}}$, c) $z = \overline{(1+\mathrm{i})} \cdot \frac{1}{2-\mathrm{i}}$.
3. Solve z from the equation
   a) $(3+4\mathrm{i})\bar{z} = 1 - 2\mathrm{i}$,
   b) $\mathrm{i}z + 2\bar{z} = 3 - \mathrm{i}$,
   c) $z^2 = -5 + 12\mathrm{i}$.
4. Prove that

$$|\operatorname{Re} z| \leq |z|, \quad |\operatorname{Im} z| \leq |z|.$$

   Show also that the equalities hold if and only if z is real or pure imaginary, respectively.
5. Prove that $|z_1 - z_2| = |1 - \overline{z_1}z_2|$, where $z_1, z_2 \in \mathbb{C}$, and $|z_1| = 1$ or $|z_2| = 1$.
6. Express $z \in \mathbb{C}$ in trigonometric form when
   a) $z = -3\mathrm{i}$, b) $z = \sqrt{3} - \mathrm{i}$, c) $z = 2 - \mathrm{i}\sqrt{12}$.
7. Evaluate $(1 - \mathrm{i}\sqrt{3})^{15}$, $(1+\mathrm{i})^{11}$, and $\frac{(1+\mathrm{i})^5}{(1-\mathrm{i}\sqrt{3})^7}$.
8. Let $z \in \mathbb{C}$, $|z| = 1$, $z \neq -1$. Prove that z can be written in the form $z = \frac{1+\mathrm{i}t}{1-\mathrm{i}t}$ for some $t \in \mathbb{R}$.
9. Solve the equations
   a) $z^4 = -1$, b) $z^6 = 1$, c) $z^3 = -\mathrm{i}$.
10. Prove that the set $\{z \in \mathbb{C} \mid |z - z_0| > r\}$ is open ($z_0 \in \mathbb{C}$, $r > 0$ are given).
11. Let $A = \{\mathrm{i}, \frac{\mathrm{i}}{2}, \frac{\mathrm{i}}{3}, \ldots\} \subset \mathbb{C}$. Determine if A is bounded, closed, or open. Find A' and $\overline{A}$.
12. Find the following limits (if they exist):
    a) $\lim_{n\to\infty} \frac{\mathrm{i}^n}{n}$, b) $\lim_{n\to\infty} \mathrm{i}^n$, c) $\lim_{n\to\infty} \frac{(1+\mathrm{i})^n}{n}$, d) $\lim_{n\to\infty} \frac{2n-\mathrm{i}n^2}{(1+\mathrm{i})n-1}$.
13. Let the sequence $(z_n) \subset \mathbb{C}$ be defined as $z_0 = 3$ and $z_{n+1} = \frac{1}{3}z_n + 2\mathrm{i}$. Show that (z_n) converges and find its limit.
14. Determine which of the following functions are bijective $D \to G$ and find $f^{-1} : G \to D$ whenever it is possible.
    a) $f(z) = \bar{z} + \mathrm{i}$, $z \in \mathbb{C}$, b) $f(z) = \frac{1}{z}$, $z \in \mathbb{C} \setminus \{0\}$,
    c) $f(z) = z^2 + \mathrm{i}$, $z \in \mathbb{C}$, d) $f(z) = z^2 + \mathrm{i}$, $0 \leq \arg z < \pi$.
15. Let $f : D \to \mathbb{C}$ be a function such that $f(z) = z^3 + \mathrm{i}$, $0 \leq \arg z < 2\pi/3$. Determine if f is bijective $D \to \mathbb{C}$. Find $f^{-1}(1)$.
16. Express the function $f(z) = f(x + \mathrm{i}y)$ in the form $f(z) = u(x,y) + \mathrm{i}v(x,y)$, $z \in D$, when
    a) $f(z) = z^3$, $z \in \mathbb{C}$, b) $f(z) = \frac{1}{z}$, $z \neq 0$, c) $f(z) = \mathrm{e}^{\mathrm{i}z}$, $z \in \mathbb{C}$.
17. Investigate the existence of the limit of $f(z)$ at the point $z = 0$, when
    a) $f(z) = \frac{\operatorname{Re} z}{z}$, b) $f(z) = \frac{z}{|z|}$, c) $f(z) = \frac{z \operatorname{Re} z}{|z|}$.

https://doi.org/10.1515/9783111632278-006

18. Find the limit $\lim_{z\to z_0} \frac{z^3+z^2+z+1}{z-z_0}$, when
 a) $z_0 = -1$, b) $z_0 = \mathrm{i}$, c) $z_0 = -\mathrm{i}$. d) Find the limit $\lim_{z\to \mathrm{i}} \frac{z^3+\mathrm{i}}{z-\mathrm{i}}$.
19. Prove using the definition of continuity that the function $f(z) = z^2 + 2z$, $z \in \mathbb{C}$ is continuous for all $z_0 \in \mathbb{C}$ but it is not continuous at $z_0 = \infty$.
20. Show that the function $f(z) = z^2$ is uniformly continuous on the set $|z - \mathrm{i}| < 2$. Is f uniformly continuous on $\mathbb{C}$?
21. Study the uniform continuity of $f(z) = \frac{1}{z}$, $z \neq 0$ on the set $|z| < 1$, $z \neq 0$.
22. Investigate if the function $f(z) = z|z|$, $z \in \mathbb{C}$ has a derivative at any $z_0 \in \mathbb{C}$.
23. Find the derivatives of the following functions (if they exist):
 a) $f(z) = \frac{z^2+1}{(z^2-1)^2}$, $z \neq \pm 1$, b) $f(z) = \mathrm{e}^{\bar{z}}$, $z \in \mathbb{C}$,
 c) $f(z) = \operatorname{Im} z$, $z \in \mathbb{C}$, d) $f(z) = z \operatorname{Im} z$, $z \in \mathbb{C}$.
24. Consider $f(z) = z^n$, $0 \le \arg z < 2\pi/n$, $n \ge 2$. Find $f'(z), z \in \mathbb{C}$, and $(f^{-1})(w)$, $w \in f(D) \setminus \{0\}$.
25. Let $f(z) = z^3$, $2\pi/3 \le \arg z < 4\pi/3$. Then $f^{-1} : \mathbb{C} \to D$ exists. Find $(f^{-1})'(\mathrm{i})$ and $(f^{-1})'(-1)$.
26. Let us assume that g is analytic in all of $\mathbb{C}$. Define the function $f : \mathbb{C} \to \mathbb{C}$ by setting
 a) $f(z) = g(\bar{z})$, $z \in \mathbb{C}$, b) $f(z) = \overline{g(\bar{z})}$, $z \in \mathbb{C}$.
 Investigate if f is analytic on $\mathbb{C}$.
27. Let $f(z) = f(x + \mathrm{i}y) = x^3 - 3xy^2 + \mathrm{i}(3x^2y - y^3)$, $z = x + \mathrm{i}y \in \mathbb{C}$. Show that f satisfies the Cauchy–Riemann conditions. Find $f'(z)$.
28. Solve

$$\mathrm{e}^z = 2 + \mathrm{i}.$$

29. Show that the function $f(z) = \frac{1}{z+\mathrm{i}}$, $z \in \mathbb{C} \setminus \{-\mathrm{i}\}$ satisfies the Cauchy–Riemann conditions.
30. Show that the function

$$f(z) = \sin z$$

 satisfies the Cauchy–Riemann conditions.
31. Prove that
 a) $\mathrm{e}^{\bar{z}} = \overline{\mathrm{e}^z}$, b) $\sin \bar{z} = \overline{\sin z}$, c) $|\mathrm{e}^z| = \mathrm{e}^x$, d) $|\cos z|^2 + |\sin z|^2 = 1 + 2\sinh^2 y$
 whenever $z \in \mathbb{C}$.
32. Find
 a) $\log(-4)$, b) $\log 3\mathrm{i}$, c) $\log(\sqrt{3} - \mathrm{i})$.
33. Find
 a) $\mathrm{i}^{2\mathrm{i}}$, b) $(-\mathrm{i})^{\mathrm{i}}$, c) $\mathrm{i}^{-\mathrm{i}}$.
34. Express the function $f(z) = \operatorname{Log} z$, $z \neq 0$, in the form $f = u + \mathrm{i}v$. Determine if it satisfies the Cauchy–Riemann conditions.
35. Find the limits
 a) $\lim_{z\to 0} \frac{\mathrm{e}^{z^2}-1}{z^2+2z}$, b) $\lim_{z\to \frac{\pi}{2}} \frac{\cos z}{z-\frac{\pi}{2}}$, c) $\lim_{z\to 0} \frac{\cos 2z-1}{\sin^2 z}$, d) $\lim_{z\to 0} \frac{\log^2(1+z)}{z^2}$.

36. Let f be analytic in a domain $A \subset \mathbb{C}$.
    a) Let us assume that $f'(z) = 0$ for all $z \in A$. Show that f is a constant function on A.
    b) Let us assume that $f = u + \mathrm{i}v$ and u is a constant function on A. Show that f is constant on A.
37. Find $\int_\gamma \bar{z}\mathrm{d}z$, where
    a) $\gamma : z(t) = t + \mathrm{i}t^2, t \in [0,1]$, b) $\gamma : z(t) = t^2 + \mathrm{i}t^4, t \in [0,1]$.
38. Find $\int_\gamma z^2\mathrm{d}z$, where γ is the line segment from i to 1 + 2i.
39. Evaluate the integral

$$\int_\gamma \frac{\mathrm{d}z}{(z-z_0)^n}, \qquad n = 2, 3, \ldots,$$

where γ is closed Jordan curve and a) z_0 is in the interior of γ, b) z_0 is in the exterior of γ.
40. Prove that

$$\int_0^{2\pi} \mathrm{e}^{\cos t}\cos(t + \sin t)\mathrm{d}t = \int_0^{2\pi} \mathrm{e}^{\cos t}\sin(t + \sin t)\mathrm{d}t = 0.$$

41. Evaluate the integral $\int_\gamma \sin^2 z\mathrm{d}z$, where γ is the line segment from 0 to i.
42. Evaluate the integrals
    a) $\int_\gamma \frac{\sin z}{z-\mathrm{i}}\mathrm{d}z$, where $\gamma : z(t) = 2\mathrm{e}^{\mathrm{i}t}, t \in [0, 2\pi]$,
    b) $\int_\gamma \frac{\sinh z}{z-\mathrm{i}\pi}\mathrm{d}z$, where $\gamma : z(t) = \mathrm{i}\pi + 2\mathrm{e}^{\mathrm{i}t}, t \in [0, 2\pi]$.
43. Evaluate

$$\int_\gamma \frac{\mathrm{e}^z}{z(z-2\mathrm{i})}\mathrm{d}z,$$

where a) $\gamma : z(t) = \mathrm{e}^{\mathrm{i}t}, t \in [0, 2\pi]$, b) $\gamma : z(t) = 3\mathrm{e}^{\mathrm{i}t}, t \in [0, 2\pi]$.
44. Evaluate

$$\frac{1}{2\pi\mathrm{i}}\int_\gamma \frac{\mathrm{e}^{az}}{z^2+1}\mathrm{d}z,$$

where $\gamma : z(t) = 3\mathrm{e}^{\mathrm{i}t}, t \in [0, 2\pi]$, and $a > 0$.
45. Evaluate

$$\frac{1}{2\pi\mathrm{i}}\int_\gamma \frac{\mathrm{e}^{az}}{(z^2+1)^2}\mathrm{d}z,$$

where γ and a are as in Exercise 44.

46. Evaluate
   a) $\int_\gamma \frac{e^{iz}}{z^3} dz$, where $\gamma : z(t) = 2e^{it}$, $t \in [0, 2\pi]$,
   b) $\int_\gamma \frac{\cos z}{(z-\pi/4)^3} dz$, where $\gamma : z(t) = e^{it}$, $t \in [0, 2\pi]$.
47. Evaluate

$$\int_\gamma \frac{e^{kz}}{z^{n+1}} dz \quad \text{and} \quad \int_\gamma \frac{\sin z}{z^{n+1}} dz,$$

where $\gamma : z(t) = e^{it}$, $t \in [0, 2\pi]$ and $k \in \mathbb{N}$.

Part II

6 Fundamental theorem of integration

The Cauchy theorem (as well as the Cauchy integral formula) allows us to prove the *fundamental theorem of integration*. Let f be analytic in the simply connected domain D. Then the integral

$$\int_{\gamma} f(\zeta)\mathrm{d}\zeta,$$

is independent on the piecewise smooth Jordan curve $\gamma \subset D$ connecting two points $z_0, z \in D$. The reason is: if we consider two different such curves γ_1 and γ_2 (both from z_0 to z), then the curve $\gamma := \gamma_1 \cup \gamma_2$ will be closed and due to the Cauchy theorem (Theorem 5.2), we have

$$0 = \int_{\gamma} f(\zeta)\mathrm{d}\zeta = \int_{\gamma_1} f(\zeta)\mathrm{d}\zeta - \int_{\gamma_2} f(\zeta)\mathrm{d}\zeta$$

or

$$\int_{\gamma_1} f(\zeta)\mathrm{d}\zeta = \int_{\gamma_2} f(\zeta)\mathrm{d}\zeta.$$

That is why the function

$$F(z) := \int_{z_0}^{z} f(\zeta)\mathrm{d}\zeta \tag{6.1}$$

is well-defined since its value is independent on the curve connecting z_0 and z. Even more is true. The function (6.1) is analytic in D and $F'(z) = f(z)$ everywhere in D. Indeed,

$$\begin{aligned}
\frac{F(z+\Delta z) - F(z)}{\Delta z} - f(z) &= \frac{1}{\Delta z}\left(\int_{z_0}^{z+\Delta z} f(\zeta)\mathrm{d}\zeta - \int_{z_0}^{z} f(\zeta)\mathrm{d}\zeta\right) - f(z) \\
&= \frac{1}{\Delta z}\int_{z}^{z+\Delta z} f(\zeta)\mathrm{d}\zeta - f(z) \\
&= \frac{1}{\Delta z}\int_{z}^{z+\Delta z} (f(\zeta) - f(z))\mathrm{d}\zeta.
\end{aligned}$$

Using the line segment from z to $z + \Delta z$ (see Figure 6.1), we obtain

https://doi.org/10.1515/9783111632278-007

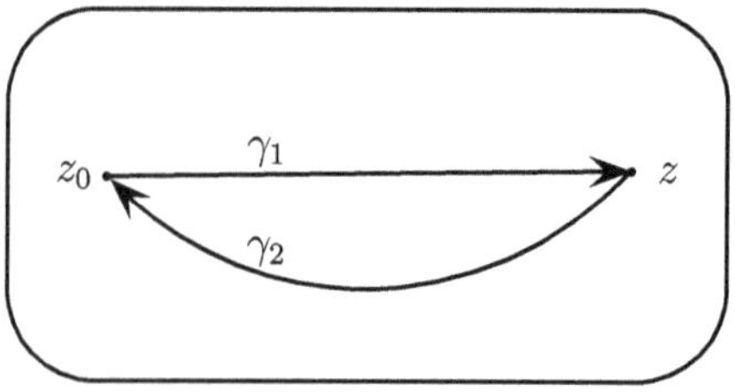

Figure 6.1: Curves connecting z_0 to z.

$$\left|\frac{F(z+\Delta z)-F(z)}{\Delta z}-f(z)\right| \leq \frac{1}{|\Delta z|}\left|\int_{z}^{z+\Delta z}|f(\zeta)-f(z)||\mathrm{d}\zeta|\right|$$
$$\leq \sup_{\zeta\in[z,z+\Delta z]}|f(\zeta)-f(z)| \to 0$$

as $\Delta z \to 0$. Hence, $F(z)$ is analytic in D and $F'(z) = f(z)$ everywhere in D. This fact justifies the following definition.

Definition 6.1. The function $\Phi(z)$ is called the *primitive* for $f(z)$ in D if $\Phi(z) \in H(D)$ and $\Phi'(z) = f(z)$.

So, if D is simply connected and f is analytic in D, then

$$F(z) = \int_{z_0}^{z} f(\zeta)\mathrm{d}\zeta$$

is a primitive for f in D. For instance, $f(z) = \bar{z}$ has no primitive.

Problem 6.2. Show that if Φ_1 and Φ_2 are primitives for f in simply connected D, then $\Phi_1(z) - \Phi_2(z) \equiv$ constant in D.

As a consequence, we have the fundamental fact: if D is simply connected, then

$$\int_{z_1}^{z_2} f(\zeta)\mathrm{d}\zeta = F(z_2) - F(z_1),$$

where F is any primitive for f. This fact is called the *fundamental theorem of complex integration* analogously to the Newton's formula for real integration.

Example 6.3. Let $D \subset \mathbb{C}$ be a simply connected domain such that $0 \notin D$ and $1 \in D$. Then $f(z) = \frac{1}{z}$ is analytic in D and

$$F(z) = \int_{1}^{z} \frac{1}{\zeta}\mathrm{d}\zeta$$

is a primitive for f in D, where D is such that any curve connecting 1 and $z \in D$ does not pass across 0. For example, D can be chosen as

$$D = \mathbb{C} \setminus \{\operatorname{Im} z = 0, \operatorname{Re} z \leq 0\}.$$

In this domain, we can take the line segment connecting 1 and z. This function $F(z)$ is said to be logarithmic function (or logarithm), i. e.,

$$\log z = \int_1^z \frac{1}{\zeta} \mathrm{d}\zeta. \tag{6.2}$$

Problem 6.4. Show that

$$\log z = \log |z| + \mathrm{i} \arg z, \quad z \in D, \tag{6.3}$$

where D is as above, i. e., $-\pi < \arg z < \pi$.

Problem 6.5. Let

$$f(z) = \frac{1}{1 + z^2}$$

and let D be simply connected such that $\pm\mathrm{i} \notin D$. Let

$$\arctan z := \int_0^z \frac{\mathrm{d}\zeta}{1 + \zeta^2},$$

where D is chosen such that any curve connecting 0 and z does not pass across $\pm\mathrm{i}$. Show that

$$\arctan z = \frac{1}{2\mathrm{i}} \log \frac{1 + \mathrm{i}z}{1 - \mathrm{i}z}.$$

The converse statement to Cauchy theorem is also true.

Theorem 6.6 (Morera's theorem). *Let f be a continuous function in a simply connected domain $D \subset \mathbb{C}$. If $\int_\gamma f(\zeta)\mathrm{d}\zeta = 0$ for every piecewise smooth closed Jordan curve in D, then f is analytic in D.*

Proof. We select a point $z_0 \in D$ and define $F(z)$ by

$$F(z) := \int_{z_0}^z f(\zeta)\mathrm{d}\zeta,$$

which is well-defined in D since the result of integration is independent on curve connecting z_0 and z in D. Since f is continuous in D, we have (choosing line segment to connect z and $z + \Delta z$)

$$\left|\frac{F(z+\Delta z)-F(z)}{\Delta z}-f(z)\right| \le \frac{1}{|\Delta z|}\left|\int_{z}^{z+\Delta z}|f(\zeta)-f(z)||d\zeta|\right|$$
$$\le \max_{\zeta\in[z,z+\Delta z]}|f(\zeta)-f(z)| \to 0$$

as $\Delta z \to 0$. Thus, $F'(z) = f(z)$, i. e., F is analytic. But since any analytic function is infinitely many times differentiable then so is f. □

7 Harmonic functions and mean value formulae

Let $u(x,y)$ be a real-valued function of two real variables x and y defined on a domain D.

Definition 7.1. If function $u(x,y)$ is twice continuously differentiable in D and satisfies the *Laplace equation*,

$$\Delta u = \partial_x^2 u(x,y) + \partial_y^2 u(x,y) = 0$$

in D, then u is said to be *harmonic* in D.

There is a close connection between harmonic and analytic functions. Indeed, if $f = u + iv$ is analytic in D, then (as we have proved) f is infinitely many times differentiable in D. So the functions u and v satisfy the Cauchy–Riemann conditions $\partial_x u = \partial_y v$ and $\partial_y u = -\partial_x v$. It follows that

$$\partial_x^2 u = \partial_{xy}^2 v, \quad \partial_y^2 u = -\partial_{xy}^2 v,$$

and hence,

$$\partial_x^2 u + \partial_y^2 u = \partial_{xy}^2 v - \partial_{xy}^2 v = 0,$$

i. e., u is harmonic. Similarly,

$$\partial_x^2 v + \partial_y^2 v = -\partial_{xy}^2 u + \partial_{xy}^2 u = 0.$$

Thus, if $f \in H(D)$, then $\operatorname{Re} f$ and $\operatorname{Im} f$ are harmonic in D.

It turns out that the converse is also true. Namely, any harmonic function is the real (or imaginary) part of some analytic function and this connection is unique up to an arbitrary constant. Let u be harmonic in a simply connected domain D. Then we may consider the differential form

$$l := -\partial_y u dx + \partial_x u dy.$$

This form is complete differential of some function v since $\partial_y(-\partial_y u) = \partial_x(\partial_x u)$ or $\Delta u = 0$, i. e.,

$$dv = -\partial_y u dx + \partial_x u dy. \tag{7.1}$$

This fact allows us to introduce function $v(x,y)$ as

$$v(x,y) = \int_{(x_0,y_0)}^{(x,y)} -\partial_y u dx + \partial_x u dy + \text{constant} \tag{7.2}$$

https://doi.org/10.1515/9783111632278-008

and this function is well-defined since the latter integral does not depend on the curve in D connecting the points (x_0, y_0) and (x, y). Even more is true, due to (7.1) and (7.2), we have

$$\partial_x v = -\partial_y u, \quad \partial_y v = \partial_x u,$$

i. e., the Cauchy–Riemann conditions are satisfied for function $f = u + \mathrm{i}v$. But u and v are twice continuously differentiable in D with Cauchy–Riemann conditions fulfilled. Thus, $f \in H(D)$ and $u = \operatorname{Re} f$. Similarly, we may construct uniquely (up to an arbitrary constant) analytic function f such that given harmonic function u is equal to $\operatorname{Im} f$.

Simultaneously, we obtained the following important result. By Corollary 5.13, we know that any analytic function is infinitely many times differentiable. Since any harmonic function is the real (or imaginary) part of some analytic function, then any harmonic function is infinitely many times differentiable.

Problem 7.2. Let $f \in H(D)$ and $f \neq 0$ everywhere in D. Prove that $\log f(z)$ is analytic in in a domain D, and $\log |f(z)|$ is harmonic in D.

Let f be analytic in a domain D containing the disk $\{z : |z - z_0| \leq R\}$. Then the Cauchy integral formula yields

$$f(z_0) = \frac{1}{2\pi \mathrm{i}} \int_{|\zeta - z_0| = R} \frac{f(\zeta)\mathrm{d}\zeta}{\zeta - z_0}.$$

If we parametrize the circle by $\zeta(t) = z_0 + Re^{\mathrm{i}t}$, $t \in [-\pi, \pi]$, then $\mathrm{d}\zeta = R\mathrm{i}e^{\mathrm{i}t}\mathrm{d}t$ and the latter integral becomes

$$f(z_0) = \frac{1}{2\pi \mathrm{i}} \int_{-\pi}^{\pi} \frac{f(z_0 + Re^{\mathrm{i}t})R\mathrm{i}e^{\mathrm{i}t}\mathrm{d}t}{Re^{\mathrm{i}t}} = \frac{1}{2\pi} \int_{-\pi}^{\pi} f(z_0 + Re^{\mathrm{i}t})\mathrm{d}t. \tag{7.3}$$

This formula is called the *mean-value formula* for analytic functions. Since any harmonic function is the real (or imaginary) part of some analytic function, then we obtain the mean-value formula also for harmonic function u as

$$\begin{aligned} u(x_0, y_0) &= \operatorname{Re} f(z_0) = \operatorname{Re} \frac{1}{2\pi} \int_{-\pi}^{\pi} f(z_0 + Re^{\mathrm{i}t})\mathrm{d}t \\ &= \frac{1}{2\pi} \int_{-\pi}^{\pi} \operatorname{Re} f(z_0 + Re^{\mathrm{i}t})\mathrm{d}t = \frac{1}{2\pi} \int_{-\pi}^{\pi} u(x_0 + R\cos t, y_0 + R\sin t)\mathrm{d}t. \end{aligned} \tag{7.4}$$

Remark. Due to periodicity, we may replace the integration from $-\pi$ to π by integration from 0 to 2π.

There is an analogue of the Cauchy formula for harmonic functions. The following theorem is valid.

Theorem 7.3 (Poisson formula). *Let real-valued function $u(re^{i\theta})$, $\theta \in [0, 2\pi]$, be harmonic in the disk $\{z : |z| \le R\}$. Then for any $0 \le r < R$, we have*

$$u(re^{i\theta}) = \frac{1}{2\pi}\int_0^{2\pi} \frac{R^2 - r^2}{R^2 - 2Rr\cos(\theta - \phi) + r^2} u(Re^{i\phi})d\phi.$$

This formula is known as the Poisson integral for the disk.

Proof. As we know from this chapter, any real-valued harmonic function is the real part of some uniquely (up to constant) determined analytic function. So, in this case $u(re^{i\theta}) = \operatorname{Re} f(z)$, where $f(z)$ is analytic in the disk $\{z : |z| \le R\}$. Moreover, we may assume without loss of generality that in this case $f(\overline{z}) = \overline{f(z)}$. This fact can be rewritten as $f(re^{-i\theta}) = u - iv$, if $f(re^{i\theta}) = u + iv$. Applying the Cauchy formula, we have

$$f(z) = f(re^{i\theta}) = u + iv = \frac{1}{2\pi i}\int_{|\zeta|=R} \frac{f(\zeta)}{\zeta - z}d\zeta = \frac{1}{2\pi}\int_0^{2\pi} \frac{f(Re^{i\phi})Re^{i\phi}}{Re^{i\phi} - re^{i\theta}}d\phi,$$

where $0 \le r < R$. Since the point $\frac{R^2}{re^{i\theta}}$ lies outside of the disk $\{z : |z| \le R\}$, then another application of the Cauchy formula leads to

$$0 = \frac{1}{2\pi i}\int_{|\zeta|=R} \frac{f(\zeta)}{\zeta - R^2r^{-1}e^{-i\theta}}d\zeta = \frac{1}{2\pi}\int_0^{2\pi} \frac{f(Re^{i\phi})Re^{i\phi}}{Re^{i\phi} - R^2r^{-1}e^{-i\theta}}d\phi.$$

Changing here ϕ to $-\phi$ and using periodicity with respect to ϕ (and our assumption about $f(z)$), we obtain

$$0 = \frac{1}{2\pi}\int_0^{2\pi} \frac{\overline{f(Re^{i\phi})}re^{i\theta}}{re^{i\theta} - Re^{i\phi}}d\phi.$$

Taking the difference between these two formulas, we can easily obtain that

$$f(re^{i\theta}) = \frac{1}{2\pi}\int_0^{2\pi} \left(\operatorname{Re} f(Re^{i\phi})\frac{Re^{i\phi} + re^{i\theta}}{Re^{i\phi} - re^{i\theta}} + i\operatorname{Im} f(Re^{i\phi})\right)d\phi.$$

Considering now the real parts, we obtain the Poisson formula. Moreover, consideration of the imaginary parts gives the following formula:

$$v(re^{i\theta}) = \frac{1}{2\pi}\int_0^{2\pi}\left(v(Re^{i\phi}) + \frac{2Rr\sin(\theta-\phi)}{R^2 - 2Rr\cos(\theta-\phi)+r^2}u(Re^{i\phi})\right)d\phi.$$

The theorem is proved. □

Remark. An analogue of the Poisson integral for the real part of the analytic function is also valid for the imaginary part of the same analytic function (compare with the latter formula in the proof of the theorem).

There is even more a general formula than Poisson integral, which is called Jensen's formula. It establishes a connection between the modulus of the zeros of the analytic function inside the disk and the average of log for this function at the boundary. More precisely, the following theorem holds.

Theorem 7.4 (Jensen's formula). *Let function $f(z)$ be analytic in the disk $\{z : |z| \le R\}$. Suppose that $a_1, a_2, \dots, a_n$ are the zeros of $f(z)$ in the interior of this disk (repeated according to their respective multiplicities), and let $f(0) \neq 0$. Then the* Jensen's formula *states*

$$\log|f(0)| = \sum_{k=1}^{n}\log\left(\frac{|a_k|}{R}\right) + \frac{1}{2\pi}\int_0^{2\pi}\log|f(Re^{i\phi})|d\phi.$$

Proof. Let us assume first that $f(z)$ has no zeros in the disk $\{z : |z| \le R\}$, then since $\log f(z)$ is analytic there (see Problem 7.2) we have by the Cauchy formula that

$$\log f(0) = \frac{1}{2\pi i}\int_{|z|=R}\frac{\log f(z)}{z}dz = \frac{1}{2\pi}\int_0^{2\pi}\log f(Re^{i\phi})d\phi.$$

Considering the real part of the latter equality, we obtain the Jensen's formula in this case, i. e.,

$$\log|f(0)| = \frac{1}{2\pi}\int_0^{2\pi}\log|f(Re^{i\phi})|d\phi.$$

Since $\log|f(z)|$ is harmonic (see Problem 7.2), then the latter formula can be seen as the generalization of the mean value property of harmonic functions.

Assume now that $f(z)$ has only one zero $a_1 = re^{i\theta}$, $r < R$, in this disk. Then applying the Cauchy formula to the function $\log(1 - zre^{-i\theta})$, we obtain

$$0 = \int_{|z|=\frac{1}{R}}\frac{\log(1-zre^{-i\theta})}{z}dz,$$

where we have used the main value of log. This equality implies that

$$\frac{1}{2\pi i}\int\limits_{|z|=\frac{1}{R}} \frac{\log(\frac{1-zre^{-i\theta}}{-zre^{-i\theta}})}{z}dz = \frac{1}{2\pi i}\int\limits_{|z|=\frac{1}{R}} \frac{\log(\frac{1}{-zre^{-i\theta}})}{z}dz$$

$$= \frac{1}{2\pi i}\int\limits_{|z|=\frac{1}{R}} \frac{\log(\frac{1}{-re^{-i\theta}})}{z}dz - \frac{1}{2\pi i}\int\limits_{|z|=\frac{1}{R}} \frac{\log z}{z}dz$$

$$= \log\left(\frac{1}{-re^{-i\theta}}\right) - \frac{1}{4\pi i}\log^2 z\Big|_{\arg z=0}^{\arg z=2\pi}$$

$$= -\log(-re^{i\theta}) - \frac{1}{4\pi i}\left(\log\frac{1}{R} + i2\pi\right)^2 + \frac{1}{4\pi i}\log^2\frac{1}{R}.$$

Considering the real part of the left and right sides of this equality, we easily obtain

$$\frac{1}{2\pi}\int\limits_0^{2\pi} \log\left|1 - \frac{R}{r}e^{i(\theta-\phi)}\right| d\phi = \log\frac{R}{r}.$$

This is the Jensen's formula for the function $f(z) = 1 - \frac{z}{a_1}$.

Now the general case can be obtained from the previous two particular cases if we represent the given function $f(z)$ with zeros $a_1, a_2, \dots, a_n$ as

$$f(z) = \left(1 - \frac{z}{a_1}\right)\left(1 - \frac{z}{a_2}\right)\cdots\left(1 - \frac{z}{a_n}\right)\varphi(z)$$

such that $\varphi(z)$ is analytic in the disk $\{z : |z| \le R\}$, has no zeros there and $\varphi(0) = f(0)$. Using this representation, the Jensen's formula is obtained now simply by adding. The theorem is completely proved. □

Corollary 7.5. *Under the conditions of the theorem above, assume in addition that besides the zeros $a_1, a_2, \dots, a_n$ the function $f(z)$ has also the poles $b_1, b_2, \dots, b_m$ (repeated also according to their respective multiplicities). Then*

$$\log|f(0)| = \sum_{k=1}^{n}\log\left(\frac{|a_k|}{R}\right) - \sum_{k=1}^{m}\log\left(\frac{|b_k|}{R}\right) + \frac{1}{2\pi}\int\limits_0^{2\pi}\log|f(Re^{i\phi})|d\phi.$$

Proof. The proof follows if we represent $f(z)$ as

$$f(z) = \frac{g(z)}{h(z)}, \quad \text{where } h(z) = \left(1 - \frac{z}{b_1}\right)\left(1 - \frac{z}{b_2}\right)\cdots\left(1 - \frac{z}{b_m}\right).$$ □

Problem 7.6. Under the conditions of Theorem 7.4, let us denote by $n(t)$ the number of zeros of $f(z)$ in the disk $\{z : |z| \le t\}$. Prove that

$$\log|f(0)| + \int_0^R \frac{n(t)}{t}\mathrm{d}t = \frac{1}{2\pi}\int_0^{2\pi} \log|f(Re^{\mathrm{i}\phi})|\mathrm{d}\phi.$$

We now prove a very important result concerning the modulus of an analytic function.

Theorem 7.7 (Maximum modulus principle). *Let f be analytic and non-constant in a connected domain D (which is not necessarily bounded). If $M := \sup_D |f(z)|$, then for any $z \in D$ we have $|f(z)| < M$, i. e., $|f(z)|$ does not attain its supremum at any point $z_0 \in D$.*

Proof. The value M cannot be equal to zero since in this case $f \equiv 0$. It contradicts with the conditions of this theorem. If $M = \infty$, then due to analyticity of f in D, we have $|f(z)| < \infty$ for every $z \in D$, i. e. $|f(z)| < M$. That is why we assume now that $0 < M < \infty$.

We will assume on the contrary that there is $z_0 \in D$ such that $M = |f(z_0)|$. The mean-value formula (7.3) leads to

$$M = |f(z_0)| = \frac{1}{2\pi}\left|\int_{-\pi}^{\pi} f(z_0 + Re^{\mathrm{i}t})\mathrm{d}t\right| \le \frac{1}{2\pi}\int_{-\pi}^{\pi} |f(z_0 + Re^{\mathrm{i}t})|\mathrm{d}t$$

for any $0 \le R \le R_0$ such that $\{z : |z - z_0| \le R\} \subset D$. Using this, we will prove that $|f(z)| = M$ for all $z \in \{z : |z - z_0| \le R\}$, $0 \le R \le R_0$. Assume again on the contrary that there is $R > 0$ with $0 \le R \le R_0$ and $t_0 \in [-\pi, \pi]$ such that $|f(z_0 + Re^{\mathrm{i}t_0})| < M$. Since $|f(z)|$ is continuous, there is $\delta > 0$ such that $|f(z_0 + Re^{\mathrm{i}t})| < M$ for any $t \in (t_0 - \delta, t_0 + \delta)$. If $t_0 = \pm\pi$, then we will consider only either $(t_0, t_0 + \delta)$ or $(t_0 - \delta, t_0)$. These assumptions lead to the following inequalities:

$$\begin{aligned} M &\le \frac{1}{2\pi}\int_{-\pi}^{\pi} |f(z_0 + Re^{\mathrm{i}t})|\mathrm{d}t = \frac{1}{2\pi}\int_{-\pi}^{t_0-\delta} |f(z_0 + Re^{\mathrm{i}t})|\mathrm{d}t \\ &\quad + \frac{1}{2\pi}\int_{t_0-\delta}^{t_0+\delta} |f(z_0 + Re^{\mathrm{i}t})|\mathrm{d}t + \frac{1}{2\pi}\int_{t_0+\delta}^{\pi} |f(z_0 + Re^{\mathrm{i}t})|\mathrm{d}t \\ &< \frac{1}{2\pi}[M(t_0 - \delta + \pi) + 2M\delta + M(\pi - t_0 - \delta)] = M. \end{aligned}$$

This contradiction shows that $|f(z)| \equiv M$ in every disk $\{z : |z - z_0| \le R\}$, $0 \le R \le R_0$. Let us show that this equality $|f(z)| = M$ holds in any point $z \in D$. In order to prove it, we join z_0 and z by a piecewise smooth Jordan curve $\gamma \subset D$ and denote by $d > 0$ the minimum distance from γ to ∂D. Next, we find consecutive points $z_0, z_1, \ldots, z_n = z$ along γ with $|z_{k+1} - z_k| \le d/2$ such that the disks $D_k = \{z : |z - z_k| \le d/2\}$, $k = 0, 1, \ldots, n-1$ are contained in D and cover γ, see Figure 7.1. Each disk D_k contains the center z_{k+1} of the next disk D_{k+1}.

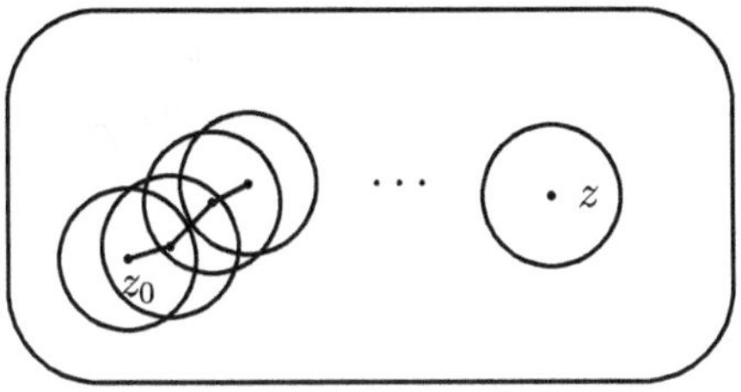

Figure 7.1: Sequence of disks D_k.

That is why it follows $|f(z)| = M$ for all $z \in D_1$, and inductively, $|f(z)| = M$ for all $z \in D_n$. Thus, $|f(z)| \equiv M$ everywhere in D.

The last step is to show that $f(z) \equiv$ constant. Indeed, since $u^2 + v^2 \equiv M^2$ then

$$\begin{cases} uu_x + vv_x = 0, \\ uu_y + vv_y = 0. \end{cases}$$

By the Cauchy–Riemann conditions, we get

$$\begin{cases} uu_x - vu_y = 0, \\ uu_y + vu_x = 0. \end{cases}$$

Hence, $u_x M^2 \equiv 0$ and $u_y M^2 \equiv 0$. Since $0 < M < \infty$, it follows that $u_x \equiv u_y \equiv 0$ in D. These two facts imply immediately that $u \equiv$ constant. Similarly, we may obtain that $v \equiv$ constant, i. e., $f \equiv$ constant. This contradiction proves the theorem completely. □

Corollary 7.8. *Let D be a bounded connected domain and let f be analytic in D and continuous in $\overline{D}$. Then either $f \equiv$ constant or $\max_{\overline{D}} |f(z)|$ achieves at the boundary ∂D.*

Proof. Since $f \in C(\overline{D})$ and $\overline{D}$ is compact set in $\mathbb{C}$, then $|f(z)|$ is continuous there, too, and by Weierstrass theorems there is $\max_{z \in \overline{D}} |f(z)|$, which is achieved at some point $z_0 \in \overline{D}$, i. e.,

$$\max_{z \in \overline{D}} |f(z)| = |f(z_0)|.$$

If $f \not\equiv$ constant, then Theorem 7.7 implies that for every $z \in D$ we have

$$\max_{z \in \overline{D}} |f(z)| > |f(z)|.$$

Thus, $z_0 \in \partial D$, i. e., $|f|$ achieves its maximum at the boundary. □

Corollary 7.9. *Let f_1 and f_2 be analytic in D and continuous in $\overline{D}$, where D is bounded. If $f_1(z) = f_2(z)$ for all $z \in \partial D$, then $f_1(z) = f_2(z)$ everywhere in $\overline{D}$.*

Proof. Let us consider $f(z) := f_1(z) - f_2(z)$. Then Corollary 7.8 implies that $\max_{z\in\overline{D}} |f(z)|$ is achieved at the boundary or $f \equiv$ constant. But $f(z) = 0$ at the boundary. That is why in both cases $f(z) \equiv 0$. □

Corollary 7.10. *Let f be analytic in D. Let us assume in addition that $f(z) \neq 0$ everywhere in D. Then either $f \equiv$ constant in $\overline{D}$ or $\inf_D |f(z)| < |f(z)|$ for all $z \in D$.*

Proof. Since $f(z) \neq 0$ and analytic in D, then $g(z) := 1/f(z)$ is well-defined and analytic in D. Theorem 7.7 implies that either $g \equiv$ constant (so is f) or for every $z \in D$ it follows that

$$|g(z)| < \sup_D |g(z)| = \frac{1}{\inf_D |f(z)|}.$$

This means that $\inf_D |f(z)| < |f(z)|$. □

As a straightforward application of maximum modulus principle, we obtain the famous *Schwarz's lemma.*

Corollary 7.11 (Schwarz's lemma). *Let $D = \{z : |z| < 1\}$ be the open unit disk in the complex plane $\mathbb{C}$ centered at the origin, and let $f : D \to \mathbb{C}$ be analytic such that $f(0) = 0$ and $|f(z)| \leq 1$ on D. Then $|f(z)| \leq |z|$ for all $z \in D$ and $|f'(0)| \leq 1$. Moreover, if $|f(z)| = |z|$ for some nonzero z or $|f'(0)| = 1$, then $f(z) = az$ for some $a \in \mathbb{C}$ with $|a| = 1$.*

Since the mean value formula holds also for harmonic functions (see (7.4)), we obtain *maximum principle* for harmonic functions.

Theorem 7.12. *Let $u(x,y)$ be real-valued, harmonic, and nonconstant in the domain D (not necessarily bounded). If $M = \sup_D u(x,y)$ and $m = \inf_D u(x,y)$, then*

$$m < u(x,y) < M \tag{7.5}$$

for any $(x,y) \in D$.

Proof. The proof literally repeats the proof of Theorem 7.7. □

Remark. In (7.5), it might be that $m = -\infty$ or $M = \infty$.

Corollary 7.13. *Let $u(x,y)$ be real-valued and harmonic in D and continuous in $\overline{D}$, where D is a bounded domain. Then either $u \equiv$ constant or for any $(x,y) \in D$, we have*

$$\min_{\overline{D}} u(x,y) < u(x,y) < \max_{\overline{D}} u(x,y),$$

i. e., $\min u(x,y)$ and $\max u(x,y)$ are achieved at the boundary ∂D.

Problem 7.14. Let $f(z) = az + b$ and $D = \{z : |z| < 1\}$. Prove that

$$\max_{|z|\le 1}|f(z)| = |a| + |b|$$

and $\max_{|z|\le 1}|f(z)| = |f(e^{i\theta_0})|$ for some real θ_0. Show also that $\theta_0 = \arg b - \arg a$.

Problem 7.15. Let $f(z) = az + b$ with $|b| > |a|$ and $D = \{z : |z| < 1\}$. Prove that $\min_{|z|\le 1}|f(z)| = |b| - |a|$ and $\min_{|z|\le 1}|f(z)| = |f(e^{i\theta_0})|$ with $\theta_0 = \arg b - \arg a + \pi$.

Problem 7.16. Prove Corollary 7.11. Hint: Consider the function $g(z) := \frac{f(z)}{z}$ if $z \neq 0$ and $g(0) := f'(0)$ by continuity and then apply maximum modulus principle to function $g(z)$.

Phragmen–Lindelöf principle

This principle is the extension of the maximum modulus principle for analytic functions to the case when the domain of analyticity is not bounded or when there is a singular point at the boundary of bounded domain. For the first case, the singular point is ∞ and we consider mostly this case. The main result is the following theorem.

Theorem 7.17 (Phragmen–Lindelöf). *Let a domain $D \subset \mathbb{C}$ be the region between two half-lines, which form angle $\frac{\pi}{\alpha}, \alpha > \frac{1}{2}$ with vertex at the origin. Assume that the function $f(z)$ is analytic on $\overline{D}$ such that*

$$|f(z)| \le M, \quad z \in \partial D,$$

and

$$|f(z)| \le M_1 e^{|z|^\beta}, \quad z \in \overline{D}$$

with some $\beta < \alpha$. Then

$$|f(z)| \le M, \quad z \in \overline{D}$$

including $z = \infty$.

Proof. We may assume without loss of generality that the domain D is symmetric with respect to the real line (see Figure 7.2).

We will denote these half-lines as $z = |z|e^{\pm i\frac{\pi}{2\alpha}}$, while the part of the circle with radius R is denoted $z = Re^{i\theta}$, $|\theta| \le \frac{\pi}{2\alpha}$. Since function z^γ for $\beta < \gamma < \alpha$ is univalent in D, we can define uniquely a new function $F(z)$ as

$$F(z) := e^{-\epsilon z^\gamma} f(z), \quad \epsilon > 0.$$

It is clear that

$$|F(z)| = e^{-\epsilon|z|^\gamma \cos(\gamma\theta)}|f(z)|, \quad |\theta| \le \frac{\pi}{2\alpha}$$

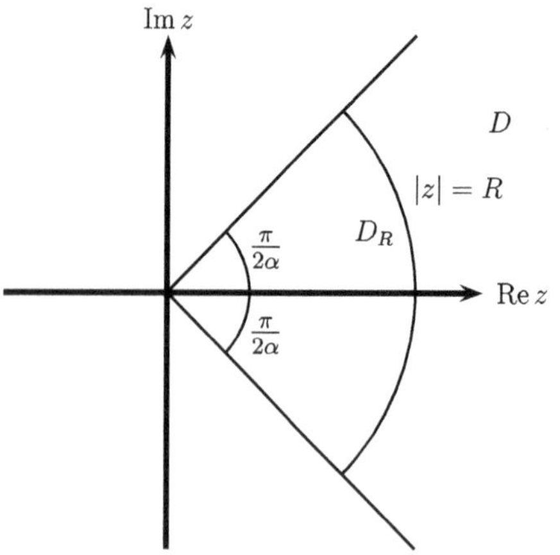

Figure 7.2: Symmetric domain D.

and, therefore, on these half-lines, i. e., for $|\theta| = \frac{\pi}{2a}$, we have the estimate

$$|F(z)| \le e^{-\epsilon|z|^{\gamma}\cos(\frac{\gamma\pi}{2a})}|f(z)| \le |f(z)| \le M.$$

Next, for the part of the circle $|z| = R$, $|\theta| \le \frac{\pi}{2a}$ we obtain that

$$|F(z)| \le M_1 e^{R^{\beta} - \epsilon R^{\gamma}\cos(\frac{\gamma\pi}{2a})}.$$

Since $\beta < \gamma < a$ and $\epsilon > 0$, then the latter inequality implies that $|F(z)| \to 0$ as $R \to +\infty$. It allows us to conclude that for $R > 0$ big enough

$$|F(z)| \le M, \quad |z| = R, \quad |\theta| \le \frac{\pi}{2a}.$$

Applying now the maximum modulus principle (see Theorem 7.7) for the bounded domain D_R (see Figure 7.2), we get that for all $z \in \overline{D_R}$,

$$|F(z)| \le M.$$

Taking into account that this estimate is independent on $R > 0$, we may conclude that the latter inequality is valid for all $z \in \overline{D}$. Hence,

$$|f(z)| \le Me^{\epsilon|z|^{\gamma}}, \quad z \in \overline{D},$$

where $\epsilon > 0$ and arbitrary small. Letting $\epsilon \to 0$ proves the theorem. □

Remark. The proof of the latter theorem shows that the half-lines can be replaced by the curves going to infinity.

Corollary 7.18. *Assume that for any $\delta > 0$ there is $M_\delta > 0$ such that*

$$|f(z)| \le M_\delta e^{\delta |z|^\alpha}, \quad z \in \overline{D},$$

and assume, as in the theorem, that on the half-lines,

$$|f(z)| \le M, \quad z \in \partial D.$$

Then

$$|f(z)| \le M, \quad z \in \overline{D}$$

including $z = \infty$.

Proof. Consider again (as in the theorem) the symmetric angle $|\theta| \le \frac{\pi}{2\alpha}$ and the function

$$F(z) := e^{-\epsilon z^\alpha} f(z), \quad z \in D, \epsilon > 0.$$

If $0 < \delta < \epsilon$ and $z \in \mathbb{R}_+$, i. e., $z = x > 0$, then we have

$$|F(z)| = e^{-\epsilon x^\alpha} |f(x)| \le M_\delta e^{-(\epsilon - \delta) x^\alpha},$$

and thus

$$\lim_{z = x \to +\infty} |F(z)| = 0.$$

Hence, for $z = x > 0$ we have that $|F(z)| \le M_1$, where $M_1 = \sup_{z=x>0} |F(z)|$. Denoting by $M_2 = \max(M, M_1)$ and applying Theorem 7.17 to each of the angles $0 \le \theta \le \frac{\pi}{2\alpha}$ and $-\frac{\pi}{2\alpha} \le \theta \le 0$, respectively, we conclude that

$$|F(z)| \le M_2, \quad z \in \overline{D}.$$

But actually $M_1 \le M$ and so $M_2 = M$. Indeed, by the definition of M_1 and continuity of $|F(z)|$ on the positive real line, we have by the Weierstrass theorem that $M_1 = |F(x_1)|$ at some point $x_1 \in \mathbb{R}_+$. Assuming that $M_1 > M$, we obtain the contradiction with the maximum modulus principle. Thus, finally

$$|F(z)| \le M, \quad z \in \overline{D}$$

and, therefore,

$$|f(z)| \le M e^{\epsilon |z|^\alpha}, \quad z \in \overline{D}.$$

In virtue of arbitrariness of $\epsilon > 0$, we deduce from here that

$$|f(z)| \le M, \quad z \in \overline{D}.$$

This completes the proof of corollary. □

Corollary 7.19. *Let function $f(z)$ be analytic and bounded in the angle mentioned in the theorem. If*

$$\lim_{z\to\infty, z=|z|e^{\pm i\frac{\pi}{2\alpha}}} f(z) = a,$$

then

$$\lim_{z\to\infty, z=|z|e^{i\theta}} f(z) = a, \quad |\theta| \le \frac{\pi}{2\alpha}$$

uniformly.

Proof. We may assume without loss of generality that $a = 0$ and $\alpha > 1$, i. e., $\frac{\pi}{\alpha} < \pi$. Consider the function

$$F(z) := \frac{z}{z+\lambda} f(z), \quad z \in D, \lambda > 0.$$

Then evidently

$$|F(z)| = \frac{|z|}{\sqrt{|z|^2 + 2\lambda|z|\cos\theta + \lambda^2}} |f(z)| \le \frac{|z|}{\sqrt{|z|^2 + \lambda^2}} |f(z)|.$$

Since $|f(z)| \le M$ in D and $\lim_{z\to\infty} f(z) = 0$ on the half-lines, then for any $\epsilon > 0$ there exists $R > 0$ such that

$$|f(z)| < \epsilon, \quad |z| > R, \quad z = |z|e^{\pm i\frac{\pi}{2\alpha}}.$$

Choosing $\lambda = \frac{RM}{\epsilon}$, we have for all $|z| \le R$ that

$$|F(z)| \le \frac{|z|}{\lambda} M \le \epsilon,$$

and we have for all $|z| > R$ on the half-lines that

$$|F(z)| \le |f(z)| < \epsilon.$$

Hence, due to Theorem 7.17 we have that

$$|F(z)| \le \epsilon, \quad z \in \overline{D},$$

and thus, for $|z| > \lambda$,

$$|f(z)| \le \left(1 + \frac{|z|}{\lambda}\right)|F(z)| < 2\epsilon.$$

This proves the corollary. □

Problem 7.20. Under the conditions of the latter corollary assume that

$$\lim_{z\to\infty, z=|z|e^{+i\frac{\pi}{2a}}} f(z) = a, \qquad \lim_{z\to\infty, z=|z|e^{-i\frac{\pi}{2a}}} f(z) = b.$$

Prove that $a = b$.

There are some interesting applications of the Phragmen–Lindelöf principle.

Proposition 7.21. *Let function $f(z)$ be analytic for all z with* $\operatorname{Re} z \ge 0$ *and*

$$|f(z)| \le M e^{k|z|}, \quad \operatorname{Re} z \ge 0,$$

for some $k > 0$. If

$$|f(z)| \le M_1 e^{-a|z|}, \quad \operatorname{Re} z = 0$$

for some $a > 0$, then $f(z) \equiv 0$ for all z, $\operatorname{Re} z \ge 0$.

Proof. Consider the function

$$F(z) := e^{-(k+ia)z} f(z), \quad \operatorname{Re} z \ge 0.$$

Then we have

$$|F(z)| = |f(z)| e^{-(k\cos\theta - a\sin\theta)|z|}, \quad |\theta| \le \frac{\pi}{2}.$$

The condition of the proposition imply

$$|F(z)| \le M, \quad \theta = 0, \quad |F(z)| \le M_1, \quad \theta = \frac{\pi}{2}.$$

Applying the Phragmen–Lindelöf principle to the angle $0 \le \theta \le \frac{\pi}{2}$, we obtain that

$$|F(z)| \le \max(M, M_1), \quad 0 \le \theta \le \frac{\pi}{2}$$

and, therefore, for this angle we have

$$|f(z)| \le \max(M, M_1) e^{(k\cos\theta - a|\sin\theta|)|z|}.$$

The same is true also for the angle $-\frac{\pi}{2} \le \theta \le 0$.

Consider now the function

$$F_1(z) := e^{Az} f(z), \quad z \in D,$$

where $A > 0$ is big enough. The estimates for $f(z)$ from above lead to the estimate

$$|F_1(z)| \le \max(M, M_1) e^{((k+A)\cos\theta - a|\sin\theta|)|z|}.$$

In particular, it yields that if $\theta = \pm\frac{\pi}{2}$ or if $\theta = \pm\arctan(\frac{k+A}{a})$, then

$$|F_1(z)| \le \max(M, M_1).$$

Applying now the Phragmen–Lindelöf principle to each of the angles: $-\frac{\pi}{2} \le \theta \le \theta_0$, $\theta_0 \le \theta \le \theta_0$, $\theta_0 \le \theta \le \frac{\pi}{2}$, where $\theta_0 = \arctan(\frac{k+A}{a})$, we get that the inequality

$$|F_1(z)| \le \max(M, M_1)$$

actually holds for the whole angle $-\frac{\pi}{2} \le \theta \le \frac{\pi}{2}$. Consequently,

$$|f(z)| \le \max(M, M_1) e^{-A\cos\theta|z|}, \quad |\theta| < \frac{\pi}{2}.$$

Letting $A \to +\infty$, we can conclude that $|f(z)| \equiv 0$ for all $z, \operatorname{Re} z \ge 0$. Proposition is proved. □

Proposition 7.22. *Let function $f(z)$ be analytic for all z, $\operatorname{Re} z \ge 0$, and*

$$|f(z)| \le M e^{k|z|}, \quad 0 < k < \pi.$$

If $f(z) = 0$ for each $z = 0, 1, 2, \ldots$, then $f(z) \equiv 0$ for all z, $\operatorname{Re} z \ge 0$.

Proof. Consider the function

$$F(z) = \frac{f(z)}{\sin \pi z}, \quad \operatorname{Re} z \ge 0.$$

We note that at the points $z = 0, 1, 2, \ldots$ this function has removable singularities (see Definition 10.9) and can be redefined as the analytic function everywhere in $\operatorname{Re} z \ge 0$. Next, the function $\frac{1}{\sin \pi z}$ is bounded on the circles $|z| = n + \frac{1}{2}$, $n = 0, 1, 2, \ldots$. Hence, we have that

$$|F(z)| \le M_1 e^{k|z|}, \quad |z| = n + \frac{1}{2}, n = 0, 1, 2, \ldots.$$

It is also true for all $z, \operatorname{Re} z = 0$ with some constant $M_1 > 0$. This implies that for all $z, n - \frac{1}{2} < |z| < n + \frac{1}{2}, n = 1, 2, \ldots$

$$|F(z)| \le M_1 e^{k(n+1/2)} \le M_2 e^{k|z|}, \quad n - \frac{1}{2} < |z| < n + \frac{1}{2}.$$

Hence, this is true for all z, $\operatorname{Re} z \geq 0$. Moreover, if $\operatorname{Re} z = 0$ we have that ($z = iy$)

$$|F(z)| \leq M e^{(k-\pi)|y|}, \quad |y| > 1.$$

The statement of this proposition follows from Proposition 7.21. □

Problem 7.23. Let function $f(z)$ be analytic for all z, $\operatorname{Re} z \geq 0$, and

$$|f(z)| \leq M_1 e^{|z|^{1-\epsilon}}$$

with $\epsilon > 0$ which is arbitrary small. Assume that

$$|f(z)| \leq M, \quad \operatorname{Re} z = 0$$

and $f(1) = 0$. Prove that

$$|f(z)| \leq M \frac{|z-1|}{|z+1|}, \quad \operatorname{Re} z > 0.$$

Hint: Consider function $\frac{1+z}{1-z} f(z)$.

Problem 7.24. Assume that function $f(z)$ is analytic in the angle $\theta_1 \leq \theta \leq \theta_2$, $\theta_2 - \theta_1 < \frac{\pi}{\rho}$, $\rho > 1$, and satisfies there the conditions

$$e^{-|z|^{\rho+\epsilon}} \leq |f(z)| \leq e^{|z|^{\rho+\epsilon}}, \quad \epsilon > 0.$$

Assume that there exist the limits

$$\lim_{z \to \infty, z = |z| e^{i\theta_j}} |z|^{-\rho} \log|f(z)| = h_j, \quad j = 1, 2.$$

If $H(\theta) = a \cos \rho\theta + b \sin \rho\theta$ with real coefficients a, b, and $H(\theta_1) = h_1$, $H(\theta_2) = h_2$, then prove that actually

$$H(\theta) \equiv \overline{\lim_{|z| \to \infty}} |z|^{-\rho} \log|f(z)|, \quad \theta_1 \leq \theta \leq \theta_2.$$

8 Liouville's theorem and the fundamental theorem of algebra

Theorem 8.1 (Cauchy's inequality). *Let f be analytic in a bounded domain D (not necessarily simply connected) and continuous in $\overline{D}$. Then for any $z_0 \in D$ and for any $n = 0, 1, \ldots,$ we have*

$$|f^{(n)}(z_0)| \le \frac{n!M}{R^n}, \tag{8.1}$$

where $M = \max_{\overline{D}} |f(z)|$ and $R = \operatorname{dist}(z_0, \partial D)$.

Proof. Let $z_0 \in D$ be arbitrary and let $R > 0$ be chosen such that we have $\{z : |z - z_0| < R\} \subset D$. Then using the Cauchy integral formula, we obtain

$$f^{(n)}(z_0) = \frac{n!}{2\pi \mathrm{i}} \int\limits_{|\zeta - z_0| = R} \frac{f(\zeta)}{(\zeta - z_0)^{n+1}} \mathrm{d}\zeta, \quad n = 0, 1, \ldots.$$

If we parametrize $|\zeta - z_0| = R$ as $\zeta = z_0 + R\mathrm{e}^{\mathrm{i}t}$, $t \in [-\pi, \pi]$, then

$$f^{(n)}(z_0) = \frac{n!}{2\pi \mathrm{i}} \int\limits_{-\pi}^{\pi} \frac{f(z_0 + R\mathrm{e}^{\mathrm{i}t})\mathrm{i}R\mathrm{e}^{\mathrm{i}t}}{R^{n+1}\mathrm{e}^{\mathrm{i}(n+1)t}} \mathrm{d}t = \frac{n!}{2\pi R^n} \int\limits_{-\pi}^{\pi} \frac{f(z_0 + R\mathrm{e}^{\mathrm{i}t})}{\mathrm{e}^{\mathrm{i}nt}} \mathrm{d}t.$$

This representation implies the inequality

$$|f^{(n)}(z_0)| \le \frac{n!}{2\pi R^n} \int\limits_{-\pi}^{\pi} \frac{|f(z_0 + R\mathrm{e}^{\mathrm{i}t})|}{|\mathrm{e}^{\mathrm{i}nt}|} \mathrm{d}t \le \frac{n!M}{2\pi R^n} 2\pi = \frac{n!M}{R^n}$$

and concludes the proof. □

Theorem 8.2 (Liouville's theorem). *Let f be analytic in the whole complex plane $\mathbb{C}$ and let $\alpha \ge 0$ be such that*

$$|f(z)| \le M(1 + |z|)^{\alpha}, \quad z \in \mathbb{C}$$

with some positive constant M. Then f is a polynomial of degree at most $[\alpha]$, where $[\alpha]$ denotes the entire part of α.

Proof. Let $n = [\alpha]$. Since f is analytic in $\mathbb{C}$, then for every $R > 0$ and every $z \in \mathbb{C}$ we have by the Cauchy integral formula that

$$f^{(n+1)}(z) = \frac{(n+1)!}{2\pi \mathrm{i}} \int\limits_{|\zeta - z| = R} \frac{f(\zeta)}{(\zeta - z)^{n+2}} \mathrm{d}\zeta.$$

That is why we have the following inequality:

https://doi.org/10.1515/9783111632278-009

$$|f^{(n+1)}(z)| \le \frac{(n+1)!}{2\pi}\int_{-\pi}^{\pi}\frac{|f(z+Re^{it})|Rdt}{R^{n+2}} \le \frac{(n+1)!}{2\pi}\frac{M(1+|z|+R)^{\alpha}}{R^{n+1}}2\pi$$
$$= M(n+1)!R^{\alpha-(n+1)}(1+(1+|z|)/R)^{\alpha} \to 0$$

as $R \to \infty$ since $n+1 > \alpha$. It means that $f^{(n+1)}(z) \equiv 0$ in $\mathbb{C}$. Thus, f is a polynomial of degree not bigger than n. □

Problem 8.3. Let f be analytic in $\mathbb{C}$ and $f^{(k)}(z) \equiv 0$ in $\mathbb{C}$ for some $k = 1, 2, \ldots$. Prove that f is a polynomial of degree not bigger than $k-1$.

Corollary 8.4. *Let f be analytic in $\mathbb{C}$ (*entire function*) and bounded in $\mathbb{C}$. Then $f \equiv$* constant.

Proof. The proof follows from Theorem 8.2 with $\alpha = 0$. □

Corollary 8.5. *Let f be analytic in $\overline{\mathbb{C}}$. Then $f \equiv$* constant.

Problem 8.6. Show that the function $f(z) = \cos z$ is not bounded.

Problem 8.7. Let f be an entire function with the property $|f(z)| \ge 1$ for all $z \in \mathbb{C}$. Show that $f \equiv$ constant.

Theorem 8.8 (The fundamental theorem of algebra). *If P is a polynomial of degree $n \ge 1$, then P has at least one zero.*

Proof. Let us assume on the contrary that this polynomial has no roots, i. e., $P(z) \ne 0$ for all $z \in \mathbb{C}$. This implies that the function

$$f(z) := \frac{1}{P(z)}$$

is an entire function, i. e., it is analytic in the whole space $\mathbb{C}$. Let us write

$$P(z) = a_n z^n + a_{n-1}z^{n-1} + \cdots + a_0, \quad a_n \ne 0$$

and consider

$$|f(z)| = \frac{1}{|P(z)|} = \frac{1}{|z|^n}\cdot\frac{1}{|a_n + a_{n-1}/z + \cdots + a_0/z^n|}. \tag{8.2}$$

For $k = 1, 2, \ldots, n$, we have $|a_{n-k}|/|z|^k \to 0$ as $|z| \to \infty$. Hence, $a_n + a_{n-1}/z + \cdots + a_0/z^n \to a_n$ as $|z| \to \infty$. Thus, (8.2) implies $|f(z)| \to 0$ as $|z| \to \infty$. In particular, there is $R > 0$ such that for all $|z| \ge R$ we have

$$|f(z)| \le 1. \tag{8.3}$$

The next step is: since f is analytic everywhere in $\mathbb{C}$, then $f(z)$ is continuous for all $z \in \mathbb{C}$. In particular, it is continuous in the closed disk $\{z : |z| \le R\}$ with R as in (8.3). By the

Weierstrass theorem for continuous functions, $|f(z)|$ is bounded in this closed disk, i. e., there is a positive number $M > 0$ such that

$$|f(z)| \leq M, \quad |z| \leq R. \tag{8.4}$$

Combining (8.3) and (8.4), we obtain that

$$|f(z)| \leq \max(1, M)$$

for all $z \in \mathbb{C}$. By Liouville's theorem, $f \equiv$ constant and so is P. This contradiction proves the theorem. □

Corollary 8.9. *Let P be a polynomial of degree $n \geq 1$. Then P has exactly n zeros, counted according to their multiplicities, and it can be represented as*

$$P(z) = a_n(z - z_1)(z - z_2)\cdots(z - z_n),$$

where $a_n \neq 0$ and $z_1, z_2, \ldots, z_n$ are the zeros of P.

Problem 8.10. Prove Corollary 8.9.

9 Representation of analytic functions via the power series

Recall that the sequence of functions $S_n(z)$, in particular, the partial sums

$$\sum_{j=0}^{n} f_j(z)$$

of some series

$$\sum_{j=0}^{\infty} f_j(z),$$

converges to $f(z)$ uniformly on a set $D \subset \mathbb{C}$ if for every $\varepsilon > 0$ there exists an integer $N_0(\varepsilon) > 0$ such that for all $n \geq N_0$ and for all $z \in D$ we have

$$|S_n(z) - f(z)| < \varepsilon,$$

in particular,

$$\left| \sum_{j=n+1}^{\infty} f_j(z) \right| < \varepsilon.$$

A useful procedure called the Weierstrass M-test can help determine whether an infinite series is uniformly convergent.

Theorem 9.1 (Weierstrass' M-test). *Suppose that the series*

$$\sum_{j=0}^{\infty} f_j(z)$$

has the property that for each $j = 0, 1, \ldots$ it holds that $|f_j(z)| \leq M_j$ for all $z \in D$. If $\sum_{j=0}^{\infty} M_j$ converges, then $\sum_{j=0}^{\infty} f_j(z)$ converges uniformly and absolutely on D.

Proof. Let

$$S_n(z) = \sum_{j=0}^{n} f_j(z)$$

be the nth partial sum of the series. If $n > m$, then for all $z \in D$ we have

$$|S_n(z) - S_m(z)| = \left| \sum_{j=m+1}^{n} f_j(z) \right| \leq \sum_{j=m+1}^{n} M_j < \varepsilon$$

https://doi.org/10.1515/9783111632278-010

for all $n > m \geq N_0(\varepsilon)$. This means that for all $z \in D$ the sequence $\{S_n(z)\}$ is a Cauchy sequence. Therefore, there is a function $f(z)$ such that

$$f(z) = \lim_{n\to\infty} S_n(z) = \sum_{j=0}^{\infty} f_j(z).$$

Moreover, this convergence is uniform and absolute on D. □

It is very important to study the particular case of the general series, when $f_j(z) = c_j(z-z_0)^j, j = 0, 1, 2, \dots$, where c_j are some complex numbers and z_0 is a fixed point of the complex plane. This type of the series is called the *power series* and they play crucial role in the theory of analytic functions. The following theorem is essential for determining the domain of convergence of power series.

Theorem 9.2 (Abel). *If a power series $\sum_{j=0}^{\infty} c_j(z-z_0)^j$ converges at some point $z_1 \neq z_0$, then it converges absolutely at any point z, which satisfies the condition $|z - z_0| < |z_1 - z_0|$.*

Proof. Take an arbitrary point z with the condition $|z - z_0| < |z_1 - z_0|$ such that $\frac{|z-z_0|}{|z_1-z_0|} =: q < 1$. By virtue of the necessary condition for convergence of the series $\sum_{j=0}^{\infty} c_j(z_1 - z_0)^j$, its terms tend to 0 as $j \to \infty$. Consequently, there exists a constant $M > 0$ such that $|c_j||z_1 - z_0|^j \leq M$. This implies that

$$\left|\sum_{j=0}^{\infty} c_j(z - z_0)^j\right| \leq \sum_{j=0}^{\infty} |c_j||z - z_0|^j \leq M \sum_{j=0}^{\infty} \left(\frac{|z - z_0|}{|z_1 - z_0|}\right)^j = M \sum_{j=0}^{\infty} q^j < \infty.$$

Thus, the theorem is proved. □

Remark. In view of the latter theorem, if we consider the least upper bound ρ of the distances $|z - z_0|$ at which the series $\sum_{j=0}^{\infty} c_j(z - z_0)^j$ converges, then at all points z' satisfying the condition $|z' - z_0| > \rho$ (under the condition $\rho < \infty$), the given power series clearly diverges. Therefore, assuming that $\rho > 0$, the disk $|z - z_0| < \rho$ is the greatest domain of convergence of this power series. The series diverges everywhere outside of this disk, and at the boundary points $|z - z_0| = \rho$ it may either converge or diverge (see, e. g., the series $\sum_{j=1}^{\infty} \frac{z^j}{j}$). Such number ρ is called the *radius of convergence* and the disk $\{z : |z - z_0| < \rho\}$ is called the *region of convergence* of the corresponding power series.

Remark. The radius of convergence ρ of the power series can be calculated as

$$\frac{1}{\rho} = \overline{\lim_{j\to\infty}} \sqrt[j]{|c_j|}$$

or

$$\frac{1}{\rho} = \lim_{j\to\infty} \left|\frac{c_{j+1}}{c_j}\right|$$

if the limit exists. These formulae are the consequences of the d'Alembert and Cauchy tests for the real number series. The first of these formulae is frequently called the *Cauchy–Hadamard formula.*

Problem 9.3. Prove the Cauchy–Hadamard formula.

Theorem 9.4. *Suppose that the power series*

$$\sum_{j=0}^{\infty} c_j(z - z_0)^j$$

has a radius of convergence $\rho > 0$. *Then, for each* $r, 0 < r < \rho$, *this series converges uniformly and absolutely on the closed disk* $\{z : |z - z_0| \le r\}$ *and defines there a continuous function.*

Proof. Given $0 < r < \rho$, choose $\zeta \in \{z : |z - z_0| < \rho\}$ such that $|\zeta - z_0| = r$. Due to the properties of the power series, we have that

$$\sum_{j=0}^{\infty} c_j(z - z_0)^j$$

converges absolutely for any $z \in \{z : |z - z_0| < \rho\}$ (see Theorem 9.2). It follows that

$$\sum_{j=0}^{\infty} |c_j(\zeta - z_0)^j| = \sum_{j=0}^{\infty} |c_j| r^j$$

converges. Moreover, for all $z \in \{z : |z - z_0| \le r\}$, we have

$$|c_j(z - z_0)^j| = |c_j||z - z_0|^j \le |c_j| r^j.$$

The conclusion now follows from the Weierstrass' M-test with $M_j = |c_j| r^j$. □

Problem 9.5. Show that the geometric series

$$\sum_{j=0}^{\infty} z^j$$

converges uniformly on the closed disk $\{z : |z| \le r\}$ with any $0 < r < 1$.

Theorem 9.6. *Suppose that the power series*

$$\sum_{j=0}^{\infty} c_j(z - z_0)^j$$

has a radius of convergence $\rho > 0$. *Then in the disk* $D_\rho = \{z : |z - z_0| < \rho\}$ *this series defines the function*

$$f(z) := \sum_{j=0}^{\infty} c_j (z - z_0)^j, \tag{9.1}$$

which is analytic in D_ρ and for each $k = 1, 2, \ldots$ it holds that

$$f^{(k)}(z) = \sum_{j=k}^{\infty} c_j j(j-1)\cdots(j-k+1)(z-z_0)^{j-k}. \tag{9.2}$$

Proof. Let $0 < r < \rho$. Then due to Theorem 9.4 in the closed disk $\overline{D_r} = \{z : |z - z_0| \le r\}$ the series (9.1) converges uniformly (and absolutely) and defines a continuous function $f(z)$. That is why we may integrate this series term by term. If $\gamma \subset D_r$ is a piecewise smooth closed Jordan curve, then

$$\int_\gamma f(z)\mathrm{d}z = \sum_{j=0}^{\infty} c_j \int_\gamma (z - z_0)^j \mathrm{d}z = 0$$

since $(z - z_0)^j$ is analytic for each $j = 0, 1, \ldots$. Applying now Morera's theorem, we conclude that f is analytic in D_ρ. Formula (9.2) follows directly by induction and it is based on the properties of the power series. □

Problem 9.7. Show that

$$\log(1 - z) = -\sum_{j=1}^{\infty} \frac{z^j}{j}$$

for all $z \in D_1 = \{z : |z| < 1\}$ or

$$\log \zeta = \sum_{j=1}^{\infty} \frac{(-1)^{j+1}}{j}(\zeta - 1)^j, \quad |\zeta - 1| < 1,$$

where log is the main branch.

Problem 9.8. Let f and g have the power series representations,

$$f(z) = \sum_{j=0}^{\infty} c_j (z - z_0)^j, \quad g(z) = \sum_{j=0}^{\infty} d_j (z - z_0)^j, \tag{9.3}$$

for $z \in D_\rho = \{z : |z - z_0| < \rho\}$. Show that

$$f(z)g(z) = \sum_{j=0}^{\infty} a_j (z - z_0)^j, \quad z \in D_\rho,$$

where $a_j = \sum_{k=0}^{j} c_k d_{j-k}$.

Problem 9.9. Let f and g have the power series representations (9.3) with $d_0 \neq 0$. Show that in some neighborhood of z_0 the function $f(z)/g(z)$ can be represented as the power series

$$\frac{f(z)}{g(z)} = \sum_{j=0}^{\infty} a_j (z - z_0)^j,$$

where a_j are uniquely determined from the equations $c_j = \sum_{k=0}^{j} a_k d_{j-k}, j = 0, 1, \ldots$.

Theorem 9.10 (Taylor's expansion). *Suppose that f is analytic in a domain D and that $D_R(z_0) = \{z : |z - z_0| < R\}$ is a disk contained in D. Then f is uniquely represented in $D_R(z_0)$ as a power series*

$$f(z) = \sum_{j=0}^{\infty} c_j (z - z_0)^j, \quad z \in D_R(z_0), \tag{9.4}$$

where

$$c_j = \frac{f^{(j)}(z_0)}{j!}.$$

Furthermore, for any $r, 0 < r < R$, the convergence is uniform on $\overline{D_r}(z_0) = \{z : |z - z_0| \le r\}$. The power series with such coefficients is called the Taylor series *for f centered at z_0.*

Proof. Let $z_0 \in D$ and let $R = \operatorname{dist}(z_0, \partial D)$ so that $D_R(z_0) = \{z : |z - z_0| < R\} \subset D$. Let $z \in D_r(z_0) = \{z : |z - z_0| < r\}$ with $0 < r < R$. The Cauchy integral formula gives that

$$f(z) = \frac{1}{2\pi i} \int_{|\zeta - z_0| = r} \frac{f(\zeta)}{\zeta - z} d\zeta = \frac{1}{2\pi i} \int_{|\zeta - z_0| = r} \frac{f(\zeta)}{(\zeta - z_0)(1 - \frac{z - z_0}{\zeta - z_0})} d\zeta.$$

Since $z \in D_r(z_0)$ and $|\zeta - z_0| = r$, we have that

$$\left| \frac{z - z_0}{\zeta - z_0} \right| = \frac{|z - z_0|}{r} < 1.$$

Therefore, we have (by geometric series)

$$\frac{1}{1 - \frac{z - z_0}{\zeta - z_0}} = \sum_{j=0}^{\infty} \left(\frac{z - z_0}{\zeta - z_0} \right)^j.$$

Moreover, for such fixed z the convergence of this series is uniform on the circle $\{\zeta : |\zeta - z_0| = r\}$. Hence, we may integrate this series term by term and obtain

$$f(z) = \frac{1}{2\pi i} \int\limits_{|\zeta - z_0| = r} \frac{f(\zeta) d\zeta}{(\zeta - z_0)(1 - \frac{z - z_0}{\zeta - z_0})}$$

$$= \frac{1}{2\pi i} \int\limits_{|\zeta - z_0| = r} \frac{f(\zeta)}{\zeta - z_0} \sum_{j=0}^{\infty} \left(\frac{z - z_0}{\zeta - z_0} \right)^j d\zeta$$

$$= \sum_{j=0}^{\infty} \left(\frac{1}{2\pi i} \int\limits_{|\zeta - z_0| = r} \frac{f(\zeta)}{(\zeta - z_0)^{j+1}} d\zeta \right) (z - z_0)^j = \sum_{j=0}^{\infty} \frac{f^{(j)}(z_0)}{j!} (z - z_0)^j.$$

Since r with $0 < r < R$ and z with $|z - z_0| < r$ are arbitrary, we may conclude that the representation (9.4) with the coefficient $c_j = f^{(j)}(z_0)/j!$ holds everywhere in $D_R(z_0)$. Even more is true: the radius of convergence of (9.4) is R and the convergence of (9.4) is uniform in $\overline{D_r}(z_0) = \{z : |z - z_0| \le r\}$ with any $r, 0 < r < R$. The latter fact follows from the properties of the power series. The uniqueness of representation (9.4) follows from the fact that necessarily $c_j = f^{(j)}(z_0)/j!$. □

Corollary 9.11 (Taylor's expansion at ∞). *Let $f(z)$ be analytic for $|z| > R$ (including $z = \infty$). Then $f(z)$ is uniquely represented in $\{z : |z| > R\}$ as the series*

$$f(z) = \sum_{j=0}^{\infty} c_j z^{-j}, \quad |z| > R,$$

where $c_j = \frac{g^{(j)}(0)}{j!}$ for $g(z) := f(1/z)$. Moreover, these coefficients are equal to

$$c_j = \frac{f^{(j)}(\infty)}{j!}, \tag{9.5}$$

where $f^{(j)}(\infty) := (f^{(j-1)})'(\infty)$.

Proof. Let us consider $g(z) := f(1/z)$. Then g is analytic in the domain $\{z : |z| < 1/R\}$. Thus, Taylor's expansion (9.4) at 0 gives

$$g(z) = \sum_{j=0}^{\infty} c_j z^j,$$

where

$$c_j = \frac{1}{2\pi i} \int\limits_{|z| = \delta} \frac{g(z) dz}{z^{j+1}} = \frac{g^{(j)}(0)}{j!}.$$

Since $f(z) = g(1/z)$, we obtain for $|z| > R$ that

$$f(z) = \sum_{j=0}^{\infty} c_j z^{-j},$$

where

$$c_j = \frac{g^{(j)}(0)}{j!} = \frac{f^{(j)}(\infty)}{j!}$$

and

$$c_j = \frac{1}{2\pi i} \int_{|z|=\delta} \frac{f(1/z)dz}{z^{j+1}} = \frac{1}{2\pi i} \int_{|z|=1/\delta} f(z)z^{j-1}dz.$$

It can be mentioned here that the definition of the derivative at ∞ leads to the fact

$$f^{(j)}(\infty) = g^{(j)}(0) = j!c_j.$$

□

Remark. The results of Problem 9.7 and Corollary 9.11 allow us to show that

$$\log\left(\frac{\zeta - 1}{\zeta}\right) = -\sum_{j=1}^{\infty} \frac{1}{j}\zeta^{-j}$$

is the Taylor's expansion of $\log(\frac{\zeta-1}{\zeta})$ at ∞.

Problem 9.12. Using Corollary 9.11 show that:
1. $f(\infty) = \lim_{z\to\infty} f(z) = c_0$,
2. $f'(\infty) = \lim_{z\to\infty} z[f(z) - f(\infty)] = c_1$,
3. $f''(\infty) = -\lim_{z\to\infty}[z^3 f'(z) + zf'(\infty)] = 2c_2$,
4. $f'''(\infty) = \lim_{z\to\infty}[z^5 f''(z) + 2z^4 f'(z) - zf''(\infty)] = 6c_3$,

where the coefficients $c_j, j = 0, 1, 2, \ldots$ are from Corollary 9.11.

Problem 9.13. Show that:
1.

$$e^z = 1 + z + \cdots + \frac{z^n}{n!} + \cdots,$$

2.

$$\sin z = z - \frac{z^3}{3!} + \cdots + (-1)^n \frac{z^{2n-1}}{(2n-1)!} + \cdots,$$

3.

$$\cos z = 1 - \frac{z^2}{2!} + \cdots + (-1)^n \frac{z^{2n}}{(2n)!} + \cdots,$$

4.

$$\sinh z = z + \frac{z^3}{3!} + \cdots + \frac{z^{2n-1}}{(2n-1)!} + \cdots,$$

5.

$$\cosh z = 1 + \frac{z^2}{2!} + \cdots + \frac{z^{2n}}{(2n)!} + \cdots$$

and all these Taylor series converge for any $|z| < \infty$.

Problem 9.14. Show that:

1.

$$\tanh z = \sum_{n=1}^{\infty} \frac{2^{2n}(2^{2n}-1)B_{2n}}{(2n)!} z^{2n-1} = z - \frac{z^3}{3} + \frac{2z^5}{15} - \cdots, \qquad |z| < \frac{\pi}{2},$$

2.

$$\coth z - \frac{1}{z} = \sum_{n=1}^{\infty} \frac{2^{2n}B_{2n}}{(2n)!} z^{2n-1} = \frac{z}{3} - \frac{z^3}{45} + \frac{2z^5}{945} - \cdots, \qquad |z| < \pi,$$

where B_{2n} denotes the Bernoulli's numbers (see Exercise 29 of Part II).

Remark. Using Problem 3.18, we obtain that

$$\tan z = -\mathrm{i}\tanh(\mathrm{i}z), \qquad \cot z = \mathrm{i}\coth(\mathrm{i}z).$$

These formulae and the results of Problem 9.14 allow us to obtain immediately the following expansions:

1.

$$\tan z = \sum_{n=1}^{\infty} \frac{(-1)^{n+1}2^{2n}(2^{2n}-1)B_{2n}}{(2n)!} z^{2n-1} = z + \frac{z^3}{3} + \frac{2z^5}{15} + \cdots, \qquad |z| < \frac{\pi}{2},$$

2.

$$\cot z - \frac{1}{z} = \sum_{n=1}^{\infty} \frac{(-1)^n 2^{2n}B_{2n}}{(2n)!} z^{2n-1} = \frac{z}{3} + \frac{z^3}{45} + \frac{2z^5}{945} + \cdots, \qquad |z| < \pi.$$

Example 9.15 (Inverse trigonometric functions). We consider the inverse functions to $\sin z$ and $\tan z$. As we know (see Example 3.32), the function $\arcsin z$ (the inverse to sin) can be defined by

$$\arcsin z = -\mathrm{i}\,\mathrm{Log}(\mathrm{i}z + \sqrt{1-z^2}),$$

which is clearly multivalued, and its derivative (for any of infinitely many branches) is equal to

$$(\arcsin z)' = \frac{1}{\sqrt{1-z^2}}, \qquad z \neq \pm 1.$$

Considering now the Taylor's expansion (so-called binomial series) for the main branch of the function $\frac{1}{\sqrt{1-z^2}}$, we obtain

$$\frac{1}{\sqrt{1-z^2}} = \sum_{n=0}^{\infty} \frac{(2n-1)!!}{2^n n!} z^{2n} = \sum_{n=0}^{\infty} \frac{(2n)!}{(2^n n!)^2} z^{2n}, \quad |z| < 1.$$

Due to the derivative of arcsin z, the main branch of arcsin z can be represented (inside of the unit circle) as

$$\arcsin z = \int_0^z \frac{1}{\sqrt{1-\zeta^2}} \mathrm{d}\zeta, \quad |z| < 1.$$

Integrating now term by term the binomial series from above, we finally get the Taylor's expansion for arcsin z,

$$\arcsin z = \sum_{n=0}^{\infty} \frac{(2n)!}{(2^n n!)^2 (2n+1)} z^{2n+1}, \quad |z| < 1.$$

With the inverse function arctan z of the function $z = \tan w$, we proceed as follows. First, we have

$$z = \tan w \equiv z = -\mathrm{i}\frac{\mathrm{e}^{\mathrm{i}w} - \mathrm{e}^{-\mathrm{i}w}}{\mathrm{e}^{\mathrm{i}w} + \mathrm{e}^{-\mathrm{i}w}} \equiv \mathrm{e}^{2\mathrm{i}w} = \frac{1+\mathrm{i}z}{1-\mathrm{i}z},$$

and, second,

$$w = \arctan z = \frac{1}{2\mathrm{i}} \log\frac{1+\mathrm{i}z}{1-\mathrm{i}z} = \frac{1}{2\mathrm{i}}(\log(1+\mathrm{i}z) - \log(1-\mathrm{i}z)),$$

where the main branch for log is considered. Taking now into account Example 9.7, we obtain

$$\arctan z = \frac{1}{2\mathrm{i}}\left(-\sum_{n=1}^{\infty} \frac{(-\mathrm{i}z)^n}{n} + \sum_{n=1}^{\infty} \frac{(\mathrm{i}z)^n}{n} \right) = \frac{1}{2\mathrm{i}} \sum_{n=1}^{\infty} \left(\frac{\mathrm{i}^n - (-\mathrm{i})^n}{n} \right) z^n.$$

The latter series leads finally to the Taylor's expansion for arctan z as

$$\arctan z = \sum_{n=1}^{\infty} \frac{(-1)^n}{2n+1} z^{2n+1}, \quad |z| < 1.$$

Similar to arcsin z and arctan z, we can obtain that arccos z and arccot z are expressed as

$$\arccos z = -\mathrm{i}\,\mathrm{Log}(z + \mathrm{i}\sqrt{1-z^2}), \quad \operatorname{arccot} z = \frac{1}{2\mathrm{i}} \mathrm{Log}\left(\frac{\mathrm{i}z - 1}{\mathrm{i}z + 1} \right).$$

Problem 9.16. Construct Taylor's expansions for $\arccos z$ and $\operatorname{arccot} z$. Hint: Take into account that

$$\arccos z = \frac{\pi}{2} - \arcsin z, \quad \operatorname{arccot} z = \arctan\frac{1}{z},$$

and hence the Taylor's expansion for $\operatorname{arccot} z$ will be at ∞.

Example 9.17. Consider function $f(z)$, which is defined by the following power series:

$$f(z) := \sum_{k=1}^{\infty} \frac{z^k}{k^2}, \quad |z| \le 1.$$

It is clear that this function is analytic for $|z| < 1$ and is continuous for $|z| \le 1$. This series is actually the Taylor's expansion for $f(z)$ at zero. Moreover, this function $f(z)$ is equal to

$$f(z) = \int_0^z \frac{1}{\zeta} \log\frac{1}{1-\zeta}\mathrm{d}\zeta, \quad |z| < 1.$$

Indeed, using Problem 9.7, we have for $|z| < 1$,

$$\begin{aligned}
\int_0^z \frac{1}{\zeta}\log\frac{1}{1-\zeta}\mathrm{d}\zeta &= -\int_0^z \frac{1}{\zeta}\log(1-\zeta)\mathrm{d}\zeta = \int_0^z \frac{1}{\zeta}\sum_{j=1}^{\infty}\frac{\zeta^j}{j}\mathrm{d}\zeta \\
&= \int_0^z \sum_{j=0}^{\infty} \frac{\zeta^j}{j+1}\mathrm{d}\zeta = \sum_{j=0}^{\infty}\frac{1}{j+1}\int_0^z \zeta^j \mathrm{d}\zeta \\
&= \sum_{j=0}^{\infty}\frac{1}{j+1}\left(\frac{\zeta^{j+1}}{j+1}\right)\Big|_0^z = \sum_{j=1}^{\infty}\frac{z^j}{j^2}.
\end{aligned}$$

Moreover, this equality can be extended by continuity to all $|z| \le 1$. Next, since

$$f'(z) = -\frac{1}{z}\log(1-z), \quad |z| < 1,$$

then it follows from here that $z = 1$ is a singular point of this analytic function while its Taylor's expansion converges uniformly for $|z| \le 1$.

Problem 9.18. Let function $f(z)$ be defined by the series

$$f(z) := \sum_{k=1}^{\infty} \frac{z^{2^k}}{k^2}, \quad |z| \le 1.$$

Show that $f(z)$ is analytic for $|z| < 1$ and is continuous for $|z| \le 1$ but every point from the unit circle $|z| = 1$ is singular for this function in a sense that there is no derivative at these points ($|z| = 1$).

The next example shows that regularity of series does not guarantee the analyticity of the corresponding function

Example 9.19. Let function $f(z)$ be defined by the series

$$f(z) := \sum_{k=1}^{\infty} \frac{z^2}{a^2 + k^2 z^2}, \qquad a \neq 0.$$

Then this function is well-defined for all $z \in \overline{\mathbb{C}}$ except for the points $z_k = \pm i\frac{a}{k}, k = 1, 2, \ldots,$ since the denominator of some fraction of this series is equal to zero at each of these points. If we consider z with $\operatorname{Im} z = 0$, i. e., $z = x + i0$, then

$$|f(z)| = \sum_{k=1}^{\infty} \frac{x^2}{a^2 + k^2 x^2} < \sum_{k=1}^{\infty} \frac{1}{k^2} < \infty.$$

Hence, $f(z)$ is bounded and continuous on the real line by the Weierstrass' M-test. But for any neighborhood of $z = 0$ on the complex plane, this function is equal to infinity at the points z_k that tend to zero. It means that there is no Taylor's expansion at $z = 0$. However, it is not so difficult to see that for $|z| > |a|$ (including $z = \infty$) this function is analytic and it has a Taylor's expansion at $z = \infty$, which is equal to

$$f(z) = \sum_{j=0}^{\infty} \left((-1)^j a^{2j} \sum_{k=1}^{\infty} \frac{1}{k^{2j+2}} \right) \frac{1}{z^{2j}}.$$

Problem 9.20. Let the coefficients of the series $\sum_{j=0}^{\infty} c_j$ be real and

$$\lim_{n \to \infty} \sum_{j=0}^{n} c_j = +\infty \text{ or } -\infty.$$

Show that $z = 1$ is a singular point of the function

$$f(z) := \sum_{j=0}^{\infty} c_j z^j.$$

Problem 9.21. Prove that if function $f(z) := \sum_{j=0}^{\infty} c_j z^j$ is bounded in the unit disk $\{z : |z| \leq 1\}$, then the series $\sum_{j=0}^{\infty} |c_j|^2$ converges.

The next problem is actually the formulation of some kind uniqueness theorem for analytic functions. The main point here is the possibility to expand these functions via the Taylor's expansion.

Problem 9.22. Let f and g be analytic in a connected domain D and $f(z) = g(z)$ on the set $E \subset D$, which has a limiting point in D. Show that $f(z) = g(z)$ for all $z \in D$.

Definition 9.23. Let f be analytic in a domain D. If

$$f(z_0) = 0, \quad f'(z_0) = 0, \quad \ldots, \quad f^{(k_0-1)}(z_0) = 0$$

but $f^{(k_0)}(z_0) \neq 0$ for some $z_0 \in D$ and $k_0 \geq 1$, then z_0 is called the *zero* of f of order k_0.

Remark. It is clear that if f is analytic in a domain D, then any possible zero of f inside of D can be only of the finite order.

Problem 9.24. Let f be analytic in a domain D and let $f \not\equiv 0$. Show that all zeros of f in D are isolated, i. e., for any bounded domain D_1 with $\overline{D_1} \subset D$ there are only at most finitely many zeros of f in D_1. Hint: Use Problem 9.22.

Problem 9.25. Let functions f and g be analytic in a connected domain D. Prove that if $f(z)g(z) \equiv 0$ in D, then either $f(z) \equiv 0$ or $g(z) \equiv 0$ in D.

Problem 9.26. Let

$$f(z) = \begin{cases} \frac{\sin z}{z}, & z \neq 0, \\ 1, & z = 0. \end{cases}$$

Show that f is analytic everywhere in $\mathbb{C}$ and find its Taylor's expansion centered at 0.

Problem 9.27. Suppose that

$$f(z) = \sum_{j=0}^{\infty} c_j z^j$$

is an entire function. Show that $\overline{f(\overline{z})}$ is also entire. When $\overline{f(\overline{z})} = f(z)$?

Problem 9.28. Let

$$f(z) = \begin{cases} e^{-1/z^2}, & z \neq 0, \\ 0, & z = 0. \end{cases}$$

Show that f is not continuous at 0 and that it has no Taylor's expansion at 0.

Problem 9.29. Let f be as in Problem 9.28. Define the Taylor's expansion for f at any point $z_0 \neq 0$.

There is an equivalent formulation of the Cauchy's inequality (see Theorem 8.1) in terms of the coefficients of the Taylor's expansions. Namely, let function $f(z)$ be represented via its Taylor's expansion in the neighborhood of some point $z_0 \neq \infty$, i. e.,

$$f(z) = \sum_{j=0}^{\infty} c_j (z - z_0)^j, \quad |z - z_0| < R.$$

If $M_r = \sup_{|z|=r} |f(z)|, r < R$, then for any $j = 0, 1, 2, \ldots$, we have *Cauchy's inequality*

$$|c_j| r^j \leq M_r. \tag{9.6}$$

Indeed, for the coefficients of the Taylor's expansion, we have (see the Cauchy formula and Theorem 9.10)

$$c_j = \frac{1}{2\pi \mathrm{i}} \int\limits_{|z-z_0|=r} \frac{f(z)}{(z-z_0)^{j+1}} \mathrm{d}z.$$

Then it follows that

$$|c_j| \leq \frac{1}{2\pi} \int\limits_{|z-z_0|=r} \frac{|f(z)|}{|z-z_0|^{j+1}} |\mathrm{d}z| \leq \frac{M_r}{r^j}.$$

Problem 9.30. Show that Cauchy's inequality (9.6) is turning to the equality if and only if $f(z) = Cz^{j_0}$ with some $j_0 = 0, 1, 2, \ldots$.

The Cauchy's inequality allows us to obtain the efficient estimate for $\operatorname{Re} f(z)$ of analytic function $f(z)$ (in particular for any harmonic function). More precisely, the following proposition holds.

Proposition 9.31. *Let function $f(z)$ be analytic in the disk $\{z : |z - z_0| < R\}$ and let $A_r = \sup_{|z|=r} \operatorname{Re} f(z), r < R$. Then the following inequality holds:*

$$|c_j| r^j \leq \max(4A_r, 0) - 2\operatorname{Re} f(0), \quad j = 1, 2, \ldots,$$

where c_j are the coefficients of the Taylor's expansion of $f(z)$ in the neighborhood of z_0.

Proof. Let $z = z_0 + r\mathrm{e}^{\mathrm{i}\theta}$, $\theta \in [0, 2\pi]$, $r < R$, and let

$$f(z) = \sum_{j=0}^{\infty} c_j (z - z_0)^j = u(r, \theta) + \mathrm{i}v(r, \theta).$$

Then

$$\operatorname{Re} f(z) = u(r, \theta) = \sum_{j=0}^{\infty} (\alpha_j \cos j\theta - \beta_j \sin j\theta) r^j, \quad c_j = \alpha_j + \mathrm{i}\beta_j.$$

Since this series converges uniformly in θ, then we can integrate it term by term (after multiplying by $\cos j\theta$ and $\sin j\theta$) and obtain for $j = 1, 2, \ldots$, respectively, that

$$\alpha_j r^j = \frac{1}{\pi} \int\limits_0^{2\pi} u(r, \theta) \cos j\theta \mathrm{d}\theta, \quad \beta_j r^j = -\frac{1}{\pi} \int\limits_0^{2\pi} u(r, \theta) \sin j\theta \mathrm{d}\theta,$$

and $a_0 = \frac{1}{2\pi}\int_0^{2\pi} u(r,\theta)\mathrm{d}\theta$. Thus, we have for $j = 1, 2, \dots$

$$c_j r^j = (\alpha_j + \mathrm{i}\beta_j)r^j = \frac{1}{\pi}\int_0^{2\pi} u(r,\theta)\mathrm{e}^{-\mathrm{i}j\theta}\mathrm{d}\theta$$

and, therefore,

$$|c_j|r^j \le \frac{1}{\pi}\int_0^{2\pi} |u(r,\theta)|\mathrm{d}\theta, \quad j = 1, 2, \dots.$$

Taking into account now formula for a_0, we obtain for $j = 1, 2, \dots$ that

$$|c_j|r^j + 2a_0 \le \frac{1}{\pi}\int_0^{2\pi} (|u(r,\theta)| + u(r,\theta))\mathrm{d}\theta \le \frac{1}{\pi}\int_0^{2\pi} 2\max(A_r, 0)\mathrm{d}\theta.$$

This proves the proposition. □

Corollary 9.32 (Liouville's theorem for $\operatorname{Re} f(z)$). *Let function $f(z)$ be an entire function and assume that there is a constant M such that $\operatorname{Re} f(z) \le M$, then $f(z) \equiv$ constant.*

Proof. If we introduce a new function $\varphi(z) := \mathrm{e}^{f(z)}$, then $|\varphi(z)| = \mathrm{e}^{\operatorname{Re} f(z)} \le \mathrm{e}^M$. The statement follows now from Liouville's theorem (see Theorem 8.2). □

Problem 9.33. Under the assumptions of Proposition 9.31, assume that

$$\sup_{|z|=r} \operatorname{Re} f(z) \le Mr^k$$

with some $M > 0$ and some $k \in \mathbb{N}_0$. Prove that $f(z)$ is a polynomial of the degree at most k.

We return now to the univalent functions in terms of Taylor's expansion in open unit disk. This consideration concerns the Taylor coefficients $c_j, j = 0, 1, 2, \dots$ of a univalent function $f(z)$, i. e., a one-to-one analytic function that maps the unit disk into the complex plane, normalized (as is always possible) so that $c_0 = 0$ and $c_1 = 1$. It is equivalent that $f(0) = 0, f'(0) = 1$. That is, we consider a function $f(z)$ defined on the open unit disk, which is analytic and injective (univalent) with the Taylor series of the form

$$f(z) = z + \sum_{j=2}^{\infty} c_j z^j, \quad |z| < 1.$$

The main statement here is the so-called *Bieberbach conjecture* (1916), which says that for such functions the following is true:

$$|c_j| \le j, \quad j = 2, 3, \dots.$$

It was finally proved by Louis de Branges in 1985. It is clear that these conditions of the Bieberbach conjecture or the Louis de Branges theorem are not sufficient (only necessary). Indeed, consider the following function with parameter $\alpha \in \mathbb{C}$:

$$f_\alpha(z) := \sum_{j=1}^{\infty} j\alpha^{j-1} z^j, \quad |z| < 1, |\alpha| \le 1.$$

This function $f_\alpha(z)$ is clearly analytic in the open unit disk and the conditions of the Bieberbach conjecture are satisfied since $|c_j| \le j$. Moreover, under the conditions for z and α, we have that actually

$$f_\alpha(z) = \frac{z}{(1-\alpha z)^2}.$$

However, if $|\alpha| = 1$, then $f_\alpha(z)$ is univalent, but if $|\alpha| < 1$, then this function $f_\alpha(z)$ is not univalent in the disk $|z| < 1/|\alpha|$.

Problem 9.34. Prove the latter statement.

This example shows that the conditions for the coefficients c_j in the Bieberbach conjecture or the Louis de Branges theorem are only necessary.

Consider the same function $f_\alpha(z)$ in the open unit disk but for $|\alpha| > 1$. Evidently, in this case, $f_\alpha(z)$ is not analytic in the unit disk since at the point $z = \frac{1}{\alpha}$ this function has a pole of order 2. In addition, this function also is not univalent.

Example 9.35. Assume that a function $f(z)$ is of the form

$$f(z) := z + \sum_{j=1}^{\infty} \frac{a_j}{z^j}, \quad |z| > 1,$$

and it is univalent and analytic in the region $|z| > 1$ except for $z = \infty$, which is a pole of order 1. We show that

$$\sum_{j=1}^{\infty} j|a_j|^2 \le 1.$$

Indeed, since function $f(z)$ is univalent in the region $|z| > 1$, then the circle $\{z : |z| = r > 1\}$ transfers (under this transformation $f(z)$) into the simple closed curve γ such that $\operatorname{int}\gamma$ has a positive Jordan measure. On this curve, $f(z) = u(\theta) + \mathrm{i}v(\theta)$, $\theta \in [0, 2\pi]$, and thus this measure is equal to (see classical Green's formula in $2D$ case)

$$\begin{aligned}\int_0^{2\pi} u(\theta)v'(\theta)d\theta &= \int_0^{2\pi} \frac{f(\theta)+\overline{f(\theta)}}{2}\,\frac{f'(\theta)-\overline{f'(\theta)}}{2\mathrm{i}}d\theta \\ &= \frac{1}{4}\int_0^{2\pi}\left(re^{\mathrm{i}\theta} + re^{-\mathrm{i}\theta} + \sum_{j=1}^{\infty}\frac{a_j e^{-\mathrm{i}j\theta} + \overline{a}_j e^{\mathrm{i}j\theta}}{r^j}\right)\end{aligned}$$

$$\times\left(re^{i\theta}+re^{-i\theta}-\sum_{j=1}^{\infty}\frac{ja_je^{-ij\theta}+j\bar{a}_je^{ij\theta}}{r^j}\right)d\theta$$

$$=\frac{\pi}{2}\left((r+\bar{a}_1r^{-1})(r-a_1r^{-1})+(r+a_1r^{-1})(r-\bar{a}_1r^{-1})-\sum_{j=2}^{\infty}\frac{2j|a_j|^2}{r^{2j}}\right)$$

$$=\pi r^2-\pi\sum_{j=1}^{\infty}\frac{j|a_j|^2}{r^{2j}}.$$

Since this value (measure) is positive, then

$$\sum_{j=1}^{\infty}\frac{j|a_j|^2}{r^{2j}}<r^2.$$

Since this inequality holds for any $r>1$, then letting $r\to 1$ yields the required result.

Problem 9.36.

1. Let function

$$f(z):=z+a_2z^2+a_3z^3+\cdots,\quad |z|<1,$$

   be univalent and analytic in the unit disk $\{z:|z|<1\}$. Show that function

$$F(z):=\sqrt{f(z^2)},$$

   where the square root is chosen to be positive for positive f, is also univalent in the unit disk $\{z:|z|<1\}$.
2. Let $\sum_{j=2}^{\infty}j|a_j|\le 1$. Show that function

$$f(z):=z+\sum_{j=2}^{\infty}a_jz^j$$

   is univalent in the unit disk $\{z:|z|<1\}$.
3. Show that function $f(z)=\frac{z}{(1-z)^3}$ is univalent in the disk $\{z:|z|<\frac{1}{2}\}$ but it is not univalent in any disk $\{z:|z|<r\}$ with $r>\frac{1}{2}$.

10 Laurent's expansions

If f is analytic in the disk $\{z : |z-z_0| < R\}$, then we have only the Taylor's representation for this function. But if f is analytic in the deleted neighborhood, i. e., the punctured disk $\{z : 0 < |z-z_0| < R\}$, then what kind of representation we may have for this function?

Let us consider the series (formally for the moment)

$$\sum_{j=-\infty}^{\infty} c_j(z-z_0)^j = \sum_{j=-\infty}^{-1} c_j(z-z_0)^j + \sum_{j=0}^{\infty} c_j(z-z_0)^j =: s_2(z) + s_1(z). \tag{10.1}$$

The first term $s_2(z)$ is called the power series with negative degrees. The second series in (10.1) defines the analytic function $s_1(z)$ in the disk $\{z : |z-z_0| < R\}$, where

$$R^{-1} = \overline{\lim_{j\to\infty}} \sqrt[j]{|c_j|} = \sup_j \sqrt[j]{|c_j|}. \tag{10.2}$$

It makes sense to consider the first series for $|z-z_0| > 0$. Thus, if we change the variables as

$$\zeta = \frac{1}{z-z_0}, \quad z = z_0 + \frac{1}{\zeta}$$

we obtain for $s_2(z)$ the representation

$$s_2(z) = s_2(z_0 + 1/\zeta) = \sum_{j=1}^{\infty} c_{-j}\zeta^j =: s_2^*(\zeta), \tag{10.3}$$

where $\zeta = 0$ corresponds to $z = \infty$. So, we have the power series with respect to positive degrees of ζ with a radius of convergence $1/r$, which satisfies (see (10.2))

$$r = \overline{\lim_{j\to\infty}} \sqrt[j]{|c_{-j}|}$$

such that $s_2^*(\zeta)$ is an analytic function (this series converges) for any $|\zeta| < 1/r$. Equivalently, $s_2(z)$ is analytic in $\{z : |z-z_0| > r\}$.

If it turns out that $r < R$, then $s(z) = s_2(z) + s_1(z)$ is analytic in the annulus $\{z : r < |z-z_0| < R\}$ centered at z_0 with radii r and R. In this case, the series (10.1) is said to be a *Laurent's expansion* for $s(z)$ in the annulus. The opposite statement also holds.

Example 10.1. Let us find three different Laurent's expansions involving powers of z for the function

$$f(z) = \frac{3}{2+z-z^2}.$$

This function has singularities at $z=-1$ and $z=2$ and is analytic in the disk $\{z : |z| < 1\}$ in the annulus $\{z : 1 < |z| < 2\}$ and in the region $\{z : |z| > 2\}$. We start by writing

https://doi.org/10.1515/9783111632278-011

$$f(z) = \frac{3}{(1+z)(2-z)} = \frac{1}{1+z} + \frac{1}{2-z} = \frac{1}{1+z} + \frac{1}{2} \cdot \frac{1}{1-z/2}.$$

We have three cases:

1. For $|z| < 1$, we have

$$\frac{1}{1+z} = \frac{1}{1-(-z)} = \sum_{j=0}^{\infty} (-z)^j = \sum_{j=0}^{\infty} (-1)^j z^j$$

   and

$$\frac{1}{2} \frac{1}{1-z/2} = \frac{1}{2} \sum_{j=0}^{\infty} \left(\frac{z}{2}\right)^j = \sum_{j=0}^{\infty} \frac{1}{2^{j+1}} z^j.$$

   Hence, we have the Taylor's expansion

$$f(z) = \sum_{j=0}^{\infty} \left((-1)^j + \frac{1}{2^{j+1}}\right) z^j.$$

2. For $1 < |z| < 2$, we have

$$\frac{1}{1+z} = \frac{1}{z} \cdot \frac{1}{1+1/z} = \frac{1}{z} \sum_{j=0}^{\infty} (-1)^j \frac{1}{z^j} = \sum_{j=1}^{\infty} \frac{(-1)^{j+1}}{z^j}$$

   and

$$\frac{1}{2} \frac{1}{1-z/2} = \frac{1}{2} \sum_{j=0}^{\infty} \left(\frac{z}{2}\right)^j = \sum_{j=0}^{\infty} \frac{1}{2^{j+1}} z^j.$$

   So,

$$f(z) = \sum_{j=1}^{\infty} \frac{(-1)^{j+1}}{z^j} + \sum_{j=0}^{\infty} \frac{1}{2^{j+1}} z^j.$$

   This is a Laurent's expansion in the annulus $1 < |z| < 2$.

3. For $|z| > 2$, we have

$$\frac{1}{1+z} = \sum_{j=1}^{\infty} \frac{(-1)^{j+1}}{z^j}$$

   and

$$\frac{1}{2} \cdot \frac{1}{1-z/2} = -\frac{1}{z} \cdot \frac{1}{1-2/z} = -\frac{1}{z} \sum_{j=0}^{\infty} \left(\frac{2}{z}\right)^j = -\sum_{j=1}^{\infty} \frac{2^{j-1}}{z^j}.$$

Therefore,

$$f(z) = \sum_{j=1}^{\infty}((-1)^{j+1} - 2^{j-1})z^{-j}.$$

This is a Laurent's expansion at ∞, and in fact, it is also a Taylor's expansion at ∞.

Problem 10.2. Find the Laurent's expansion for e^{-1/z^2} centered at $z_0 = 0$.

Theorem 10.3. *Suppose that f is analytic in the annulus $\{z : r < |z - z_0| < R\}$ with $0 \le r < R$. Then for every $z \in \{z : r < |z - z_0| < R\}$, we have*

$$f(z) = \sum_{j=-\infty}^{\infty} c_j(z - z_0)^j, \tag{10.4}$$

where the coefficients c_j are uniquely determined by

$$c_j = \frac{1}{2\pi i}\int_{\gamma} \frac{f(\zeta)d\zeta}{(\zeta - z_0)^{j+1}}, \quad j = 0, \pm 1, \pm 2, \dots \tag{10.5}$$

with a piecewise smooth closed Jordan curve $\gamma \subset \{z : r < |z - z_0| < R\}$ and $z_0 \in \operatorname{int}\gamma$. Moreover, the convergence in (10.4) is uniform on any closed subannulus $\{z : r < r_1 \le |z - z_0| \le R_1 < R\}$.

Proof. Let $z \in \{z : r < |z - z_0| < R\}$. Then we can find $r_1 > r$ and $R_1 < R$ such that $z \in \{z : r_1 < |z - z_0| < R_1\}$. Using the Cauchy integral formula for the multiply connected domain, we obtain

$$\begin{aligned} f(z) &= \frac{1}{2\pi i}\int_{|\zeta - z_0| = R_1} \frac{f(\zeta)d\zeta}{\zeta - z} - \frac{1}{2\pi i}\int_{|\zeta - z_0| = r_1} \frac{f(\zeta)d\zeta}{\zeta - z} \\ &= \frac{1}{2\pi i}\int_{|\zeta - z_0| = R_1} \frac{f(\zeta)}{\zeta - z_0}\frac{d\zeta}{1 - \frac{z - z_0}{\zeta - z_0}} + \frac{1}{2\pi i}\int_{|\zeta - z_0| = r_1} \frac{f(\zeta)}{z - z_0}\frac{d\zeta}{1 - \frac{\zeta - z_0}{z - z_0}}. \end{aligned}$$

Since

$$\frac{1}{1 - \frac{z - z_0}{\zeta - z_0}} = \sum_{j=0}^{\infty}\frac{(z - z_0)^j}{(\zeta - z_0)^j}, \quad |z - z_0| < R_1$$

and

$$\frac{1}{1 - \frac{\zeta - z_0}{z - z_0}} = \sum_{j=0}^{\infty}\frac{(\zeta - z_0)^j}{(z - z_0)^j}, \quad |z - z_0| > r_1,$$

we may integrate in these series term by term (since these series converge uniformly on the circles $|\zeta - z_0| = R_1$ and $|\zeta - z_0| = r_1$, respectively) and obtain

$$\begin{aligned} f(z) &= \sum_{j=0}^{\infty} (z - z_0)^j \frac{1}{2\pi i} \int\limits_{|\zeta - z_0| = R_1} \frac{f(\zeta) d\zeta}{(\zeta - z_0)^{j+1}} \\ &\quad + \sum_{j=0}^{\infty} \frac{1}{(z - z_0)^{j+1}} \frac{1}{2\pi i} \int\limits_{|\zeta - z_0| = r_1} \frac{f(\zeta) d\zeta}{(\zeta - z_0)^{-j}} \\ &= \sum_{j=-\infty}^{-1} (z - z_0)^j \frac{1}{2\pi i} \int\limits_{|\zeta - z_0| = r_1} \frac{f(\zeta) d\zeta}{(\zeta - z_0)^{j+1}} \\ &\quad + \sum_{j=0}^{\infty} (z - z_0)^j \frac{1}{2\pi i} \int\limits_{|\zeta - z_0| = R_1} \frac{f(\zeta) d\zeta}{(\zeta - z_0)^{j+1}} \\ &= \sum_{j=-\infty}^{\infty} (z - z_0)^j \frac{1}{2\pi i} \int\limits_{\gamma} \frac{f(\zeta) d\zeta}{(\zeta - z_0)^{j+1}}, \end{aligned}$$

where the integrals are considered for an arbitrary piecewise smooth closed Jordan curve $\gamma \subset \{z : r < |z - z_0| < R\}$ and $z_0 \in \operatorname{int} \gamma$. We have used the fact that these integrals are independent on such curves due to the Cauchy theorem for multiply connected domains.

Thus, we proved the Laurent's expansions (10.4)–(10.5). It is evident that this representation is unique since we obtain necessarily (10.5). Uniform convergence of (10.4) for $z \in \{z : r < r_1 \le |z - z_0| \le R_1 < R\}$ follows from the arbitrariness of r_1 and R_1 in the preceding considerations. □

Definition 10.4. The series (10.4) with the coefficients (10.5) is called the *Laurent's expansion* (representation) of the analytic function f in the annulus $\{z : r < |z - z_0| < R\}$ and

$$f_1(z) = \sum_{j=0}^{\infty} c_j (z - z_0)^j, \quad f_2(z) = \sum_{j=1}^{\infty} c_{-j} (z - z_0)^{-j}$$

are called the *regular* and *main parts* of this expansion, respectively.

If f is analytic in the annulus $\{z : 0 < |z - z_0| < r\}$ with some $r > 0$, then z_0 is said to be an *isolated singular point* of f. Then Theorem 10.3 says that in this annulus $f(z)$ can be represented via the Laurent series

$$f(z) = \sum_{j=-\infty}^{\infty} c_j (z - z_0)^j,$$

where c_j are calculated by (10.5).

Example 10.5. Let us find the Laurent's expansion for

$$f(z) = \frac{\cos z - 1}{z^4}$$

that involves powers of z. Since

$$\cos z - 1 = -\frac{z^2}{2!} + \frac{z^4}{4!} - \frac{z^6}{6!} + \cdots$$

and this representation is valid for all $|z| < \infty$, then

$$f(z) = \frac{\cos z - 1}{z^4} = -\frac{1}{z^2 2!} + \frac{1}{4!} - \frac{z^2}{6!} + \cdots = -\frac{1}{2} \cdot \frac{1}{z^2} + \sum_{j=0}^{\infty} \frac{(-1)^j z^{2j}}{(2j+4)!}.$$

This is the Laurent's expansion in the neighborhood of $z = 0$ and $z = \infty$ both.

Problem 10.6. Find the Laurent's expansion for

$$f(z) = \frac{\sin 2z}{z^4}$$

that involves powers of z.

Problem 10.7. Find three Laurent's expansions for

$$f(z) = \frac{1}{z^2 - 5z + 6}$$

centered at $z_0 = 0$.

Problem 10.8. Find two Laurent's expansions for

$$\frac{1}{z(4-z)^2}$$

that involve powers of z.

Definition 10.9. If the number of nonzero coefficients (10.5) for $j < 0$ is:
1. empty,
2. finite, or
3. infinite,

then z_0 is called a *removable singular point* (in short, removable singularity), a *pole*, and an *essentially singular point* (in short, essential singularity) for f, respectively.

Let z_0 be removable for f. Then its Laurent's expansion has the form

$$f(z) = \sum_{j=0}^{\infty} c_j (z - z_0)^j,$$

where $z \in \{z : 0 < |z - z_0| < r\}$. But this series, as a power series, converges in the whole disk $\{z : |z - z_0| < r\}$ and it is equal to $f(z)$ for all $z \neq z_0$. If we define f at the point z_0 as

$$f(z_0) := c_0 = \lim_{z \to z_0} \sum_{j=0}^{\infty} c_j (z - z_0)^j,$$

then we obtain a new function in the whole disk $\{z : |z - z_0| < r\}$, which is analytic there. In particular, f is bounded in the closed disk $\{z : |z - z_0| \leq r_1\}$ with $r_1 < r$. The opposite property is also true. The following theorem holds.

Theorem 10.10. *Let f be analytic in the annulus $\{z : 0 < |z - z_0| < r\}$ for some $r > 0$. Then z_0 is a removable singular point of f if and only if f is bounded in the deleted neighborhood of z_0.*

Proof. We have proved the necessary condition above. It remains to prove this theorem only in the opposite direction. Let us assume that f is bounded in some deleted neighborhood of z_0, i. e., there is $M > 0$ such that

$$|f(z)| \leq M, \quad 0 < |z - z_0| < \delta.$$

Due to Theorem 10.3, we have that for all $z \in \{z : 0 < |z - z_0| < r\}$ it holds that

$$f(z) = \sum_{j=-\infty}^{\infty} c_j (z - z_0)^j,$$

where

$$c_j = \frac{1}{2\pi \mathrm{i}} \int_{|\zeta - z_0| = \delta} \frac{f(\zeta)}{(\zeta - z_0)^{j+1}} \mathrm{d}\zeta, \quad 0 < \delta < r.$$

Thus,

$$|c_j| \leq \frac{1}{2\pi} \max_{|\zeta - z_0| = \delta} |f(\zeta)| \delta^{-j} 2\pi \leq M\delta^{-j}, \quad j = 0, \pm 1, \pm 2, \ldots. \tag{10.6}$$

But for $j < 0$, it follows that $c_j = 0$ because we may let $\delta \to 0$ in these estimates. Hence, z_0 is a removable singular point. □

Remark. The estimate (10.6) has independent interest.

If f is analytic in some domain $D \subset \mathbb{C}$, then $z_0 \in D$ is a zero of order m of f if in some neighborhood $U_\delta(z_0) \subset D$ the function f admits the representation

$$f(z) = (z - z_0)^m \varphi(z), \tag{10.7}$$

where $\varphi(z)$ is analytic in $U_\delta(z_0)$ and $\varphi(z_0) \neq 0$. It is equivalent to

$$f(z_0) = 0, \quad f'(z_0) = 0, \quad \dots, \quad f^{(m-1)}(z_0) = 0, \quad f^{(m)}(z_0) \neq 0.$$

Another fact is: if $f \not\equiv 0$ is analytic in $D \subset \mathbb{C}$ and $f(z_0) = 0$, $z_0 \in D$, then the order of the zero is always finite, i. e., there is $m \in \mathbb{N}$ such that (10.7) holds. Indeed, if we assume on the contrary that $f^{(k)}(z_0) = 0$, $k = 0, 1, \dots$, then by the Taylor's expansion we have

$$f(z) = \sum_{j=0}^{\infty} \frac{f^{(j)}(z_0)}{j!}(z - z_0)^j \equiv 0$$

for all $z \in U_\delta(z_0)$. Further, using the procedure of continuation (see the proof of Theorem 7.7) for any $z_1 \in D$, we may obtain $f(z_1) = 0$, i. e., $f(z) \equiv 0$ in D. This contradiction proves the fact.

Another consequence is: if $f(z_n) = 0$ and $z_n \to z_0$, $z_n \neq z_0$ with $z_0, z_n \in D$, then $f \equiv 0$ in D.

Let us assume that z_0 is a pole of f. Then f has the Laurent's expansion of the form

$$f(z) = \sum_{j=0}^{\infty} c_j(z - z_0)^j + \sum_{j=1}^{m} c_{-j}(z - z_0)^{-j}, \tag{10.8}$$

where $c_{-m} \neq 0$. Then we say that z_0 is a *pole* of order m.

Theorem 10.11. *Let f be analytic in the annulus $\{z : 0 < |z - z_0| < r\}$. Then f has a pole of some order m at z_0 if and only if $\lim_{z\to z_0} |f(z)| = \infty$.*

Proof. Let us assume first that z_0 is a pole of order $m \in \mathbb{N}$. Then the following representation holds:

$$f(z) = \sum_{j=0}^{\infty} c_j(z - z_0)^j + \sum_{j=1}^{m} c_{-j}(z - z_0)^{-j},$$

where $c_{-m} \neq 0$. Then for the function

$$F(z) := (z - z_0)^m f(z),$$

we have

$$F(z) = c_{-m} + c_{-m+1}(z - z_0) + \dots + c_{-1}(z - z_0)^{m-1} + c_0(z - z_0)^m + \dots,$$

i. e., $F(z)$ has a removable singularity at z_0. Moreover, there exists

$$\lim_{z\to z_0} F(z) = c_{-m} \neq 0.$$

This fact implies that there is $0 < \delta < r$ such that

$$|F(z)| > \frac{|c_{-m}|}{2} > 0$$

for all $0 < |z - z_0| < \delta$ and, therefore,

$$|f(z)| > \frac{|c_{-m}|}{2}|z - z_0|^{-m},$$

i. e.,

$$\lim_{z\to z_0}|f(z)| = \infty. \tag{10.9}$$

Conversely, if (10.9) holds, then

$$\lim_{z\to z_0}\frac{1}{|f(z)|} = \lim_{z\to z_0}\frac{1}{f(z)} = 0.$$

This fact can be interpreted as follows: a new function

$$g(z) := \frac{1}{f(z)}$$

is analytic in $\{z : 0 < |z - z_0| < \delta\}$ and the function g can be extended as an analytic function everywhere in $\{z : |z - z_0| < \delta\}$ and z_0 is a zero of analytic function g. But the zero of a analytic function (if it is not identically zero) is of finite order, say m. That is why

$$g(z) = (z - z_0)^m\varphi(z),$$

where $\varphi(z)$ is analytic and $\varphi(z_0) \neq 0$. Hence,

$$f(z) = \frac{1}{(z - z_0)^m}\frac{1}{\varphi(z)},$$

where $1/\varphi(z)$ is analytic in the neighborhood of z_0 and $1/\varphi(z_0) \neq 0$. This condition allows us to represent $1/\varphi(z)$ via its Taylor's expansion for $|z - z_0| < \delta$ as

$$\frac{1}{\varphi(z)} = \sum_{j=0}^{\infty} a_j(z - z_0)^j,$$

where $a_0 = 1/\varphi(z_0) \neq 0$. This implies that the Laurent's expansion for f is

$$f(z) = \frac{a_0}{(z - z_0)^m} + \cdots + \frac{a_{m-1}}{(z - z_0)} + \sum_{j=0}^{\infty} a_{m+j}(z - z_0)^j.$$

This means that z_0 is a pole of order m for f. □

Corollary 10.12. *A point z_0 is a pole of order m for function f, which is analytic in the annulus $\{z : 0 < |z - z_0| < \delta\}$ if and only if z_0 is a zero of $1/f$ of order m and this function is analytic in $\{z : |z - z_0| < \delta\}$.*

Theorem 10.13. *Let f be analytic in the annulus $\{z : 0 < |z - z_0| < r\}$. Then f has an essentially singular point at z_0 if and only if there is no $\lim_{z \to z_0} f(z)$ (finite or infinite).*

Proof. If we assume on the contrary that there is $\lim_{z \to z_0} f(z)$ finite or infinite, then in the first case z_0 is a removable singularity, and in the second case, it is a pole of some order. This contradiction proves the theorem. □

As the consequence of this fact, we can obtain that there exist two different sequences z_n' and z_n'' converging to z_0 such that $f(z_n')$ is bounded and $\lim_{n \to \infty} f(z_n'') = \infty$. Even more is true.

Theorem 10.14 (Casorati–Sokhotski–Weierstrass). *Let $z_0 \neq \infty$ be an essential singularity of $f(z)$. Let E be the set of all values of $f(z)$ in the deleted neighborhood of z_0. Then E is* dense *in $\mathbb{C}$, i. e., for any $\varepsilon > 0$, $\delta > 0$ (or $M > 0$) and complex number w, there exists a complex number z with $0 < |z - z_0| < \delta$ such that $|f(z) - w| < \varepsilon$ (or $|f(z)| > M$), in short $\overline{\{f(U(z_0) \setminus z_0)\}} = \overline{\mathbb{C}}$.*

Proof. Let a be an arbitrary point of $\mathbb{C}$. Let us assume that for any

$$z \in \{z \in \mathbb{C} : 0 < |z - z_0| < \delta\}$$

we have $f(z) \neq a$. Otherwise, there is

$$z' \in \{z \in \mathbb{C} : 0 < |z - z_0| < \delta\}$$

such that $f(z') = a$ and everything is proved.

Then in the deleted neighborhood of z_0 the function

$$g(z) := \frac{1}{f(z) - a}$$

is well-defined. For this function $g(z)$, the point z_0 will be also essential singularity. Indeed, if z_0 is a pole or removable singularity for $g(z)$, then z_0 is a zero or removable singularity for $f(z) - a$, respectively. In both cases, we have a contradiction with essential singularity at z_0 for $f(z)$. Thus, $g(z)$ cannot be bounded in this neighborhood and, therefore, there is a sequence z_n converging to z_0 such that $\lim_{n \to \infty} g(z_n) = \infty$ or $\lim_{n \to \infty} f(z_n) = a$. It means that $a \in \overline{E}$ and the theorem is proved. □

Remark. If $z_0 = \infty$ is an essential singularity for $f(z)$, then (by definition) zero is the essential singularity for $\varphi(z) = f(1/z)$. Then Theorem 10.14 holds for $\varphi(z)$ in the deleted neighborhood of zero, which is equivalent to the fact that Theorem 10.14 holds for $f(z)$ in the neighborhood of ∞.

There is a substantial strengthening of Theorem 10.14, which only guarantees that the range of $f(z)$ is dense in $\mathbb{C}$. Namely, the following *great Picard's theorem* holds. We give it without proof.

Theorem 10.15 (Picard). *If analytic function $f(z)$ has an essential singularity at z_0, then on any deleted neighborhood of z_0 the function $f(z)$ takes on all possible complex values, with at most a single exception, infinitely often.*

Example 10.16.
1. The function $f(z) = e^{1/z}$ has an essential singularity at $z = 0$ but still never attains the value 0.
2. The function

$$f(z) = \frac{1}{1 - e^{1/z}}$$

has an essential singularity at $z = 0$ and attains the value ∞ infinitely often in any neighborhood of 0 ($z_n = i/2\pi n$, $n = \pm 1, \pm 2, \ldots$). However, it does not attain the values 0 or 1, since $e^{1/z} \neq 0$. It must be mentioned here that $z = 0$ is not isolated.

Example 10.17.
1. Consider the function

$$f(z) = \frac{\sin z}{z}.$$

Since

$$\frac{\sin z}{z} = \frac{1}{z}\left(z - \frac{z^3}{3!} + \frac{z^5}{5!} - \cdots\right) = 1 - \frac{z^2}{3!} + \frac{z^4}{5!} - \cdots$$

for $|z| > 0$, then we can remove the singularity at $z = 0$ if we define $f(0) = 1$ since then f will be analytic at $z = 0$.
2. Consider the function

$$g(z) = \frac{\cos z - 1}{z^2}.$$

Since again for all $|z| > 0$, we have

$$\frac{\cos z - 1}{z^2} = \frac{1}{z^2}\left(-\frac{z^2}{2!} + \frac{z^4}{4!} - \cdots\right) = -\frac{1}{2} + \frac{z^2}{4!} - \frac{z^4}{6!} + \cdots$$

then defining $g(0) = -1/2$ we obtain the function that is analytic for all z.

Example 10.18. Consider the function

$$f(z) = \frac{\sin z}{z^3}.$$

Since for all $|z| > 0$, we have

$$\frac{\sin z}{z^3} = \frac{1}{z^3}\left(z - \frac{z^3}{3!} + \frac{z^5}{5!} - \cdots\right) = \frac{1}{z^2} - \frac{1}{3!} + \frac{z^2}{5!} - \cdots,$$

then $c_{-2} = 1 \neq 0$. Therefore, $f(z)$ has a pole of order 2 at 0.

Example 10.19. Consider the function

$$f(z) = z^2 \sin\frac{1}{z}.$$

Since for all $|z| > 0$, we have

$$z^2 \sin\frac{1}{z} = z^2\left(\frac{1}{z} - \frac{1}{3!z^3} + \frac{1}{5!z^5} - \cdots\right) = z - \frac{1}{3!z} + \frac{1}{5!z^3} - \cdots,$$

then the Laurent's expansion has infinitely many negative powers of z. Hence, $z = 0$ is essentially singular point for f.

Problem 10.20. Suppose that f has a removable singularity at z_0. Show that the function $1/f$ has either a removable singularity or a pole at z_0.

Problem 10.21.
1. Let f be analytic and have a zero of order k at z_0. Show that f' has a zero of order $k-1$ at z_0.
2. Let f be analytic and have a zero of order k at z_0. Show that f'/f has a *simple pole* (pole of order 1) at z_0.
3. Let f have a pole of order k at z_0. Show that f' has a pole of order $k+1$ at z_0.

Problem 10.22. Find and classify the singularities of

$$f(z) = \frac{1}{\sin\frac{1}{z}}.$$

Let f be analytic in the region $\{z : |z| > R\}$. Then the function

$$\varphi(z) := f(1/z)$$

is analytic in the annulus $\{z : 0 < |z| < 1/R\}$. Hence, $z = 0$ might be an isolated singular point for φ. The Laurent's expansion for φ gives

$$\varphi(z) = \sum_{j=-\infty}^{\infty} c_j z^j, \quad 0 < |z| < \frac{1}{R}.$$

Thus, we have the following expansion for f:

$$f(z) = \varphi(1/z) = \sum_{j=-\infty}^{\infty} c_j z^{-j} = \sum_{j=-\infty}^{\infty} c_{-j} z^j, \quad |z| > R. \tag{10.10}$$

Definition 10.23. If $z = 0$ is a removable singularity, a pole, or an essential singularity for $\varphi(z) = f(1/z)$, then $z = \infty$ is called a removable singularity, a pole, or an essential singularity for $f(z)$, respectively.

Remark. This definition implies that if the number of coefficients in (10.10) for $j > 0$ is empty, finite, or infinite, then $z = \infty$ is a removable singularity, a pole, or an essential singularity, respectively.

Remark. There is even a more general observation. Namely, let z_0 be an isolated singular point of the function $f(z)$ in some domain $D \subset \mathbb{C}$ and let the analytic function $\zeta = \psi(z)$ establish a reciprocal one-to-one correspondence between D and D', where the inverse function $z = \varphi(\zeta)$ is well-defined. Then the point $\zeta_0 = \psi(z_0)$ is an isolated singular point of the analytic function $f(\varphi(\zeta))$, and the type of this singularity is the same as that of the point z_0.

Example 10.24.
1. Let f be a polynomial of degree n, i. e.,
$$f(z) = a_0 z^n + a_1 z^{n-1} + \cdots + a_n, \quad a_0 \neq 0.$$
Then $z = \infty$ is a pole of order n.
2. Let f be analytic in the whole space $\mathbb{C}$. If $z = \infty$ is a removable singularity, then $f \equiv$ constant, and if $z = \infty$ is a pole of order n, then f is a polynomial of degree n.

Problem 10.25. Consider the function
$$f(z) = z^2 \sin \frac{1}{z}.$$
Show that f has a pole of order 1 at $z = \infty$. Compare this result with the second part of Example 10.24.

Example 10.26. It is clear that the function $f(z) = z + \frac{1}{z}$ is analytic for all z such that $0 < |z| < \infty$ and $f(z)$ has a pole of order 1 at $z = 0$ and $z = \infty$. What is more, the expression $z + \frac{1}{z}$ is the Laurent's expansion for $f(z)$ at $z = 0$ and at $z = \infty$ as well. We find now the Laurent's expansions for the functions $f(z) = e^{z+1/z}$ and $\sin(z + \frac{1}{z})$ at $z = 0$ and $z = \infty$ at the same time due to symmetry. Let us consider the first function. Since we have that
$$e^z = \sum_{l=0}^{\infty} \frac{z^l}{l!}, \quad e^{1/z} = \sum_{k=0}^{\infty} \frac{1}{k! z^k}, \quad 0 < |z| < \infty,$$
then
$$e^{z+1/z} = \sum_{l,k=0}^{\infty} \frac{z^{l-k}}{l!k!} = \sum_{n=-\infty}^{\infty} z^n \sum_{l,k=0, l-k=n}^{\infty} \frac{1}{l!k!}.$$

Due to the symmetry of this function with respect to z and $\frac{1}{z}$, the coefficients in the Laurent's expansion in front of z^n and z^{-n} are equal. That is why we have

$$e^{z+1/z} = \sum_{n=0}^{\infty}\left(\sum_{k=0}^{\infty}\frac{1}{k!(n+k)!}\right)z^n + \sum_{n=1}^{\infty}\left(\sum_{k=0}^{\infty}\frac{1}{k!(n+k)!}\right)z^{-n}.$$

Based on Theorem 10.3 (see formula (10.5)), we may calculate the coefficients $c_n, n \in \mathbb{N}_0$ of this expansion. Indeed,

$$c_n = \frac{1}{2\pi i}\int_{|\zeta|=1}\frac{e^{\zeta+1/\zeta}}{\zeta^{n+1}}d\zeta = \frac{1}{2\pi}\int_{-\pi}^{\pi}\frac{e^{2\cos\theta}}{e^{i\theta(n+1)}}e^{i\theta}d\theta = \frac{1}{2\pi}\int_{-\pi}^{\pi}e^{2\cos\theta}\cos(n\theta)d\theta.$$

In particular, we obtain an equality of independent interest

$$\frac{1}{2\pi}\int_{-\pi}^{\pi}e^{2\cos\theta}\cos(n\theta)d\theta = \sum_{k=0}^{\infty}\frac{1}{k!(n+k)!}.$$

Considering the second function, i. e., $f(z) = \sin(z+1/z)$, we obtain similar to the previous function that (again due to the symmetry)

$$\sin\left(z+\frac{1}{z}\right) = \sum_{n=0}^{\infty}c_n z^n + \sum_{n=1}^{\infty}c_n z^{-n},$$

where the coefficients $c_n, n \in \mathbb{N}_0$ are equal to

$$c_n = \frac{1}{2\pi}\int_{-\pi}^{\pi}\sin(2\cos\theta)\cos(n\theta)d\theta.$$

Example 10.27. We want to now find the Laurent's expansion for the function $f(z) = e^{z-1/z}$. As in the previous example, we can write

$$e^{z-1/z} = \sum_{l,k=0}^{\infty}\frac{(-1)^k z^{l-k}}{l!k!} = \sum_{n=-\infty}^{\infty}z^n\sum_{l,k=0,l-k=n}^{\infty}\frac{(-1)^k}{l!k!}.$$

Due to the symmetrical property of this function if we change z to $-\frac{1}{z}$, the coefficients c_n in the Laurent's expansion in front of z^n and z^{-n} are connected to each other as follows:

$$c_n = (-1)^n c_{-n}, \quad n \in \mathbb{Z}.$$

That is why we have

$$e^{z-1/z} = \sum_{n=0}^{\infty}\left(\sum_{k=0}^{\infty}\frac{(-1)^k}{k!(n+k)!}\right)z^n + \sum_{n=1}^{\infty}(-1)^n\left(\sum_{k=0}^{\infty}\frac{(-1)^k}{k!(n+k)!}\right)z^{-n}.$$

As above, we may calculate the coefficients c_n as

$$
\begin{aligned}
c_n &= \frac{1}{2\pi i}\int_{|\zeta|=1} \frac{e^{\zeta-1/\zeta}}{\zeta^{n+1}}d\zeta = \frac{1}{2\pi}\int_{-\pi}^{\pi}\frac{e^{2i\sin\theta}}{e^{i\theta(n+1)}}e^{i\theta}d\theta \\
&= \frac{1}{2\pi}\int_{-\pi}^{\pi} e^{i(2\cos\theta-n\theta)}d\theta = \frac{1}{2\pi}\int_{-\pi}^{\pi}\cos(2\sin\theta-n\theta)d\theta.
\end{aligned}
$$

First of all, we can see that really $c_n = (-1)^n c_{-n}$ and, second, for $n \in \mathbb{N}_0$ we have the equality

$$
\sum_{k=0}^{\infty}\frac{(-1)^k}{k!(n+k)!} = \frac{1}{2\pi}\int_{-\pi}^{\pi}\cos(2\sin\theta-n\theta)d\theta,
$$

which has independent interest.

Problem 10.28. Find the Laurent's expansion for the function $\cos(z+\frac{1}{z})$.

Example 10.29. Consider one more example of the Laurent's expansion for the function $f(z) = \sin z \cdot \sin(1/z)$. Using the Laurent's expansion for $\sin(\cdot)$, we have

$$
\begin{aligned}
\sin z \cdot \sin(1/z) &= \sum_{l=0}^{\infty}(-1)^l\frac{z^{2l+1}}{(2l+1)!}\sum_{k=0}^{\infty}(-1)^k\frac{z^{-2k-1}}{(2k+1)!} \\
&= \sum_{n=-\infty}^{\infty} z^{2n}\sum_{k,l=0,l-k=n}^{\infty}\frac{(-1)^{l+k}}{(2l+1)!(2k+1)!}.
\end{aligned}
$$

Due to the symmetry of this function with respect to z and $\frac{1}{z}$, the coefficients in the Laurent's expansion in front of z^{2n} and z^{-2n} are equal. That is why we have

$$
\begin{aligned}
\sin z \cdot \sin(1/z) &= \sum_{n=0}^{\infty}(-1)^n\left(\sum_{k=0}^{\infty}\frac{1}{(2k+1)!(2n+2k+1)!}\right)z^{2n} \\
&\quad + \sum_{n=1}^{\infty}(-1)^n\left(\sum_{k=0}^{\infty}\frac{1}{(2k+1)!(2n+2k+1)!}\right)z^{-2n}.
\end{aligned}
$$

Further, since

$$
\sin(e^{i\theta})\cdot\sin(e^{-i\theta}) = \frac{1}{2}(\cosh(2\sin\theta)-\cos(2\cos\theta)), \quad \theta \in [-\pi,\pi],
$$

then we obtain that the coefficients $c_{2n}, n \in \mathbb{N}_0$ are equal to

$$
c_{2n} = \frac{1}{4\pi i}\int_{-\pi}^{\pi}(\cosh(2\sin\theta)-\cos(2\cos\theta))ie^{i\theta}e^{-i(2n+1)\theta}d\theta.
$$

In particular, we have that (since sine is odd)

$$(-1)^n \sum_{k=0}^{\infty} \frac{1}{(2k+1)!(2n+2k+1)!}$$
$$= \frac{1}{4\pi} \int_{-\pi}^{\pi} (\cosh(2\sin\theta) - \cos(2\cos\theta))\cos(2n\theta)d\theta.$$

Problem 10.30. Find the Laurent's expansions at the points $z = 0$ and $z = \infty$ of the functions

$$\cos z \cdot \cos(1/z), \quad \sin z \cdot \cos(1/z), \quad \cos z \cdot \sin(1/z).$$

Problem 10.31. Using the Laurent's expansions of the functions $\sin(z + \frac{1}{z})$ and $\cos(z + \frac{1}{z})$ (see Example 10.26), find the Laurent's expansions for the functions $\sinh(z - \frac{1}{z})$ and $\cosh(z - \frac{1}{z})$, respectively.

Example 10.32. Consider the main branch of multivalued function

$$f(z) = \frac{1}{\sqrt{1+z^2}},$$

which is equal to the positive square root when $z = x + i0$, $x \in \mathbb{R}$. In the annulus $1 < |z| < \infty$, this function has no singularities. Thus, we will investigate it in the neighborhood of ∞. For this purpose, consider the function

$$\varphi(z) = f\left(\frac{1}{z}\right) = \frac{z}{\sqrt{1+z^2}}, \quad |z| < 1.$$

Using the binomial series in the neighborhood of zero (see Example 9.15), we obtain

$$\varphi(z) = z \sum_{n=0}^{\infty} \frac{(2n)!}{(2^n n!)^2} (-z^2)^n = \sum_{n=0}^{\infty} (-1)^n \frac{(2n)!}{(2^n n!)^2} z^{2n+1}.$$

It means that the Laurent's expansion for function $f(z)$ in the annulus $1 < |z| < \infty$ is given by

$$f(z) = \sum_{n=0}^{\infty} (-1)^n \frac{(2n)!}{(2^n n!)^2} \frac{1}{z^{2n+1}}.$$

This representation shows that $z = \infty$ is a removable singular point for this function. Moreover, this series can be considered as the Taylor's expansion of this branch at ∞. It also means that this branch of $f(z)$ is analytic at ∞.

Example 10.33. Let function $f(z)$ be defined by the series (see, for comparison, Example 9.19)

$$f(z) := \sum_{j=0}^{\infty} \frac{1}{j!} \frac{1}{1 + a^{2j} z^2}, \qquad a > 1.$$

This function is well-defined everywhere on the complex plane $\overline{\mathbb{C}}$ (including $z = \infty$) except for the points $z_k = \pm \frac{\mathrm{i}}{a^k}$, $k = 0, 1, 2, \ldots$. These points are clearly singular for this function and they accumulate to $z = 0$. That is why there is no Laurent's expansion for $f(z)$ in the neighborhood of $z = 0$. But it is clear that $f(z)$ is analytic in the annulus $\{z : 1 < |z| < \infty\}$. Indeed, for $|z| \geq r > 1$, we have

$$|f(z)| \leq \sum_{j=0}^{\infty} \frac{1}{j!} \frac{1}{r^2 a^{2j} - 1} < \infty.$$

It means that this series converges uniformly for $|z| \geq r > 1$ and, therefore, $f(z)$ is continuous for $|z| > 1$. Moreover, since for $|z| \geq r > 1$, we have

$$|f'(z)| \leq \sum_{j=0}^{\infty} \frac{1}{j!} \frac{2a^{2j} r}{(r^2 a^{2j} - 1)^2} < \infty,$$

then $f(z)$ is analytic there. Next, $\lim_{z \to \infty} f(z) = 0$ and we can represent this function via the Laurent's expansion as follows ($f(z)$ is even):

$$f(z) = \sum_{j=1}^{\infty} \frac{c_{2j}}{z^{2j}},$$

where the coefficients c_{2j} are equal to

$$c_{2j} = \frac{1}{2\pi \mathrm{i}} \int_{|z|=R>1} \zeta^{2j-1} \left(\sum_{k=0}^{\infty} \frac{1}{k!} \frac{1}{1 + a^{2k} \zeta^2} \right) \mathrm{d}\zeta.$$

The latter series in the integral can be evaluated as

$$\sum_{k=0}^{\infty} \frac{1}{k!} \frac{1}{1 + a^{2k} \zeta^2} = \sum_{k=0}^{\infty} \frac{\zeta^{-2} a^{-2k}}{k!} \sum_{m=0}^{\infty} (-1)^m \zeta^{-2m} a^{-2km}.$$

Since this series converges absolutely and uniformly for $|\zeta| \geq r > 1$, then we can calculate the coefficients c_{2j} as

$$c_{2j} = \frac{1}{2\pi \mathrm{i}} \sum_{k=0}^{\infty} \sum_{m=0}^{\infty} (-1)^m \int_{|\zeta|=R>1} \frac{\zeta^{2j-2m-3} a^{-2k(m+1)}}{k!} \mathrm{d}\zeta.$$

The only member of this series that gives a nonzero integral is the member with ζ^{-1} (index $m = j - 1$), i. e., the term with the coefficient

$$(-1)^{j-1}\sum_{k=0}^{\infty}\frac{a^{-2kj}}{k!} = (-1)^{j-1}e^{a^{-2j}}.$$

Hence,

$$c_{2j} = \frac{1}{2\pi i}\int\limits_{|z|=R>1}(-1)^{j-1}e^{a^{-2j}}\frac{1}{\zeta}d\zeta = (-1)^{j-1}e^{a^{-2j}}$$

and, therefore, the Laurent's expansion for $f(z)$ is equal to

$$f(z) = \sum_{j=1}^{\infty}(-1)^{j-1}\frac{e^{a^{-2j}}}{z^{2j}}.$$

It can be mentioned that actually this series is Taylor's expansion of $f(z)$ at $z = \infty$. Besides this, $z = \infty$ is a zero of order two for this function.

Problem 10.34. Let function $f(z)$ be defined via the following series:

$$f(z) = \sum_{n=1}^{\infty}\frac{z^{n-1}}{(1-z^n)(1-z^{n+1})}.$$

Prove that

$$f(z) = \begin{cases} \frac{1}{(1-z)^2}, & |z| < 1, \\ \frac{1}{z(1-z)^2}, & |z| > 1 \end{cases}$$

and that $f(z)$ is analytic for all $|z| \neq 1$ (including $z = \infty$). Prove also that $f'(0) = 1$ and $f'(\infty) = 0$.

Problem 10.35. Let function $f(z)$ be defined via the following series:

$$f(z) = \sum_{n=0}^{\infty}\frac{1}{z^n + z^{-n}}.$$

Prove that

$$f(z) = \begin{cases} \sum_{k=0}^{\infty}\frac{(-1)^k}{1-z^{2k+1}}, & |z| < 1, \\ \sum_{k=0}^{\infty}\frac{(-1)^k z^{2k+1}}{z^{2k+1}-1}, & |z| > 1 \end{cases}$$

and that $f(z)$ is analytic for all $|z| \neq 1$ (including $z = \infty$). Prove also that $f'(0) = 1$ and $f'(\infty) = 1$.

Problem 10.36. Show that

$$e^{z+1/(2z^2)} = \sum_{n=-\infty}^{\infty} c_n z^n,$$

where

$$c_n = \frac{1}{2\pi\sqrt{e}} \int_{-\pi}^{\pi} e^{\cos\theta + \cos^2\theta} \cos(\sin\theta(1 - \cos\theta) - n\theta)\,d\theta.$$

Hint: Use the same procedure as in Examples 10.26 and 10.27.

11 Residues and their calculus

Recall that if a piecewise smooth closed Jordan curve $\gamma : z(t), a \leq t \leq b$ is parametrized so that $\operatorname{int}\gamma$ is kept on the left as $z(t)$ moves around γ, then we say that γ is *oriented positively*. Otherwise, γ is said to be *oriented negatively* (see Figure 11.1).

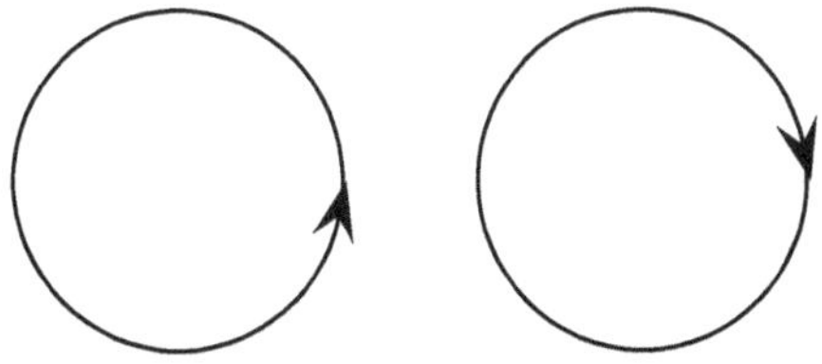

Figure 11.1: Orientation illustrated with circles.

Let z_0 be an isolated singular point for a function f, i. e., $f(z)$ is analytic in the annulus $\{z : 0 < |z - z_0| < r\}$ if $z_0 \neq \infty$ and in the region $\{z : |z| > R\}$ if $z_0 = \infty$.

Definition 11.1. The *residue* of f at the point z_0 is defined by

$$\operatorname*{Res}_{z=z_0} f := \frac{1}{2\pi i} \int_{\gamma} f(\zeta) d\zeta, \tag{11.1}$$

where $z_0 \in \operatorname{int}\gamma$, $\gamma \subset \{z : 0 < |z - z_0| < \delta\}$, and γ is positively oriented if $z_0 \neq \infty$ and $0 \in \operatorname{int}\gamma$, $\gamma \subset \{z : |z| > R\}$, and γ is negatively oriented if $z_0 = \infty$ (but it is positively oriented with respect to $z = \infty$).

Remark. Due to the Cauchy theorem, the integral in (11.1) is independent on the corresponding curve, and thus the residue can be rewritten as

$$\begin{aligned} \operatorname*{Res}_{z=z_0\neq\infty} f &= \frac{1}{2\pi i} \int_{|\zeta - z_0| = \delta} f(\zeta) d\zeta, \\ \operatorname*{Res}_{z=\infty} f &= -\frac{1}{2\pi i} \int_{|\zeta| = R} f(\zeta) d\zeta. \end{aligned} \tag{11.2}$$

If $z_0 \neq \infty$, then the formulas (10.5) show us that

$$\operatorname*{Res}_{z=z_0} f = \frac{1}{2\pi i} \int_{|\zeta - z_0| = \delta} f(\zeta) d\zeta = c_{-1}, \tag{11.3}$$

https://doi.org/10.1515/9783111632278-012

where c_{-1} is the coefficient in front of $(z-z_0)^{-1}$ of the Laurent's expansion for f. If $z_0 = \infty$, then

$$\operatorname*{Res}_{z=\infty} f = -\frac{1}{2\pi i} \int\limits_{|\zeta|=R} f(\zeta)d\zeta = -c_1, \tag{11.4}$$

where c_1 is the coefficient in front of z^{-1} of the Laurent's expansion for f.

Example 11.2. Let us find $\operatorname{Res}_{z=0} g$ of

$$g(z) = \frac{3}{2z + z^2 - z^3}.$$

Since

$$\frac{3}{2z + z^2 - z^3} = \frac{3/2}{z} \cdot \frac{1}{1 + z/2 - z^2/2}$$

and

$$g_1(z) = \frac{1}{1 + z/2 - z^2/2}$$

is analytic in the neighborhood of $z = 0$ and such that $g_1(0) = 1$, then the Laurent's expansion for g has the form

$$g(z) = \frac{3/2}{z} + \sum_{j=0}^{\infty} c_j z^j.$$

Thus,

$$\operatorname*{Res}_{z=0} g = 3/2.$$

Example 11.3. If $f(z) = e^{2/z}$, then the Laurent's expansion of f about the point 0 has the form

$$e^{2/z} = 1 + \frac{2}{z} + \frac{2^2}{2!z^2} + \cdots$$

and $\operatorname{Res}_{z=0} f = 2$. At the same time (by definition), $\operatorname{Res}_{z=\infty} f = -2$.

Theorem 11.4 (Residues at poles). *If f has a pole of order k at $z_0 \neq \infty$, then*

$$\operatorname*{Res}_{z=z_0} f = \frac{1}{(k-1)!} \lim_{z\to z_0} \frac{d^{k-1}}{dz^{k-1}}((z-z_0)^k f(z)). \tag{11.5}$$

Proof. Suppose that f has a pole of order k at $z_0 \neq \infty$. Then f can be written as

$$f(z) = \frac{c_{-k}}{(z-z_0)^k} + \frac{c_{-k+1}}{(z-z_0)^{k-1}} + \cdots + \frac{c_{-1}}{(z-z_0)} + \sum_{j=0}^{\infty} c_j (z-z_0)^j, \quad c_{-k} \neq 0.$$

Multiplying both sides by $(z-z_0)^k$ give

$$(z-z_0)^k f(z) = c_{-k} + \cdots + c_{-1}(z-z_0)^{k-1} + \sum_{j=0}^{\infty} c_j (z-z_0)^{j+k}.$$

If we differentiate both sides $k-1$ times, we get

$$\frac{d^{k-1}}{dz^{k-1}}((z-z_0)^k f(z)) = (k-1)! c_{-1} + \sum_{j=0}^{\infty} c_j (j+k)\cdots(j+2)(z-z_0)^{j+1}.$$

Letting $z \to z_0$, the result is

$$\lim_{z\to z_0} \frac{d^{k-1}}{dz^{k-1}}((z-z_0)^k f(z)) = (k-1)! c_{-1}.$$

By (11.3), this leads to (11.5). □

Corollary 11.5. *Let $f = \varphi/\psi$ be such that*

$$\varphi(z_0) \neq 0, \quad \psi(z_0) = 0, \quad \psi'(z_0) \neq 0.$$

Then f has a pole of order 1 at z_0 and

$$\operatorname*{Res}_{z=z_0} f = \frac{\varphi(z_0)}{\psi'(z_0)}. \tag{11.6}$$

Proof. The conditions for φ and ψ show that z_0 is a pole of order 1 for $f = \varphi/\psi$. Hence,

$$\operatorname*{Res}_{z=z_0} f = \lim_{z\to z_0}\left((z-z_0)\frac{\varphi(z)}{\psi(z)}\right) = \lim_{z\to z_0} \frac{\varphi(z)}{\frac{\psi(z)-\psi(z_0)}{z-z_0}} = \frac{\varphi(z_0)}{\psi'(z_0)}$$

by Theorem 11.4. □

Corollary 11.6. *Let $f = \varphi/\psi$ be such that*

$$\varphi(z_0) \neq 0, \quad \psi(z_0) = \psi'(z_0) = 0, \quad \psi''(z_0) \neq 0.$$

Then

$$\operatorname*{Res}_{z=z_0} f = \frac{2\varphi'(z_0)}{\psi''(z_0)} - \frac{2\varphi(z_0)\psi'''(z_0)}{3(\psi''(z_0))^2}.$$

Corollary 11.7. *If f has a pole of order k at $z_0 = \infty$, then*

$$\operatorname*{Res}_{z=\infty} f = -\frac{1}{(k+1)!}\lim_{z\to 0}\frac{d^{k+1}}{dz^{k+1}}(z^k f(1/z)). \tag{11.7}$$

Problem 11.8. Prove Corollaries 11.6 and 11.7.

Problem 11.9. Find the residue of

$$f(z) = \frac{\pi\cot(\pi z)}{z^2}$$

at $z_0 = 0$.

Theorem 11.10 (Cauchy's residue theorem). *Let $D \subset \mathbb{C}$ be a simply connected domain and let γ be a piecewise smooth closed Jordan curve, which is positively oriented and lies in D. If f is analytic in D except for the points $z_1, z_2, \ldots, z_n \in \operatorname{int}\gamma$, then*

$$\int_\gamma f(\zeta)d\zeta = 2\pi i \sum_{j=1}^{n} \operatorname*{Res}_{z=z_j} f. \tag{11.8}$$

Proof. Since there are finitely many singular points in $\operatorname{int}\gamma$, there exists $r > 0$ such that the positively oriented circles $\gamma_j := \{z : |z - z_j| = r\}, j = 1, 2, \ldots, n$ are mutually disjoint and all lie in $\operatorname{int}\gamma$. Applying the Cauchy theorem for the multiply connected domain, we obtain

$$\int_\gamma f(\zeta)d\zeta + \sum_{j=1}^{n}\int_{-\gamma_j} f(\zeta)d\zeta = 0$$

or

$$\int_\gamma f(\zeta)d\zeta = \sum_{j=1}^{n}\int_{\gamma_j} f(\zeta)d\zeta = 2\pi i\sum_{j=1}^{n}\operatorname*{Res}_{z=z_j} f.$$ □

Corollary 11.11. *Let $D \subset \mathbb{C}$ be a multiply connected bounded domain with the boundary ∂D, which is a combination of finitely many disjoint piecewise smooth closed Jordan curves. If f is analytic in D and continuous in $\overline{D}$ except for the points $z_1, z_2, \ldots, z_n \in D$, then*

$$\int_{\partial D} f(\zeta)d\zeta = 2\pi i\sum_{j=1}^{n}\operatorname*{Res}_{z=z_j} f,$$

where the integration holds over positively oriented curves.

Corollary 11.12. *Let f be analytic in $\mathbb{C}$ except for $z_1, z_2, \ldots, z_n, z_0 = \infty$. Then*

$$\sum_{j=0}^{n}\operatorname*{Res}_{z=z_j} f = 0. \tag{11.9}$$

Proof. Let $R > 0$ be chosen so that $z_1, z_2, \ldots, z_n \in \{z : |z| < R\}$. Theorem 11.10 gives that

$$\frac{1}{2\pi i} \int_{|\zeta|=R} f(\zeta)d\zeta = \sum_{j=1}^{n} \operatorname*{Res}_{z=z_j} f,$$

where the circle $\{\zeta : |\zeta| = R\}$ is positively oriented. But

$$\frac{1}{2\pi i} \int_{|\zeta|=R} f(\zeta)d\zeta = -\operatorname*{Res}_{z=\infty} f$$

by (11.2). This clearly proves the corollary. □

Example 11.13. Let us find the isolated singular points and the residues at these points for

$$f(z) = \frac{e^z}{1 - \cos z}.$$

Since $e^z \neq 0$ for all $z \in \mathbb{C}$, then the singular points of f may appear only when $1 - \cos z = 0$ or $e^{iz} = 1$. So, the singular points are

$$z_n = 2\pi n, \quad n = 0, \pm 1, \pm 2, \ldots.$$

At the same time, we have

$$(1 - \cos z)'|_{z=z_n} = \sin z|_{z=z_n} = 0, \quad (1 - \cos z)''|_{z=z_n} = \cos z|_{z=z_n} = 1.$$

It means that all these points z_n are zeros of order 2 of the denominator. Therefore, all these points z_n are poles of order 2 for $f(z)$. From these considerations, it follows also that $z = \infty$ is not an isolated singular point (it is not classified). By Theorem 11.4, we have that

$$\begin{aligned}
\operatorname*{Res}_{z=z_n} f &= \lim_{z\to z_n} \frac{d}{dz}\left((z - z_n)^2 \frac{e^z}{1 - \cos z}\right) = \lim_{\zeta\to 0} \frac{d}{d\zeta}\left(\zeta^2 \frac{e^{\zeta+2\pi n}}{1 - \cos\zeta}\right) \\
&= \lim_{\zeta\to 0} \frac{d}{d\zeta}\left(\zeta^2 \frac{e^{\zeta+2\pi n}}{\zeta^2/2! - \zeta^4/4! + \cdots}\right) = e^{2\pi n} \lim_{\zeta\to 0} \frac{d}{d\zeta}\left(\frac{e^\zeta}{1/2 - \zeta^2/4! - \cdots}\right) \\
&= e^{2\pi n} \lim_{\zeta\to 0}\left(\frac{e^\zeta}{1/2 - \zeta^2/4! + \cdots} - \frac{e^\zeta(-2\zeta/4! + 4\zeta^3/6! - \cdots)}{1/2 - \zeta^2/4! + \zeta^4/6! - \cdots}\right) = 2e^{2\pi n}
\end{aligned}$$

for $n = 0, \pm 1, \pm 2, \ldots$. This can be proved also using Corollary 11.6.

Problem 11.14. Show that if $n \in \mathbb{N}$ is even then

$$\operatorname*{Res}_{z=\frac{1}{2}} (\tan(\pi z))^{n-1} = \frac{(-1)^{n/2}}{\pi}.$$

There is one more quite simple but very practical example.

Example 11.15. Let P be a polynomial of degree at most 2. Let us show that if a, b, and c are distinct complex numbers, then

$$f(z) = \frac{P(z)}{(z-a)(z-b)(z-c)} = \frac{A}{z-a} + \frac{B}{z-b} + \frac{C}{z-c},$$

where

$$A = \frac{P(a)}{(a-b)(a-c)} = \operatorname*{Res}_{z=a} f,$$
$$B = \frac{P(b)}{(b-a)(b-c)} = \operatorname*{Res}_{z=b} f$$

and

$$C = \frac{P(c)}{(c-a)(c-b)} = \operatorname*{Res}_{z=c} f.$$

Indeed, since

$$\frac{P(z)}{(z-a)(z-b)(z-c)} = \frac{A}{z-a} + \frac{B}{z-b} + \frac{C}{z-c}$$

then $z = a$, $z = b$, and $z = c$ are singular points of f (if, of course, a, b, and c are not zeros of P). That is why the terms

$$\frac{A}{z-a}, \quad \frac{B}{z-b}, \quad \frac{C}{z-c}$$

are the main parts of the Laurent's expansion for f around a, b, and c, respectively. Thus,

$$A = \operatorname*{Res}_{z=a} f, \quad B = \operatorname*{Res}_{z=b} f, \quad C = \operatorname*{Res}_{z=c} f$$

and

$$A = \lim_{z\to a} \frac{P(z)(z-a)}{(z-a)(z-b)(z-c)} = \frac{P(a)}{(a-b)(a-c)},$$
$$B = \lim_{z\to b} \frac{P(z)(z-b)}{(z-a)(z-b)(z-c)} = \frac{P(b)}{(b-a)(b-c)},$$
$$C = \lim_{z\to c} \frac{P(z)(z-c)}{(z-a)(z-b)(z-c)} = \frac{P(c)}{(c-a)(c-b)}.$$

Problem 11.16. Show that if P has degree of at most 3, then

$$f(z) = \frac{P(z)}{(z-a)^2(z-b)} = \frac{A}{(z-a)^2} + \frac{B}{z-a} + \frac{C}{z-b}, \quad a \neq b,$$

where

$$A = \operatorname*{Res}_{z=a}((z-a)f), \quad B = \operatorname*{Res}_{z=a} f, \quad C = \operatorname*{Res}_{z=b} f.$$

Problem 11.17. Let γ be a piecewise smooth closed Jordan curve and let f be analytic in $\operatorname{int}\gamma$. Let $z_0 \in \operatorname{int}\gamma$ be the only zero of f and of order k. Show that

$$\frac{1}{2\pi \mathrm{i}} \int_{\gamma} \frac{f'(\zeta)}{f(\zeta)} \mathrm{d}\zeta = k = \operatorname*{Res}_{z=z_0} \frac{f'}{f}.$$

A function that is analytic everywhere on the complex plane, except for simple poles, admits a special type of representation there.

Proposition 11.18. *Let function $f(z)$ be analytic on the complex plane $\mathbb{C}$ except for the points $z_1, z_2, \ldots, z_n, \ldots$, which are the poles of order 1 with the residues at these points $a_1, a_2, \ldots, a_n, \ldots$, respectively. Assume that there exists a sequence of closed curves γ_n such that $\operatorname{int}\gamma_n$ includes $z_1, z_2, \ldots, z_n$ and only them, and the following conditions are fulfilled as $n \to \infty$:*

$$R_n := \operatorname{dist}(0, \gamma_n) \to \infty, \quad |\gamma_n| = O(R_n), \quad f(z) = o(R_n), \quad z \in \gamma_n.$$

Then

$$f(z) = f(0) + z \sum_{n=1}^{\infty} \frac{a_n}{z_n(z - z_n)}, \qquad z \neq z_n.$$

Proof. Consider the following integral:

$$I_n(z) := \frac{1}{2\pi \mathrm{i}} \int_{\gamma_n} \frac{f(\zeta)}{\zeta(\zeta - z)} \mathrm{d}\zeta, \quad z \in \operatorname{int}\gamma_n.$$

Applying Cauchy's residue theorem (see Theorem 11.10) to the function $\frac{f(\zeta)}{\zeta(\zeta - z)}$, we obtain

$$\begin{aligned} I_n(z) &= \sum_{k=1}^{n} \operatorname*{Res}_{\zeta=z_k} \frac{f(\zeta)}{\zeta(\zeta - z)} + \operatorname*{Res}_{\zeta=z} \frac{f(\zeta)}{\zeta(\zeta - z)} + \operatorname*{Res}_{\zeta=0} \frac{f(\zeta)}{\zeta(\zeta - z)} \\ &= \sum_{k=1}^{n} \frac{a_k}{z_k(z_k - z)} + \frac{f(z)}{z} - \frac{f(0)}{z}. \end{aligned}$$

This integral $I_n(z)$ can be estimated as

$$|I_n(z)| \le \frac{1}{2\pi} \int_{\gamma_n} \frac{|f(\zeta)|}{|\zeta||\zeta - z|} |\mathrm{d}\zeta| \le \frac{1}{2\pi} \frac{\max_{\gamma_n} |f(\zeta)||\gamma_n|}{R_n(R_n - |z|)}.$$

The latter estimate and the conditions of proposition imply that $I_n(z) \to 0$ as $n \to \infty$ if $z \neq z_k, k = 1, 2, \dots$. This proves the proposition. □

This proposition allows us to get some special representation for the trigonometric functions.

Example 11.19. Let function $f(z)$ be defined as

$$f(z) = \frac{1}{\sin z} - \frac{1}{z}, \quad z \neq 0.$$

It is clear that $f(z)$ has poles of order 1 at the points $z_n = \pi n$, $n = \pm 1, \pm 2, \dots$. The residues at these points are equal to

$$\begin{aligned}\operatorname*{Res}_{z=z_n} f(z) &= \lim_{z\to\pi n}(z-\pi n)\left(\frac{1}{\sin z} - \frac{1}{z}\right) = \lim_{\zeta\to 0}\zeta\left(\frac{1}{\sin(\zeta+\pi n)} - \frac{1}{\zeta+\pi n}\right)\\ &= \lim_{\zeta\to 0}\frac{\zeta}{(-1)^n \sin\zeta} = (-1)^n.\end{aligned}$$

Moreover, the point $z = 0$ is a removable singularity for this function since due to L'Hôpital's rule (see Proposition 3.37), we have that

$$\lim_{z\to 0}\left(\frac{1}{\sin z} - \frac{1}{z}\right) = \lim_{z\to 0}\frac{z - \sin z}{z \sin z} = 0.$$

Considering now the closed curves γ_n, which are equal to the boundary of quadrants with the vertices at the points $(n + 1/2)(\pm 1 \pm \mathrm{i})\pi$, we can easily obtain that function $f(z)$ is uniformly bounded on γ_n. Application of Proposition 11.18 yields

$$\begin{aligned}\frac{1}{\sin z} &= \frac{1}{z} + z\sum_{n=-\infty, n\neq 0}^{\infty}\frac{(-1)^n}{\pi n(z-\pi n)} = \frac{1}{z} + z\sum_{n=1}^{\infty}\frac{(-1)^n}{\pi n(z-\pi n)} + z\sum_{n=1}^{\infty}\frac{(-1)^{n+1}}{\pi n(z+\pi n)}\\ &= \frac{1}{z} + 2z\sum_{n=1}^{\infty}\frac{(-1)^n}{z^2-(\pi n)^2}.\end{aligned}$$

Problem 11.20. Show that:

1. $\frac{1}{\cos z} = 4\pi\sum_{n=0}^{\infty}\frac{(-1)^n(2n+1)}{((2n+1)\pi)^2-4z^2}$,
2. $\frac{1}{e^z-1} = \frac{1}{z} - \frac{1}{2} + 2z\sum_{n=1}^{\infty}\frac{1}{z^2+(2\pi n)^2}$,
3. $\frac{1}{\sin^2 z} = \sum_{n=-\infty}^{\infty}\frac{1}{(z-\pi n)^2}$,
4. $\tan z = 8z\sum_{n=0}^{\infty}\frac{1}{((2n+1)\pi)^2-4z^2}$,
5. $\cot z = \frac{1}{z} + 2z\sum_{n=1}^{\infty}\frac{1}{z^2-(\pi n)^2}$,
6. $\frac{1}{\cosh z} = 4\pi\sum_{n=0}^{\infty}\frac{(-1)^n(2n+1)}{((2n+1)\pi)^2+4z^2}$,
7. $\frac{1}{\sinh z} = \frac{1}{z} + 2z\sum_{n=1}^{\infty}\frac{(-1)^n}{z^2+(\pi n)^2}$,

8. $\coth z = \frac{1}{z} + 2z \sum_{n=1}^{\infty} \frac{1}{z^2+(\pi n)^2}$,
9. $\tanh z = 8z \sum_{n=0}^{\infty} \frac{1}{((2n+1)\pi)^2+4z^2}$.

There is quite close connection (due to the Cauchy's residue theorem) between some series and some integrals. We illustrate this by the following example.

Example 11.21. Let us show that for all $|z| < 1$ and for any number $a \neq n, n \in \mathbb{N}$, the following is true:

$$-\sum_{n=1}^{\infty} \frac{(-z)^n}{n-a} = \frac{\pi z^a}{\sin(\pi a)} + \frac{z}{2\mathrm{i}\sin(\pi a)} \int_{|\zeta|=1} \frac{\zeta^{a-1} - z^{a-1}}{\zeta - z} \mathrm{d}\zeta.$$

Since $|z| < 1$, then the above integral can be represented as

$$\int_{|\zeta|=1} \frac{\zeta^{a-1} - z^{a-1}}{\zeta} \sum_{n=0}^{\infty} \frac{z^n}{\zeta^n} \mathrm{d}\zeta = \sum_{n=0}^{\infty} z^n \int_{|\zeta|=1} \zeta^{a-2-n} \mathrm{d}\zeta - \sum_{n=0}^{\infty} z^{n+a-1} \int_{|\zeta|=1} \zeta^{-n-1} \mathrm{d}\zeta.$$

Both integrals on the right-hand side of the latter equality can be calculated straightforwardly. Indeed,

$$\int_{|\zeta|=1} \zeta^{a-2-n} \mathrm{d}\zeta = \int_{-\pi}^{\pi} \mathrm{e}^{\mathrm{i}\theta(a-2-n)} \mathrm{i}\mathrm{e}^{\mathrm{i}\theta} \mathrm{d}\theta = \frac{(-1)^n 2\mathrm{i}\sin(\pi a)}{n+1-a}$$

and this value is well-defined due to the condition on a. The second integral is well known and it is equal to

$$\int_{|\zeta|=1} \zeta^{-1-n} \mathrm{d}\zeta = \begin{cases} 2\pi\mathrm{i}, & n = 0, \\ 0, & n = 1, 2, \dots. \end{cases}$$

Combining these two integrals, we obtain

$$\int_{|\zeta|=1} \frac{\zeta^{a-1} - z^{a-1}}{\zeta - z} \mathrm{d}\zeta = \sum_{n=0}^{\infty} z^n \frac{2\mathrm{i}(-1)^n}{n+1-a} \sin(\pi a) - z^{a-1} 2\pi\mathrm{i}.$$

Multiplying the latter equality by z and rearranging the terms, one can obtain the needed result.

Based on this formula and using L'Hôpital's rule ($a \to 0$), it can be shown that

$$\frac{\log(1+z)}{z} = \frac{1}{2\pi\mathrm{i}} \int_{|\zeta|=1} \frac{\log \zeta}{\zeta(\zeta - z)} \mathrm{d}\zeta.$$

12 The principle of the argument and Rouche's theorem

Let G be a domain on the complex plane and let D be a bounded subdomain of G such that $\overline{D} \subset G$. The domain D needs not be simply connected but the boundary ∂D of this domain is a combination of finitely many disjoint piecewise smooth closed Jordan curves. Let f be an analytic function on G. Consequently, f is analytic on the closed domain $\overline{D}$.

Proposition 12.1. *Let the domains D and G be as above and let f be analytic on G, except for a finite number of poles $z_k \in D$ of order μ_k for $k = 1, 2, \ldots, n$. Let us assume in addition that $f(z) \neq 0$ on $\overline{D}$ except for a finite number of zeros $w_k \in D$ of order λ_k, $k = 1, 2, \ldots, m$. Then the function*

$$\frac{f'(z)}{f(z)}$$

is analytic on $\overline{D}$ except for the points $\{z_k\}_{k=1}^{n}$ and $\{w_k\}_{k=1}^{m}$ (which are poles of order 1 for f'/f) and

$$\frac{1}{2\pi \mathrm{i}} \int_{\partial D} \frac{f'(\zeta) d\zeta}{f(\zeta)} = N - P, \tag{12.1}$$

where $N = \sum_{k=1}^{m} \lambda_k$ and $P = \sum_{k=1}^{n} \mu_k$.

Proof. Consider the function f'/f in the neighborhood of a pole z_k. Then $f(z)$ can be represented there as

$$f(z) = (z - z_k)^{-\mu_k} f_1(z),$$

where $f_1(z)$ is analytic in this neighborhood and $f_1(z_k) \neq 0$. This implies that

$$\frac{f'(z)}{f(z)} = \frac{-\mu_k (z - z_k)^{-\mu_k - 1} f_1(z) + (z - z_k)^{-\mu_k} f_1'(z)}{(z - z_k)^{-\mu_k} f_1(z)} = -\frac{\mu_k}{z - z_k} + \frac{f_1'(z)}{f_1(z)}, \tag{12.2}$$

where the second term $f_1'(z)/f_1(z)$ in the latter sum is analytic in this neighborhood of z_k since $f_1(z_k) \neq 0$. The representation (12.2) shows that z_k is a pole of order 1 for f'/f and

$$\operatorname*{Res}_{z=z_k} \frac{f'(z)}{f(z)} = -\mu_k. \tag{12.3}$$

Consider now the function f'/f in the neighborhood of a zero w_k. Then we have that

$$f(z) = (z - w_k)^{\lambda_k} f_2(z),$$

where $f_2(z)$ is analytic in this neighborhood and $f_2(w_k) \neq 0$. Thus, we have

https://doi.org/10.1515/9783111632278-013

$$\frac{f'(z)}{f(z)} = \frac{\lambda_k(z-w_k)^{\lambda_k-1}f_2(z)+(z-w_k)^{\lambda_k}f_2'(z)}{(z-w_k)^{\lambda_k}f_2(z)} = \frac{\lambda_k}{z-w_k}+\frac{f_2'(z)}{f_2(z)}, \tag{12.4}$$

where the second term in the latter sum is analytic in this neighborhood since $f_2(w_k) \neq 0$. The representation (12.4) shows also that w_k is a pole of order 1 for f'/f and

$$\operatorname*{Res}_{z=w_k} \frac{f'(z)}{f(z)} = \lambda_k. \tag{12.5}$$

Since the function f'/f is analytic on $\overline{D}$ except for the points $\{z_k\}_{k=1}^n$, $\{w_k\}_{k=1}^m$ (where it has the simple poles), then applying the Cauchy's residue theorem (see Theorem 11.10) we obtain (see (12.3) and (12.5))

$$\frac{1}{2\pi \mathrm{i}} \int\limits_{\partial D} \frac{f'(\zeta)\mathrm{d}\zeta}{f(\zeta)} = \sum_{k=1}^{n} \operatorname*{Res}_{z=z_k} \frac{f'(z)}{f(z)} + \sum_{k=1}^{m} \operatorname*{Res}_{z=w_k} \frac{f'(z)}{f(z)} = -\sum_{k=1}^{n} \mu_k + \sum_{k=1}^{m} \lambda_k = N - P.$$

This completes the proof. □

Corollary 12.2. *Suppose that $f(z)$ is analytic on $\overline{D}$ and $f(z) \neq 0$ on $\overline{D}$ except for the zeros $w_k \in D$ of order λ_k, $k = 1, 2, \dots, m$. Then*

$$\frac{1}{2\pi \mathrm{i}} \int\limits_{\partial D} \frac{f'(\zeta)\mathrm{d}\zeta}{f(\zeta)} = N. \tag{12.6}$$

Let γ be a piecewise smooth closed Jordan curve and let $f(z)$ be analytic on $\overline{\operatorname{int}\gamma}$.

Definition 12.3. Let ζ_0 be a point of γ and $\varphi_0 = \operatorname{Arg} f(z)$ at ζ_0. Let also $\varphi_1 = \operatorname{Arg} f(z)$ at ζ_0 after going around once along this curve from ζ_0 to ζ_0 in a positive direction. Then the value $\varphi_1 - \varphi_0$ is called the *variation* of $\operatorname{Arg} f(z)$ along curve γ and it is denoted by

$$\varphi_1 - \varphi_0 = \operatorname*{Var}_{\gamma} \operatorname{Arg} f.$$

Theorem 12.4 (The principle of argument). *Let f be analytic on $\overline{\operatorname{int}\gamma}$, where γ is a piecewise smooth closed Jordan curve, except for the poles $\{z_k\}_{k=1}^n \subset \operatorname{int}\gamma$ of order μ_k. Assume that $f(z) \neq 0$ on $\overline{\operatorname{int}\gamma}$ except for the zeros $\{w_k\}_{k=1}^m \subset \operatorname{int}\gamma$ of order λ_k. Then*

$$\frac{1}{2\pi \mathrm{i}} \int\limits_{\gamma} \frac{f'(z)}{f(z)}\mathrm{d}z = \frac{1}{2\pi} \operatorname*{Var}_{\gamma} \operatorname{Arg} f(z) = N - P, \tag{12.7}$$

where $N = \sum_{k=1}^m \lambda_k$ and $P = \sum_{k=1}^n \mu_k$.

Proof. Since $f(z) \neq 0$ on γ, we may consider the multivalued function

$$\operatorname{Log} f(z) = \log|f(z)| + \mathrm{i} \operatorname{Arg} f(z).$$

Moreover, this function is analytic in the neighborhood of γ and

$$(\operatorname{Log} f(z))' = \frac{f'(z)}{f(z)}.$$

Proposition 12.1 says that

$$\frac{1}{2\pi \mathrm{i}} \int_{\gamma} (\operatorname{Log} f(\zeta))' \mathrm{d}\zeta = N - P.$$

It is equivalent to the changes of $\operatorname{Log} f(\zeta)$ after going around once along γ from ζ_0 to ζ_0, i. e.,

$$\begin{aligned} N - P &= \frac{1}{2\pi \mathrm{i}} [\operatorname{Log} f(\zeta)]_{\zeta=\zeta_0}^{\zeta=\zeta_0} = \frac{1}{2\pi \mathrm{i}} [\log|f(\zeta)| + \mathrm{i} \operatorname{Arg} f(\zeta)]_{\zeta=\zeta_0}^{\zeta=\zeta_0} \\ &= \frac{\operatorname{Arg} f(\zeta)}{2\pi}\bigg|_{\zeta=\zeta_0}^{\zeta=\zeta_0} = \frac{\operatorname{Var}_\gamma \operatorname{Arg} f(\zeta)}{2\pi}. \end{aligned}$$

□

Theorem 12.5 (Rouche). *Assume that G is a simply connected domain, γ is a piecewise smooth closed Jordan curve in G, and f and g are analytic functions on G except for the finitely many poles, which are located in* $\operatorname{int} \gamma$. *If $|f(\zeta)| > |g(\zeta)|$ on γ, then*

$$N_{f+g} - P_{f+g} = N_f - P_f, \tag{12.8}$$

where N_f, N_{f+g}, P_f, and P_{f+g} denote the number of zeros or poles (taking into account their multiplicity) for functions f and $f + g$, respectively.

Proof. The conditions for f and g on γ show that $|f(\zeta)| > 0$ and $|f + g| \geq |f| - |g| > 0$ on γ, i. e., f and $f + g$ are not equal to zero on γ. That is why we may apply Theorem 12.4 and obtain

$$\frac{1}{2\pi} \operatorname*{Var}_{\gamma} \operatorname{Arg}(f + g) - \frac{1}{2\pi} \operatorname*{Var}_{\gamma} \operatorname{Arg}(f) = (N_{f+g} - P_{f+g}) - (N_f - P_f).$$

But the left-hand side of the latter equality is equal to (see the proof of Theorem 12.4)

$$\frac{1}{2\pi} \operatorname*{Var}_{\gamma} \operatorname{Arg} \frac{f + g}{f} = \frac{1}{2\pi} \operatorname*{Var}_{\gamma} \operatorname{Arg}(1 + g/f).$$

We will show now that this value is equal to zero. Indeed, since on γ we have

$$|g/f + 1 - 1| = |g/f| < 1$$

then the value of $g/f + 1$ on γ changes inside the circle $\{w : |w - 1| < 1\}$ such that $w = 0$ does not belong to this set (see Figure 12.1).

Since it does not go around zero along γ, then $\operatorname{Var}_\gamma \operatorname{Arg}(1 + g/f) = 0$. Hence, the equality (12.8) holds. □

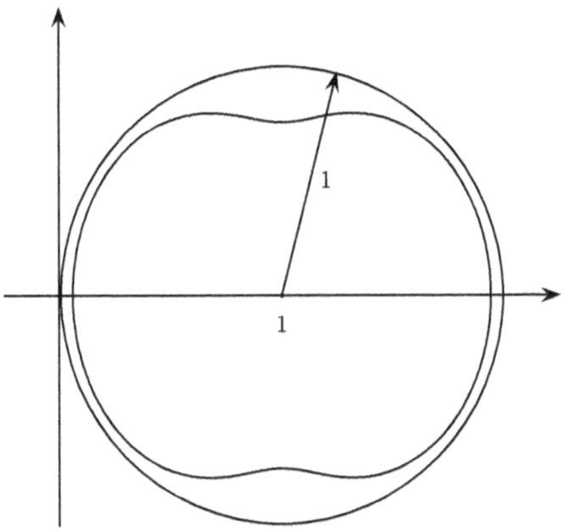

Figure 12.1: Image of $g/f + 1$ on γ contained in the unit circle.

Corollary 12.6. *Suppose that f and g are analytic. Then under the conditions of Theorem* 12.5, *we have that*

$$N_{f+g} = N_f. \tag{12.9}$$

Example 12.7. Let $P(z) = z^{10} - 5z^7 + 2$. The fundamental theorem of algebra says that this polynomial has exactly 10 roots (taking into account their multiplicities). The question now is: how many of these roots are located in the unit disk $\{z : |z| < 1\}$? Indeed, if we denote $g(z) = z^{10} + 2$ and $f(z) = -5z^7$, then $P(z) = f(z) + g(z)$. The function f has 7 roots in this disk and for $|z| = 1$ we have that

$$|g(z)| = |z^{10} + 2| \leq |z|^{10} + 2 = 3 < 5 = |f(z)| = 5|z|^7.$$

By Rouche's theorem, we obtain $N_{f+g} = N_f = 7$.

Problem 12.8. Prove the fundamental theorem of algebra using Corollary 12.6.

Problem 12.9. Show that the equation

$$a_0 + a_1 \cos\varphi + a_2 \cos 2\varphi + \cdots + a_n \cos n\varphi = 0,$$

where $0 \leq a_0 < a_1 < \cdots < a_n$ has $2n$ simple roots on the interval $(0, 2\pi)$.

Problem 12.10. Let γ be a closed Jordan curve. If

$$|a_k z^k| > |a_0 + a_1 z + \cdots + a_{k-1} z^{k-1} + a_{k+1} z^{k+1} + \cdots + a_n z^n|$$

for all $z \in \gamma$, then prove that there are exactly k roots of the polynomial

$$a_0 + a_1 z + \cdots + a_n z^n$$

in $\operatorname{int} \gamma$.

Problem 12.11. Show that for any small $\epsilon > 0$ there exists $n \in \mathbb{N}$ such that all zeros of the function

$$f(z) = \sum_{k=0}^{n} \frac{1}{k!z^k}$$

are in the disk $\{z : |z| \leq \epsilon\}$.

Problem 12.12. Let $a > \mathrm{e}$. Show that the equation

$$\mathrm{e}^z = az^n, \quad n \in \mathbb{N}_0,$$

has exactly n zeros in the unit disk. Hint: Use Rouche's theorem with $f(z) = \mathrm{e}^z$ and $g(z) = az^n$.

Problem 12.13. Find the number of roots of the equation

$$z^6 + 6z + 10 = 0$$

in each quadrant of the complex plane $\mathbb{C}$.

Problem 12.14. Find the number of roots of the polynomial of degree 8,

$$P_8(z) := z^8 - 5z^5 - 2z + 1$$

inside the unit disk $\{z : |z| < 1\}$. Hint: Use Rouche's theorem for $f(z) = -5z^5 + 1$ and $g(z) = z^8 - 2z$.

Problem 12.15. Show that the roots of the polynomial of degree 4,

$$P_4(z) := z^4 + z^3 + 4z^2 + 2z + 3$$

are not real and not purely imaginary, and two of them are in the second quadrant and other two (complex conjugate) are in the third quadrant. Hint: Show that for $R > 0$ big enough the variation of $\operatorname{Arg} P_4(z)$ along the closed curve $\gamma := \{z : |z| = R\} \cap \{\operatorname{Re} z = 0, \operatorname{Im} z \geq 0\} \cap \{\operatorname{Im} = 0, \operatorname{Re} \geq 0\}$ in the first quadrant is equal to zero.

Problem 12.16. Show that the polynomial of degree 4,

$$P_4(z) := z^4 + az^3 + b, \quad a, b > 0,$$

has two (conjugate) roots with a positive real part.

Problem 12.17. Show that if $f(z)$ is analytic and univalent in the domain D, then $f'(z) \neq 0$ for all $z \in D$. Hint: Assuming on the contrary that there is $z_0 \in D$ such that $f'(z_0) = 0$, obtain the contradiction with the univalent function.

13 Calculation of integrals by residue theory

13.1 Trigonometric integrals

Suppose that we want to evaluate an integral of the form

$$\int_{0}^{2\pi} R(\cos t, \sin t)\mathrm{d}t, \tag{13.1}$$

where $R(u, v)$ is a rational function of two variables u and v, i. e.,

$$R(u, v) = \frac{\sum_{k,l} a_{kl} u^k v^l}{\sum_{m,n} b_{mn} u^n v^m}$$

and the summation in both sums is finite. Due to periodicity, (13.1) is equal to

$$\int_{-\pi}^{\pi} R(\cos t, \sin t)\mathrm{d}t. \tag{13.2}$$

Consider the unit circle $\{z : |z| = 1\}$, which is parameterized as (positive orientation) $\gamma : z(t) = \mathrm{e}^{\mathrm{i}t}, t \in [-\pi, \pi]$. Then

$$\cos t = \frac{\mathrm{e}^{\mathrm{i}t} + \mathrm{e}^{-\mathrm{i}t}}{2} = \frac{z + 1/z}{2} = \frac{z^2 + 1}{2z},$$
$$\sin t = \frac{\mathrm{e}^{\mathrm{i}t} - \mathrm{e}^{-\mathrm{i}t}}{2\mathrm{i}} = \frac{z - 1/z}{2\mathrm{i}} = \frac{z^2 - 1}{2\mathrm{i}z}$$

and

$$\mathrm{d}z = \mathrm{d}(\mathrm{e}^{\mathrm{i}t}) = \mathrm{e}^{\mathrm{i}t}\mathrm{i}\mathrm{d}t$$

or

$$\mathrm{d}t = \frac{\mathrm{d}z}{\mathrm{i}\mathrm{e}^{\mathrm{i}t}} = \frac{\mathrm{d}z}{\mathrm{i}z}.$$

The integral (13.1) transforms to the curve integral

$$\int_{-\pi}^{\pi} R(\cos t, \sin t)\mathrm{d}t = \int_{\gamma} R\left(\frac{z^2 + 1}{2z}, \frac{z^2 - 1}{2\mathrm{i}z}\right)\frac{\mathrm{d}z}{\mathrm{i}z} = \int_{\gamma} \widetilde{R}(z)\mathrm{d}z, \tag{13.3}$$

where

$$\widetilde{R}(z) = \frac{1}{\mathrm{i}z} R\left(\frac{z^2 + 1}{2z}, \frac{z^2 - 1}{2\mathrm{i}z}\right)$$

https://doi.org/10.1515/9783111632278-014

is a rational function of only one variable z. This rational function $\widetilde{R}$ may have only poles (zeros of the denominator of $\widetilde{R}$).

Let us consider the poles of $\widetilde{R}$, which are located inside the unit disk $\{z : |z| < 1\}$ and denote them as $z_1, z_2, \ldots, z_m$. The residue theorem gives

$$\int_{-\pi}^{\pi} R(\cos t, \sin t)\mathrm{d}t = \int_{\gamma} \widetilde{R}(z)\mathrm{d}z = 2\pi\mathrm{i}\sum_{j=1}^{m} \operatorname*{Res}_{z=z_j} \widetilde{R}. \tag{13.4}$$

Example 13.1. Let us evaluate the integral

$$\int_{0}^{2\pi} \frac{1}{3 + 2\sin t}\mathrm{d}t.$$

Due to (13.3), we have

$$\int_{0}^{2\pi} \frac{1}{3 + 2\sin t}\mathrm{d}t = \int_{\gamma} \frac{1}{\mathrm{i}z}\frac{1}{3 + 2\frac{z^2-1}{2\mathrm{i}z}}\mathrm{d}z = \int_{\gamma} \frac{\mathrm{d}z}{z^2 + 3\mathrm{i}z - 1},$$

where γ is the unit circle. The zeros of the denominator are

$$z_{1,2} = \frac{-3\mathrm{i} \mp \mathrm{i}\sqrt{5}}{2}.$$

It is easy to see that $|z_1| > 1$ and $|z_2| < 1$. By (13.4), we get

$$\int_{0}^{2\pi} \frac{1}{3 + 2\sin t}\mathrm{d}t = 2\pi\mathrm{i}\operatorname*{Res}_{z=z_2} \frac{1}{z^2 + 3\mathrm{i}z - 1} = 2\pi\mathrm{i}\frac{1}{2z_2 + 3\mathrm{i}} = \frac{2\pi}{\sqrt{5}}$$

after using also (11.6).

Example 13.2. Let us evaluate the integral

$$I := \int_{0}^{2\pi} \frac{1}{1 + 3\cos^2 t}\mathrm{d}t.$$

Repeating the same procedure as above, we obtain

$$I = \int_{\gamma} \frac{1}{\mathrm{i}z}\frac{1}{1 + 3(\frac{z^2+1}{2z})^2}\mathrm{d}z = \frac{1}{\mathrm{i}}\int_{\gamma} \frac{4z\mathrm{d}z}{3z^4 + 10z^2 + 3}.$$

The zeros of the denominator are

$$z_1 = \mathrm{i}\sqrt{3}, \quad z_2 = -\mathrm{i}\sqrt{3}, \quad z_3 = \mathrm{i}/\sqrt{3}, \quad z_4 = -\mathrm{i}/\sqrt{3}.$$

It is clear that $|z_1|, |z_2| > 1$, and $|z_3|, |z_4| < 1$. That is why

$$I = 2\pi\left(\operatorname*{Res}_{z=z_3}\frac{4z}{3z^4+10z^2+3} + \operatorname*{Res}_{z=z_4}\frac{4z}{3z^4+10z^2+3}\right)$$
$$= 2\pi\left(\frac{4z_3}{12z_3^3+20z_3} + \frac{4z_4}{12z_4^3+20z_4}\right)$$
$$= 2\pi\left(\frac{i/\sqrt{3}}{3(i/\sqrt{3})^3+5i/\sqrt{3}} - \frac{i/\sqrt{3}}{3(-i/\sqrt{3})^3-5i/\sqrt{3}}\right) = \pi.$$

Problem 13.3. Evaluate

$$\int_0^{2\pi}\frac{\cos(2t)}{5-4\cos t}\,dt.$$

Problem 13.4. Evaluate

$$\int_0^{2\pi}\frac{\sin^2 t}{5+4\cos t}\,dt.$$

Problem 13.5. Show that if $w \in \mathbb{C}\setminus\{w \in \mathbb{R} : |w| \geq 1\}$, then

$$\int_0^{\pi}\frac{dt}{1-w\cos t} = \frac{\pi}{\sqrt{1-w^2}}.$$

Explain the choice of the complex number w.

Problem 13.6. Show that if $a > b > 0$, then

1.
$$\int_0^{2\pi}\frac{dt}{(a+b\cos t)^2} = \frac{2\pi a}{\sqrt{(a^2-b^2)^3}},$$

2.
$$\int_0^{2\pi}\frac{dt}{(a+b\cos^2 t)^2} = \frac{\pi(2a+b)}{\sqrt{(a^2+ab)^3}}.$$

The next example shows that the above technique cannot be used for the evaluation of the following integral.

Example 13.7. We will show that

$$\int_0^{\pi}\frac{x\sin x}{1-2a\cos x+a^2}\,dx = \begin{cases}\frac{\pi}{a}\log(1+a), & 0 < a < 1,\\ \frac{\pi}{a}\log(1+1/a), & a > 1.\end{cases}$$

Let us assume first that $0 < a < 1$. Consider the function

$$f(z) := \frac{z\mathrm{e}^{\mathrm{i}z}}{a\mathrm{e}^{\mathrm{i}z} - 1}, \qquad z \in \{z : |\operatorname{Re} z| \le \pi, \operatorname{Im} z \ge 0\}.$$

Since $0 < a < 1$, then this function is analytic in the specified region. Applying now the Cauchy theorem for the rectangle with the vertices at the points $-\pi, \pi, \pi + \mathrm{i}n, -\pi + \mathrm{i}n, n \in \mathbb{N}$, we obtain

$$\int_{-\pi}^{\pi} f(x)\mathrm{d}x + \int_{0}^{n} f(\pi + \mathrm{i}y)\mathrm{i}\mathrm{d}y + \int_{\pi}^{-\pi} f(x + \mathrm{i}n)\mathrm{d}x + \int_{n}^{0} f(-\pi + \mathrm{i}y)\mathrm{i}\mathrm{d}y = 0.$$

The third integral tends to zero as $n \to \infty$ since it is equal to

$$-\int_{-\pi}^{\pi} \frac{(x + \mathrm{i}n)\mathrm{e}^{\mathrm{i}x}}{a\mathrm{e}^{\mathrm{i}x} - \mathrm{e}^{n}}\mathrm{d}x \to 0, \qquad n \to \infty.$$

The sum of the second and the fourth integrals is equal to

$$\mathrm{i}\int_{0}^{n} \frac{(\pi + \mathrm{i}y)\mathrm{e}^{\mathrm{i}(\pi+\mathrm{i}y)}}{a\mathrm{e}^{\mathrm{i}(\pi+\mathrm{i}y)} - 1}\mathrm{d}y + \mathrm{i}\int_{0}^{n} \frac{(\pi - \mathrm{i}y)\mathrm{e}^{\mathrm{i}(-\pi+\mathrm{i}y)}}{a\mathrm{e}^{\mathrm{i}(-\pi+\mathrm{i}y)} - 1}\mathrm{d}y = 2\pi\mathrm{i}\int_{0}^{n} \frac{\mathrm{d}y}{a + \mathrm{e}^{y}}\mathrm{d}y.$$

Letting now $n \to \infty$, we get

$$\int_{-\pi}^{\pi} \frac{x\mathrm{e}^{\mathrm{i}x}}{a\mathrm{e}^{\mathrm{i}x} - 1}\mathrm{d}x + 2\pi\mathrm{i}\int_{0}^{\infty} \frac{\mathrm{d}y}{a + \mathrm{e}^{y}}\mathrm{d}y = 0.$$

Equating the imaginary part of this equality to zero, we obtain that

$$\int_{-\pi}^{\pi} \frac{x \sin x}{1 - 2a\cos x + a^2}\mathrm{d}x = 2\pi\int_{0}^{\infty} \frac{\mathrm{d}y}{a + \mathrm{e}^{y}}.$$

But the improper integral on the right-hand side of this equality can be easily evaluated and it is equal to $\frac{1}{a}\log(1+a)$. Taking into account that the integrand on the left-hand side is even, we finally obtain the needed result in the case $0 < a < 1$. The case $a > 1$ can be reduced to the previous case if we change a to $\frac{1}{a}$. Moreover, the obtained result can be easily extended (by continuity with respect to parameter a) for all $0 \le a < \infty$.

Problem 13.8. Show that for $|a| < 1$ we have

$$\int_{0}^{2\pi} \frac{\mathrm{d}t}{1 - 2a\cos t + a^2} = \frac{2\pi}{1 - a^2}.$$

What can one say about this integral for $|a| > 1$?

Problem 13.9. Prove that for $n \in \mathbb{N}$,

$$\int_0^{2\pi} \mathrm{e}^{\cos t} \cos(nt - \sin t)\mathrm{d}t = \frac{2\pi}{n!}, \quad \int_0^{2\pi} \mathrm{e}^{\cos t} \sin(nt - \sin t)\mathrm{d}t = 0.$$

Hint: Use Euler's formula.

13.2 Improper integrals of the form $\int_{-\infty}^{\infty} f(x)\mathrm{d}x$

Let $f(x)$ be a continuous real-valued function of $x \in \mathbb{R}$. The *Cauchy principal value of the integral*

$$\int_{-\infty}^{\infty} f(x)\mathrm{d}x$$

is defined by

$$\text{p.v.} \int_{-\infty}^{\infty} f(x)\mathrm{d}x = \lim_{R\to\infty} \int_{-R}^{R} f(x)\mathrm{d}x$$

provided the limit exists. By this definition, we obtain

$$\text{p.v.} \int_{-\infty}^{\infty} f(x)\mathrm{d}x = 0$$

if f is odd and

$$\text{p.v.} \int_{-\infty}^{\infty} f(x)\mathrm{d}x = 2\int_{0}^{\infty} f(x)\mathrm{d}x$$

if f is even.

Theorem 13.10. *Let f be analytic for $\operatorname{Im} z > 0$ and continuous for $\operatorname{Im} z \geq 0$ except for the singular points $z_1, z_2, \ldots, z_n$ with $\operatorname{Im} z_j > 0$ for all $j = 1, 2, \ldots, n$. If $f(z) = o(1/|z|)$ for $z \to \infty$, $\operatorname{Im} z > 0$, then*

$$\text{p.v.} \int_{-\infty}^{\infty} f(x)\mathrm{d}x = 2\pi\mathrm{i} \sum_{j=1}^{n} \operatorname*{Res}_{z=z_j} f. \tag{13.5}$$

Proof. Let $R > 0$ be chosen such that all points $z_1, z_2, \ldots, z_n$ belong to the region $\{z : |z| < R, \operatorname{Im} z > 0\}$. Let γ_R be the union of the line segment $[-R, R]$ and the upper semicircle Γ_R^+ (see Figure 13.1). The residue theorem gives that

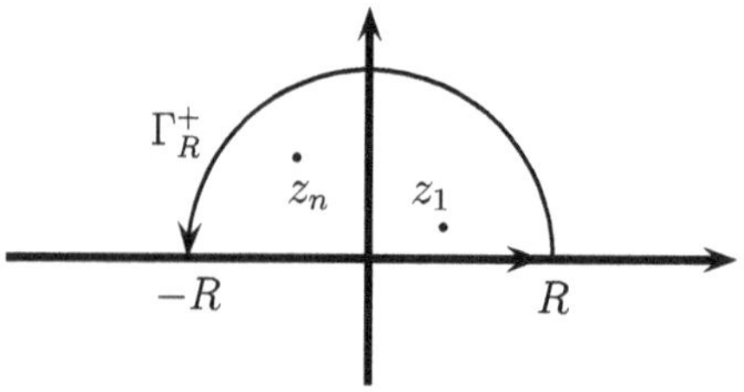

Figure 13.1: Semicircle Γ_R^+ containing singularities.

$$\int_{-R}^{R} f(x)\mathrm{d}x + \int_{\Gamma_R^+} f(z)\mathrm{d}z = 2\pi\mathrm{i}\sum_{j=1}^{n} \operatorname*{Res}_{z=z_j} f.$$

But

$$\left|\int_{\Gamma_R^+} f(z)\mathrm{d}z\right| = \left|\int_0^{\pi} f(R\mathrm{e}^{\mathrm{i}t})R\mathrm{e}^{\mathrm{i}t}\mathrm{i}\mathrm{d}t\right| \le \int_0^{\pi} |f(R\mathrm{e}^{\mathrm{i}t})|R\mathrm{d}t$$

$$= \int_0^{\pi} o(1/R)R\mathrm{d}t = o_R(1)\pi \to 0$$

as $R \to \infty$. That is why

$$\lim_{R\to\infty} \int_{-R}^{R} f(x)\mathrm{d}x = 2\pi\mathrm{i}\sum_{j=1}^{n} \operatorname*{Res}_{z=z_j} f$$

and the proof is concluded. □

Example 13.11. Let us evaluate the integral

$$\int_0^{\infty} \frac{1}{x^4+1}\mathrm{d}x = \frac{1}{2}\int_{-\infty}^{\infty} \frac{1}{x^4+1}\mathrm{d}x.$$

The singular points of

$$\frac{1}{z^4+1}$$

are

$$z_0 = \mathrm{e}^{\mathrm{i}\pi/4}, \quad z_1 = \mathrm{e}^{\mathrm{i}3\pi/4}, \quad z_2 = \mathrm{e}^{\mathrm{i}5\pi/4}, \quad z_3 = \mathrm{e}^{\mathrm{i}7\pi/4}.$$

It is clear also that $\operatorname{Im} z_0, \operatorname{Im} z_1 > 0$, and $\operatorname{Im} z_2, \operatorname{Im} z_3 < 0$. Hence,

$$\begin{aligned}\int_0^\infty \frac{1}{x^4+1}dx &= \pi i\left(\operatorname*{Res}_{z=z_0}\frac{1}{z^4+1} + \operatorname*{Res}_{z=z_1}\frac{1}{z^4+1}\right) = \pi i\left(\frac{1}{4z_0^3} + \frac{1}{4z_1^3}\right)\\ &= \frac{\pi i}{4}(e^{-3i\pi/4} + e^{-9i\pi/4})\\ &= \frac{\pi i}{4}\left(\cos\frac{3\pi}{4} - i\sin\frac{3\pi}{4} + \cos\frac{9\pi}{4} - i\sin\frac{9\pi}{4}\right)\\ &= \frac{\pi i}{4}\left(-2i\sin\frac{\pi}{4}\right) = \frac{\pi\sqrt{2}}{4}.\end{aligned}$$

Example 13.12. Let us evaluate the integral

$$\int_0^\infty \frac{x^4}{x^6+1}dx.$$

The singular points of

$$\frac{z^4}{z^6+1}$$

are

$$z_k = e^{i(\pi/6+2\pi k/6)}, \quad k = 0, 1, \dots, 5.$$

It is clear that only z_0, z_1 and z_2 belong to the upper half-plane. Thus,

$$\begin{aligned}\int_0^\infty \frac{x^4}{x^6+1}dx &= \pi i\sum_{j=0}^{2}\operatorname*{Res}_{z=z_j}\frac{z^4}{z^6+1} = \pi i\left(\frac{z_0^4}{6z_0^5} + \frac{z_1^4}{6z_1^5} + \frac{z_2^4}{6z_2^5}\right)\\ &= \frac{\pi i}{6}\left(\frac{1}{z_0} + \frac{1}{z_1} + \frac{1}{z_2}\right) = \frac{\pi i}{6}(e^{-i\pi/6} + e^{-i\pi/2} + e^{-i5\pi/6})\\ &= \frac{\pi i}{6}\left(\cos\frac{\pi}{6} - i\sin\frac{\pi}{6} - i + \cos\frac{5\pi}{6} - i\sin\frac{5\pi}{6}\right)\\ &= \frac{\pi i}{6}\left(-2i\sin\frac{\pi}{6} - i\right) = \frac{\pi}{3}.\end{aligned}$$

Problem 13.13. Evaluate the integral

$$\int_{-\infty}^\infty \frac{x^2}{(x^2+4)^2}dx.$$

Problem 13.14. Evaluate the integral

$$\int_{-\infty}^{\infty} \frac{1}{(x^4+1)^2} dx.$$

Problem 13.15. Evaluate the integral

$$\text{p.v.} \int_{-\infty}^{\infty} \frac{1}{x(x^2+1)} dx.$$

Problem 13.16. Prove that:

1.

$$\int_{-\infty}^{\infty} \frac{dx}{(x^2+a^2)(x^2+b^2)^2} = \frac{\pi(2|a|+|b|)}{2a^2|ab|(|a|+|b|)^2},$$

2.

$$\int_{0}^{\infty} \frac{dx}{(a+bx^2)^n} = \pi\sqrt{a/b}\frac{(2n-3)!!}{(2a)^n(n-1)!}, \qquad n \in \mathbb{N},$$

under the condition $ab > 0$. Here, $(2n-3)!!$ for $n = 1$ must be substituted by 1.

13.3 Improper integrals of the form $\int_{-\infty}^{\infty} e^{iax} f(x)dx$

Theorem 13.17 (Jordan's lemma). *Let us assume that* f *is continuous in the region* $\{z : |z| > R, \operatorname{Im} z > 0\}$ *for some* $R > 0$. *If*

$$\lim_{z\to\infty} f(z) = 0, \quad \operatorname{Im} z > 0,$$

then

$$\lim_{R\to\infty} \int_{|\zeta|=R, \operatorname{Im}\zeta>0} e^{ia\zeta} f(\zeta)d\zeta = 0 \tag{13.6}$$

for any $a > 0$.

Proof. Under the conditions for f, we have that for any $\varepsilon > 0$ there exists $R > 0$ such that

$$|f(z)| < \varepsilon, \quad |z| > R, \operatorname{Im} z > 0.$$

We parametrize the semicircle as $\gamma : \zeta(t) = Re^{it}$, $t \in (0, \pi)$. In that case, we obtain

$$\begin{aligned}\left|\int_\gamma e^{ia\zeta} f(\zeta)d\zeta\right| &\le \int_\gamma |e^{ia\zeta}||f(\zeta)||d\zeta| < \varepsilon \int_0^\pi |e^{iaR(\cos t + i\sin t)}|R dt \\ &= \varepsilon R \int_0^\pi e^{-aR\sin t} dt = 2\varepsilon R \int_0^{\pi/2} e^{-aR\sin t} dt \\ &< 2\varepsilon R \int_0^{\pi/2} e^{-aR2t/\pi} dt\end{aligned}$$

since $\sin t > 2t/\pi$ for $0 < t < \pi/2$ and $a > 0$. The latter integral can be evaluated precisely and, therefore,

$$\left|\int_\gamma e^{ia\zeta} f(\zeta)d\zeta\right| < \frac{\pi\varepsilon}{a}(1 - e^{-aR}) < \frac{\pi\varepsilon}{a}.$$

Since $\varepsilon > 0$ was arbitrary, we obtain (13.6). □

Corollary 13.18. *Let us assume that f is continuous in the region $\{z : |z| > R, \operatorname{Im} z < 0\}$ for some $R > 0$. If*

$$\lim_{z\to\infty} f(z) = 0, \quad \operatorname{Im} z < 0,$$

then

$$\lim_{R\to\infty} \int_{|\zeta|=R, \operatorname{Im}\zeta<0} e^{ia\zeta} f(\zeta)d\zeta = 0 \tag{13.7}$$

for any $a < 0$.

Corollary 13.19. *Let us assume that f is continuous in the regions $\{z : |z| > R, \operatorname{Re} z < 0\}$, or $\{z : |z| > R, \operatorname{Re} z > 0\}$ for some $R > 0$. If*

$$\lim_{z\to\infty \operatorname{Re} z<0} f(z) = 0 \quad \text{or} \quad \lim_{z\to\infty \operatorname{Re} z>0} f(z) = 0,$$

then

$$\lim_{R\to\infty} \int_{|\zeta|=R, \operatorname{Re}\zeta<0} e^{a\zeta} f(\zeta)d\zeta = 0 \tag{13.8}$$

or

$$\lim_{R\to\infty} \int_{|\zeta|=R, \operatorname{Re}\zeta>0} e^{a\zeta} f(\zeta)d\zeta = 0 \tag{13.9}$$

for any $a > 0$ or $a < 0$, respectively.

Theorem 13.20. *Let f be analytic for $\operatorname{Im} z > 0$ and continuous for $\operatorname{Im} z \geq 0$ except at the singular points $z_1, z_2, \dots, z_n$ with $\operatorname{Im} z_j > 0$ for all $j = 1, 2, \dots, n$. If $f(z) = o(1)$ for $z \to \infty$, $\operatorname{Im} z > 0$, then*

$$\text{p.v.} \int_{-\infty}^{\infty} e^{iax} f(x) dx = 2\pi i \sum_{j=1}^{n} \operatorname*{Res}_{z=z_j}(e^{iaz} f(z)) \tag{13.10}$$

for $a > 0$.

Proof. Let $R > 0$ be chosen such that all singular points $z_1, z_2, \dots, z_n$ belong to the region $\{z : |z| < R, \operatorname{Im} z > 0\}$. Let γ_R be the union of the line segment $[-R, R]$ with the upper semicircle Γ_R^+. The residue theorem gives that

$$\int_{-R}^{R} e^{iax} f(x) dx + \int_{\Gamma_R^+} e^{iaz} f(z) dz = 2\pi i \sum_{j=1}^{n} \operatorname*{Res}_{z=z_j}(e^{iaz} f).$$

Jordan's lemma (see (13.6)) implies that for $a > 0$ the integral over Γ_R^+ tends to zero as $R \to \infty$. Hence, letting $R \to \infty$ we obtain (13.10). □

Example 13.21. Let us evaluate the integral

$$\int_{0}^{\infty} \frac{x \sin x}{x^2 + 4} dx.$$

Indeed, we have

$$\begin{aligned} \int_{0}^{\infty} \frac{x \sin x}{x^2 + 4} dx &= \frac{1}{2} \text{p.v.} \int_{-\infty}^{\infty} \frac{x \sin x}{x^2 + 4} dx = \frac{1}{2} \operatorname{Im}\left(\text{p.v.} \int_{-\infty}^{\infty} \frac{x e^{ix}}{x^2 + 4} dx \right) \\ &= \frac{1}{2} \operatorname{Im}\left(2\pi i \operatorname*{Res}_{z=2i} \frac{e^{iz} z}{z^2 + 4} \right) = \frac{1}{2} \operatorname{Im}\left(2\pi i \frac{e^{i2i} 2i}{2 \cdot 2i} \right) \\ &= \frac{1}{2} \operatorname{Im}\left(\frac{e^{-2}}{2} 2\pi i \right) = \frac{\pi}{2} e^{-2}. \end{aligned}$$

Example 13.22. Let us evaluate the integral ($a > 0$),

$$\int_{0}^{\infty} \frac{\cos(ax)}{x^2 + 1} dx.$$

Indeed, we have using Theorem 13.20 that

$$\int_0^\infty \frac{\cos(ax)}{x^2+1}dx = \frac{1}{2}\,\text{p.v.}\int_{-\infty}^{\infty}\frac{\cos(ax)}{x^2+1}dx = \frac{1}{2}\operatorname{Re}\left(\text{p.v.}\int_{-\infty}^{\infty}\frac{e^{iax}}{x^2+1}dx\right)$$
$$= \frac{1}{2}\operatorname{Re}\left(2\pi i \operatorname*{Res}_{z=i}\frac{e^{iaz}}{z^2+1}\right) = \frac{1}{2}\operatorname{Re}\left(2\pi i\frac{e^{iai}}{2i}\right) = \frac{\pi}{2}e^{-a}.$$

This result can be easily generalized for any $a \in \mathbb{R}$.

Problem 13.23. Prove that

$$\int_0^\infty \frac{\cos x - \cos ax}{x}dx = \log a$$

for $a > 0$. Hint: Integrate function $f(z) = \frac{e^{iz}-e^{iaz}}{z}$ over the boundary of the quarter part of the disk with radius R and let then $R \to \infty$, and use Problem 5.20.

Generalize this example to the case when $A_1 + A_2 + \cdots + A_n = 0$ and $a_1, a_2, \ldots, a_n > 0$, to obtain that

$$\int_0^\infty \frac{A_1\cos a_1x + A_2\cos a_2x + \cdots + A_n\cos a_nx}{x}dx$$
$$= -A_1\log a_1 - A_2\log a_2 - \cdots - A_n\log a_n.$$

Problem 13.24. Let $k > 0$. Prove that

$$\lim_{R\to+\infty}\int_{-R}^{R}\frac{e^{ixt}}{k+it}dt = e^{-xk}\begin{cases}2, & x>0,\\ 1, & x=0,\\ 0, & x<0.\end{cases}$$

This integral is known as the *discontinuous Cauchy multiplier*.

Problem 13.25. Assume that x is not an integer and $0 < \nu < 1$. Show that

$$\frac{1}{2\pi i}\lim_{n\to\infty}\sum_{k=-n}^{n}\frac{e^{i2kn\nu}}{k-x} = \frac{e^{i2\pi x\nu}}{1-e^{i2\pi x}}.$$

In particular,

$$\lim_{n\to\infty}\sum_{k=-n}^{n}\frac{e^{ikn}}{k-x} = -\frac{\pi}{\sin\pi x}.$$

Hint: Integrate function $f(z) = \frac{e^{i(2\nu-1)\pi z}}{(z-x)\sin\pi z}$ over the circle with radius $R = n + 1/2$.

Definition 13.26. Let f be a continuous real-valued function of $x \in [a,b]$ except for possibly the point $c \in (a,b)$. The principal value of the integral

$$\int_a^b f(x)\mathrm{d}x$$

is defined as

$$\text{p.v.}\int_a^b f(x)\mathrm{d}x := \lim_{\varepsilon\to+0}\left[\int_a^{c-\varepsilon} f(x)\mathrm{d}x + \int_{c+\varepsilon}^b f(x)\mathrm{d}x\right]$$

if the limit exists.

Example 13.27. Let us evaluate the principal value integral

$$\text{p.v.}\int_a^b \frac{1}{x-c}\mathrm{d}x, \quad a < c < b.$$

By definition, we have

$$\text{p.v.}\int_a^b \frac{\mathrm{d}x}{x-c} = \lim_{\varepsilon\to+0}\left[\int_a^{c-\varepsilon}\frac{\mathrm{d}x}{x-c} + \int_{c+\varepsilon}^b \frac{\mathrm{d}x}{x-c}\right]$$
$$= \lim_{\varepsilon\to+0}[\log|-\varepsilon| - \log|a-c| + \log|b-c| - \log|\varepsilon|] = \log\frac{b-c}{c-a}.$$

Example 13.28. Let us evaluate the integral

$$\int_0^\infty \frac{\sin x}{x}\mathrm{d}x.$$

We have

$$\int_0^\infty \frac{\sin x}{x}\mathrm{d}x = \frac{1}{2}\,\text{p.v.}\int_{-\infty}^\infty \frac{\sin x}{x}\mathrm{d}x,$$

where principal value integral is considered with respect to ∞ and 0. We have

$$\text{p.v.}\int_{-\infty}^\infty \frac{\sin x}{x}\mathrm{d}x = \frac{1}{\mathrm{i}}\,\text{p.v.}\int_{-\infty}^\infty \frac{\mathrm{e}^{\mathrm{i}x}}{x}\mathrm{d}x = \frac{1}{\mathrm{i}}\lim_{\substack{R\to\infty\\ \varepsilon\to0}}\left(\int_{-R}^{-\varepsilon}\frac{\mathrm{e}^{\mathrm{i}x}}{x}\mathrm{d}x + \int_\varepsilon^R \frac{\mathrm{e}^{\mathrm{i}x}}{x}\mathrm{d}x\right).$$

Here, we have used the fact that

$$\text{p.v.}\int_{-\infty}^\infty \frac{\cos x}{x}\mathrm{d}x = 0.$$

Consider the function

$$f(z) = \frac{e^{iz}}{z}.$$

It has only one singular point $z = 0$. That is why we consider the closed curve (Figure 13.2)

$$\gamma = [-R, -\varepsilon] \cup \Gamma_\varepsilon^- \cup [\varepsilon, R] \cup \Gamma_R^+.$$

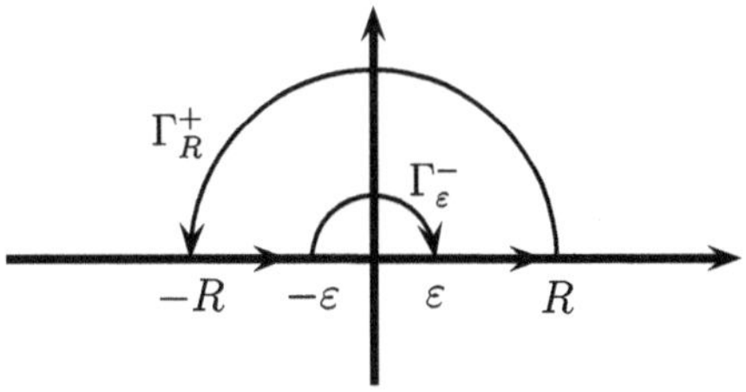

Figure 13.2: Semicircles Γ_R^+ and Γ_ε^-.

Inside of γ, the function f is analytic and continuous up to the curve γ. Using the Cauchy theorem, we have

$$0 = \int_\gamma \frac{e^{iz}}{z} dz = \int_{-R}^{-\varepsilon} \frac{e^{ix}}{x} dx + \int_{\Gamma_\varepsilon^-} \frac{e^{iz}}{z} dz + \int_\varepsilon^R \frac{e^{ix}}{x} dx + \int_{\Gamma_R^+} \frac{e^{iz}}{z} dz. \tag{13.11}$$

The integral over Γ_R^+ tends to 0 as $R \to \infty$ due to Jordan's lemma. The integral over Γ_ε^- can be evaluated as

$$\int_{\Gamma_\varepsilon^-} \frac{e^{iz}}{z} dz = -\int_{\Gamma_\varepsilon^+} \frac{e^{iz}}{z} dz = -\int_0^\pi \frac{e^{i\varepsilon e^{it}} i\varepsilon e^{it}}{\varepsilon e^{it}} dt = -i \int_0^\pi e^{i\varepsilon \cos t} e^{-\varepsilon \sin t} dt.$$

But the last integral tends to $-i\pi$ as $\varepsilon \to 0$ due to continuity of the functions $e^{i\varepsilon \cos t}$ and $e^{-\varepsilon \sin t}$ with respect to ε and $t \in [0, \pi]$.

Letting now $\varepsilon \to 0$ and $R \to \infty$ in (13.11), we obtain

$$0 = \lim_{R\to\infty, \varepsilon\to 0} \left(\int_{-R}^{-\varepsilon} \frac{e^{ix}}{x} dx + \int_\varepsilon^R \frac{e^{ix}}{x} dx \right) - i\pi$$

or

$$\text{p.v.} \int_{-\infty}^{\infty} \frac{e^{ix}}{x} dx = i\pi.$$

Therefore,

$$\int_0^\infty \frac{\sin x}{x}\,dx = \frac{1}{2i}\,\text{p.v.}\int_{-\infty}^\infty \frac{e^{ix}}{x}\,dx = \frac{\pi}{2}.$$

This integral is called the *Dirichlet integral.*

Problem 13.29. Show that (see, for comparison, Example 13.28)

$$\int_0^\infty \frac{\sin^2 x}{x^2}\,dx = \frac{\pi}{2}.$$

Problem 13.30. Show that for any $a \in \mathbb{R}$, we have

$$\int_0^\infty \frac{\sin ax}{e^{2\pi x} - 1}\,dx = \operatorname{sgn} a\left(\frac{1}{4}\frac{e^a + 1}{e^a - 1} - \frac{1}{2a}\right),$$

where

$$\operatorname{sgn} a = \begin{cases} 1, & a > 0, \\ 0, & a = 0, \\ -1, & a < 0. \end{cases}$$

Hint: Integrate function $f(z) = \frac{e^{iaz}}{e^{2\pi z}-1}$ over the rectangle with the vertices at the points 0, R, $R + i$, i.

Problem 13.31. Prove that

$$\int_0^\infty \frac{\sin(xb)}{x(x^2 + a^2)}\,dx = \frac{\pi \operatorname{sgn} b}{2a^2}\left(1 - e^{-|a||b|}\right),$$

where a and b are real and $a \neq 0$. Hint: Use the same procedure as in Example 13.28.

Problem 13.32. Prove that

$$\int_0^\infty \frac{\sin(xb)}{x(x^2 + a^2)^2}\,dx = \operatorname{sgn} b\left(\frac{\pi}{2a^4} - \frac{\pi}{4a^3}e^{-|a||b|}\left(|b| + 2/|a|\right)\right),$$

where a and b are real and $a \neq 0$. Hint: Use the same procedure as in Example 13.28.

Problem 13.33. Evaluate the integrals

$$\int_0^\infty e^{-x^2}\cos(ax^2)dx, \quad \int_0^\infty e^{-x^2}\sin(ax^2)dx$$

for real a. Hint: Use Problem 5.21.

13.4 Improper integrals for multivalued functions

In this section, we study improper integrals of the form

$$\int_0^\infty x^{a-1}f(x)dx, \quad \int_0^1 (1-x)^{-a}x^{a-1}f(x)dx, \quad \int_0^\infty f(x)\log x dx.$$

Theorem 13.34. *Let $f(z)$ be a single-valued analytic function for $z \in \mathbb{C}$ except for the singular points $z_1, z_2, \ldots, z_n$, which are not lying on $\mathbb{R}_+$ for all $j = 1, 2, \ldots, n$. If $0 < a < 1$, and if $f(z)$ has a removable singularity at $z = 0$ and $f(z) = O(\frac{1}{|z|})$ for $z \to \infty$, then*

$$\int_0^\infty x^{a-1}f(x)dx = \frac{2\pi i}{1-e^{i2\pi a}}\sum_{j=1}^{n}\operatorname*{Res}_{z=z_j}(z^{a-1}f(z)). \tag{13.12}$$

Proof. Consider the domain $D_{\epsilon,R}$, which is a disk with radius $R > 0$ big enough and without a disk with radius $\epsilon > 0$ small enough, and with a cut along the positive real axis from ϵ to R. In this domain, the multivalued (in general) function $g(z) := z^{a-1}f(z)$ is a single-valued analytic function except for the singular points $z_1, z_2, \ldots, z_n$, which are inside of the domain $D_{\epsilon,R}$. Thus, by the residue theorem, we have that

$$\begin{aligned}
2\pi i\sum_{j=1}^{n}\operatorname*{Res}_{z=z_j} g(z) &= \int_{\partial D_{\epsilon,R}} g(z)dz \\
&= \int_\epsilon^R x^{a-1}f(x)dx + \int_{|z|=R} z^{a-1}f(z)dz \\
&\quad + \int_R^\epsilon z^{a-1}f(z)dz - \int_{|z|=\epsilon} z^{a-1}f(z)dz =: I_1 + I_2 + I_3 + I_4,
\end{aligned}$$

where the integration over the circles $|z| = R$ and $|z| = \epsilon$ are taken in positive direction with respect to the origin. Consider now each of the terms $I_j, j = 1, 2, 3, 4$. By hypothesis in the neighborhood of the infinity, we have

$$|I_2| \leq \left| \int_0^{2\pi} R^{\alpha-1} \mathrm{e}^{\mathrm{i}(\alpha-1)t} f(R\mathrm{e}^{\mathrm{i}t}) \mathrm{i}R\mathrm{e}^{\mathrm{i}t} \mathrm{d}t \right| \leq 2\pi M R^{\alpha-1} \to 0, \quad R \to +\infty,$$

since $0 < \alpha < 1$. Similarly, using the conditions in the neighborhood of the origin, we obtain

$$|I_4| \leq \left| \int_0^{2\pi} \epsilon^{\alpha-1} \mathrm{e}^{\mathrm{i}(\alpha-1)t} f(\epsilon\mathrm{e}^{\mathrm{i}t}) \mathrm{i}\epsilon\mathrm{e}^{\mathrm{i}t} \mathrm{d}t \right| \leq 2\pi M_1 \epsilon^{\alpha} \to 0, \quad \epsilon \to 0,$$

since $0 < \alpha < 1$. Here, M, M_1 are some positive constants, which correspond to the hypotheses of the theorem. The term I_3 is the integral over the lower lip of the cut where $\arg z = 2\pi$. Therefore, we have

$$I_3 = -\int_{\epsilon}^{R} x^{\alpha-1} \mathrm{e}^{\mathrm{i}2\pi(\alpha-1)} f(x) \mathrm{d}x = -\mathrm{e}^{\mathrm{i}2\pi(\alpha-1)} I_1.$$

Taking now the limits in $I_j, j = 1, 2, 3, 4$ as $R \to +\infty$ and $\epsilon \to 0$, we finally get

$$\int_0^{\infty} x^{\alpha-1} f(x) \mathrm{d}x = \frac{2\pi\mathrm{i}}{1 - \mathrm{e}^{\mathrm{i}2\pi\alpha}} \sum_{j=1}^{n} \operatorname*{Res}_{z=z_j} (z^{\alpha-1} f(z)).$$

Thus, the theorem is proved. □

Example 13.35. Let us evaluate the integral

$$\int_0^{\infty} \frac{x^{-\alpha}}{1+x} \mathrm{d}x, \quad 0 < \alpha < 1.$$

The integrand satisfies all conditions of Theorem 13.34 and has only one singular point $z = -1$. Thus, (13.12) implies that

$$\int_0^{\infty} \frac{x^{-\alpha}}{1+x} \mathrm{d}x = \frac{2\pi\mathrm{i}}{1 - \mathrm{e}^{-2\pi\mathrm{i}\alpha}} \operatorname*{Res}_{z=-1} \frac{z^{-\alpha}}{1+z} = 2\pi\mathrm{i} \frac{\mathrm{e}^{-\pi\mathrm{i}\alpha}}{1 - \mathrm{e}^{-2\pi\mathrm{i}\alpha}} = \frac{\pi}{\sin(\alpha\pi)}.$$

Problem 13.36. Show that for $-1 < \alpha < 3$,

$$\int_0^{\infty} \frac{x^{\alpha}}{(1+x^2)^2} \mathrm{d}x = \frac{\pi(1-\alpha)}{4\cos(\pi\alpha/2)}.$$

Theorem 13.37. *Let $f(z)$ be a single-valued analytic function for $z \in \mathbb{C}$ except for the singular points $z_1, z_2, \ldots, z_n$, which are not lying on the interval $[0, 1]$ for all $j = 1, 2, \ldots, n$. If $0 < \alpha < 1$ and $f(z)$ has a removable singularity at $z = \infty$, then*

$$\int_0^1 x^{a-1}(1-x)^{-a}f(x)\mathrm{d}x = \frac{\pi a_0}{\sin(a\pi)} + \frac{2\pi\mathrm{i}}{1-\mathrm{e}^{\mathrm{i}2\pi a}}\sum_{j=1}^{n}\operatorname*{Res}_{z=z_j}(z^{a-1}(1-z)^{-a}f(z)), \tag{13.13}$$

where $a_0 = \lim_{z\to\infty} f(z)$.

Proof. Consider the domain $D_{\epsilon,R}$, which is a disk with radius $R > 0$ big enough and without two disks with radius $\epsilon > 0$ small enough around $z = 0$, $z = 1$, and with a cut along the positive real axes from ϵ to $1-\epsilon$. In this domain, the multivalued (in general) function $g(z) := z^{a-1}(1-z)^{-a}f(z)$ is a single-valued analytic function except the singular points $z_1, z_2, \dots, z_n$, which are inside of the domain $D_{\epsilon,R}$. Thus, by the residue theorem, we have that

$$\begin{aligned}
2\pi\mathrm{i}\sum_{j=1}^{n}\operatorname*{Res}_{z=z_j} g(z) &= \int_{\partial D_{\epsilon,R}} g(z)\mathrm{d}z \\
&= \int_{\epsilon}^{1-\epsilon} x^{a-1}(1-x)^{-\epsilon}f(x)\mathrm{d}x \\
&\quad + \int_{|z|=R} z^{a-1}(1-z)^{-a}f(z)\mathrm{d}z + \int_{1-\epsilon}^{\epsilon} z^{a-1}(1-z)^{-a}f(z)\mathrm{d}z \\
&\quad - \int_{|z|=\epsilon} z^{a-1}(1-z)^{-a}f(z)\mathrm{d}z - \int_{|z-1|=\epsilon} z^{a-1}(1-z)^{-a}f(z)\mathrm{d}z \\
&=: I_1 + I_2 + I_3 + I_4 + I_5,
\end{aligned}$$

where the integration over the circles $|z| = R$ and $|z| = \epsilon$, $|z-1| = \epsilon$ are taken in positive direction with respect to $z = 0$ and $z = 1$, respectively. By hypothesis in the neighborhood of the infinity, we have that $f(z) = a_0 + \frac{a_1}{z} + \cdots$ and, therefore,

$$I_2 = \int_0^{2\pi} R^{a-1}\mathrm{e}^{\mathrm{i}(a-1)t}(1-R\mathrm{e}^{\mathrm{i}t})^{-a}\left(a_0 + \frac{a_1}{R\mathrm{e}^{\mathrm{i}t}} + \cdots\right)\mathrm{i}R\mathrm{e}^{\mathrm{i}t}\mathrm{d}t \to 2\pi\mathrm{i}a_0\mathrm{e}^{\mathrm{i}\pi a}$$

as $R \to +\infty$. We have used here the fact that $-1 = \mathrm{e}^{-\mathrm{i}\pi}$. Next,

$$I_4 = -\int_0^{2\pi} \epsilon^{a-1}\mathrm{e}^{\mathrm{i}(a-1)t}(1-\epsilon\mathrm{e}^{\mathrm{i}t})^{-a}f(\epsilon\mathrm{e}^{\mathrm{i}t})\mathrm{i}\epsilon\mathrm{e}^{\mathrm{i}t}\mathrm{d}t \to 0, \quad \epsilon \to 0,$$

since $0 < \epsilon < 1$ and $f(z)$ has no singularities at $z = 0$. Similarly, we get that

$$I_5 = -\int_0^{2\pi} (1+\epsilon\mathrm{e}^{\mathrm{i}t})^{a-1}(-\epsilon\mathrm{e}^{\mathrm{i}t})^{-a}f(1+\epsilon\mathrm{e}^{\mathrm{i}t})\mathrm{i}\epsilon\mathrm{e}^{\mathrm{i}t}\mathrm{d}t \to 0, \quad \epsilon \to 0.$$

The term I_3 is the integral over the lower lip of the cut where $\arg z = 2\pi$. Therefore, we have

$$I_3 = \int_{1-\epsilon}^{\epsilon} x^{a-1} e^{i2\pi(a-1)} (1-x)^{-\epsilon} f(x) dx = -e^{i2\pi a} I_1.$$

Taking now the limits in $I_j, j = 1, 2, 3, 4, 5$ as $R \to +\infty$ and $\epsilon \to 0$ and combining results for I_j, we finally get (13.13). The theorem is completely proved. □

Example 13.38. Immediate application of Theorem 13.37 with $f \equiv 1$ yields that

$$\int_0^1 x^{a-1}(1-x)^{-a} dx = \frac{\pi}{\sin(a\pi)}, \qquad 0 < a < 1.$$

Note that the integral at hand is a particular case of Euler's beta function (see Theorem 17.27 and Example 13.35).

Theorem 13.39. *Let f be analytic for $\operatorname{Im} z > 0$ and continuous for $\operatorname{Im} z \geq 0$ except for the singular points $z_1, z_2, \ldots, z_n$ with $\operatorname{Im} z_j > 0$ for all $j = 1, 2, \ldots, n$. Assume that the restriction of $f(z)$ to the real line $\mathbb{R}$ is even. If $f(z) = o(\frac{1}{|z|\log|z|})$ for $z \to \infty$, $\operatorname{Im} z > 0$, then*

$$\int_0^{\infty} f(x) \log x dx = i\pi \sum_{j=1}^{n} \operatorname*{Res}_{z=z_j} \left(f(z) \left(\log z - \frac{i\pi}{2} \right) \right). \tag{13.14}$$

Proof. Consider the domain $D_{R,\epsilon}$ in the upper half-plane such that its boundary is the curve consisting of the segments $[-R, -\epsilon]$, $[\epsilon, R]$ of the real axis and the semicircles $\Gamma_R^+ = \{z : |z| = R, \operatorname{Im} z > 0\}$, and $\gamma_\epsilon^+ = \{z : |z| = \epsilon, \operatorname{Im} z > 0\}$ connecting them. The function $g(z) := f(z) \log z$, which is the branch of the analytic multivalued function, coincides with $f(x) \log x$ on the positive real semiaxis, on the negative real semiaxis, i. e., for $z = -x = xe^{i\pi}$, $x > 0$, and takes the value $g(-x) = f(-x)\log(-x) = f(x)(\log x + i\pi)$. Moreover, this function $g(z)$ is analytic in the domain $D_{R,\epsilon}$ except for the singular points $z_j, j = 1, 2, \ldots n$ of the function $f(z)$. Hence, by the residue theorem we have that

$$\begin{aligned}
2\pi i \sum_{j=1}^{n} \operatorname*{Res}_{z=z_j} g(z) &= \int_{\partial D_{\epsilon,R}} g(z) dz \\
&= \int_{\epsilon}^{R} f(x) \log x dx + \int_{\Gamma_R^+} f(z) \log z dz \\
&\quad - \int_{R}^{\epsilon} f(x)(\log x + i\pi) dx - \int_{\gamma_\epsilon^+} f(z) \log z dz =: I_1 + I_2 + I_3 + I_4,
\end{aligned}$$

where integration over semicircles are taken in the positive direction with respect to the origin. By hypothesis in the neighborhood of the infinity, we have that for R big enough

$$|I_2| \leq \int_0^{\pi} |f(Re^{it})|(\log R + \pi)R\mathrm{d}t \to 0, \quad R \to +\infty.$$

Combining I_1 and I_3, we get

$$I_1 + I_3 = \int_{\epsilon}^{R} f(x)(2\log x + i\pi)\mathrm{d}x \to \int_0^{\infty} f(x)(2\log x + i\pi)\mathrm{d}x$$

as $R \to \infty$, $\epsilon \to 0$. Similar to I_2, we can estimate I_4 as follows:

$$|I_2| \leq \int_0^{\pi} |f(\epsilon e^{it})|(|\log \epsilon| + \pi)\epsilon\mathrm{d}t \to 0, \quad \epsilon \to 0,$$

if we take into account that $f(z)$ has no singularity at 0 and $\lim_{v\to 0} v\log v = 0$ for real-valued log. Next, since $f(x)$ is even and satisfies all conditions of Theorem 13.10, then

$$\int_0^{\infty} f(x)\mathrm{d}x = \pi i \sum_{j=1}^{n} \operatorname*{Res}_{z=z_j} f(z).$$

Taking now the limits in $I_j, j = 1, 2, 3, 4$ as $R \to +\infty$ and $\epsilon \to 0$ and combining results for I_j, we get

$$2\pi i \sum_{j=1}^{n} \operatorname*{Res}_{z=z_j}(f(z)\log z) = 2\int_0^{\infty} f(x)\log x\mathrm{d}x + (i\pi)^2 \sum_{j=1}^{n} \operatorname*{Res}_{z=z_j} f(z).$$

This rearranges to (13.14). The theorem is finally proved. □

Example 13.40. Let us evaluate the integral

$$\int_0^{\infty} \frac{\log x}{(a^2 + x^2)^2}\mathrm{d}x, \quad a > 0.$$

Applying Theorem 13.39, we easily get ($z = ia$ is a pole of order 2 in the upper half-plane)

$$\int_0^{\infty} \frac{\log x}{(a^2 + x^2)^2}\mathrm{d}x = \pi i \operatorname*{Res}_{z=ia} \frac{\log z - \frac{i\pi}{2}}{(a^2 + z^2)^2} = \frac{\pi(\log a - 1)}{4a^3}.$$

In particular,

$$\int_0^\infty \frac{\log x}{(e^2+x^2)^2}\,dx = 0.$$

Problem 13.41. Evaluate the *Fresnel integrals*

$$\int_0^\infty \cos(x^2)\,dx \quad \text{and} \quad \int_0^\infty \sin(x^2)\,dx.$$

Problem 13.42. Prove that

$$\int_0^\infty \frac{\log x}{x^2+a^2}\,dx = \frac{\pi \log a}{2a}$$

for any $a > 0$. In particular,

$$\int_0^\infty \frac{\log x}{1+x^2}\,dx = 0.$$

Problem 13.43. Using the result of Problem 13.42, prove that

$$\int_0^\infty \frac{\log^2 x}{x^2+a^2}\,dx = \frac{\pi}{2a}\log^2 a + \frac{\pi^3}{8a}.$$

Problem 13.44. Show that

$$\int_0^\infty \frac{x^{\alpha-1}}{x+\lambda}\,dx = \lambda^{\alpha-1}\frac{\pi}{\sin(\alpha\pi)}$$

for $0 < \alpha < 1$ and $\lambda > 0$.

Problem 13.45. Show that

$$\int_0^\infty \sin(x^\alpha)\,dx = \Gamma\left(\frac{\alpha+1}{\alpha}\right)\sin\frac{\pi}{2\alpha}, \qquad \alpha > 1,$$

where Γ is Euler's gamma function (see the definition in Chapter 17).

Problem 13.46. Show that

$$\int_0^\infty x^m e^{ix^n}\,dx = \frac{1}{n}\Gamma\left(\frac{m+1}{n}\right)e^{i\frac{\pi(m+1)}{2n}}, \qquad -1 < m < n-1, \qquad n > 0,$$

where Γ is Euler's gamma function (see the definition in Chapter 17). In particular, if $a > 0$ and $0 < p < 1$, then

$$\int_0^\infty t^{p-1}\cos at\,dt = \frac{\Gamma(p)\cos\frac{\pi p}{2}}{a^p}, \quad \int_0^\infty t^{p-1}\sin at\,dt = \frac{\Gamma(p)\sin\frac{\pi p}{2}}{a^p}.$$

Problem 13.47.

1. Prove that for $0 < \sigma < 1$, we have

$$\int_0^\infty \frac{\log(1+x)}{x^{\sigma+1}}\,dx = \frac{\pi}{\sigma\sin\sigma\pi}, \quad \text{p.v.}\int_0^\infty \frac{\log|1-x|}{x^{\sigma+1}}\,dx = \frac{\pi}{\sigma\tan\sigma\pi}.$$

2. Show that for $0 < a < 1$, we have

$$\text{p.v.}\int_0^\infty \frac{x^{a-1}}{1-x}\,dx = \pi\cot(a\pi).$$

3. Show that for $0 < b < 1$ and $-\pi < a < \pi$, we have

$$\int_0^\infty \frac{x^{b-1}}{x+e^{ia}}\,dx = \frac{\pi e^{ia(b-1)}}{\sin\pi b}.$$

Problem 13.48. Prove that

$$\int_0^\infty \frac{x^{-a}}{1+2x\cos\lambda+x^2}\,dx = \frac{\pi}{\sin a\pi}\,\frac{\sin a\lambda}{\sin\lambda}$$

for $-1 < a < 1$ and $-\pi < \lambda < \pi$.

14 Calculation of series by residue theory

There are two results, which may work in applications to the calculation of number series by residue theory.

Theorem 14.1. *Let $f(z)$ be analytic in $\mathbb{C}$ except for the finite number of points $\{z_j\}_{j=1}^m$ with $\operatorname{Im} z_j \neq 0$. Let us assume in addition that $f(z) \to 0$ as $|z| \to \infty$. Then*

$$\sum_{k=-\infty}^{\infty} (-1)^k f(k) = -\sum_{j=1}^{m} \operatorname*{Res}_{z=z_j} \frac{\pi f(z)}{\sin \pi z}. \tag{14.1}$$

Proof. For any $n \in \mathbb{Z}$ large enough and for $R > 0$, let us consider the curve (rectangle)

$$\begin{aligned}\Gamma_{n,R} = \{z \in \mathbb{C} : {} & x + \mathrm{i}R, x \in [-n - 1/2, n + 1/2],\\ & x - \mathrm{i}R, x \in [-n - 1/2, n + 1/2], -n - 1/2 + \mathrm{i}y, y \in [-R, R],\\ & n + 1/2 + \mathrm{i}y, y \in [-R, R]\}\end{aligned}$$

such that all singular points of $f(z)$ belong to $\operatorname{int} \Gamma_{n,R}$. Then the function

$$\frac{\pi f(z)}{\sin \pi z}$$

has the singular points

$$\{z_j\}_{j=1}^m, \quad \widetilde{z_k} = k, k = 0, \pm 1, \pm 2, \dots, \pm n$$

inside $\operatorname{int} \Gamma_{n,R}$. Using now the Cauchy residue theorem for this special domain $\operatorname{int} \Gamma_{n,R}$, we obtain

$$\begin{aligned}\int_{\Gamma_{n,R}} \frac{\pi f(z)}{\sin \pi z} \mathrm{d}z &= 2\pi \mathrm{i} \sum_{k=-n}^{n} \operatorname*{Res}_{z=k} \frac{\pi f(z)}{\sin \pi z} + 2\pi \mathrm{i} \sum_{j=1}^{m} \operatorname*{Res}_{z=z_j} \frac{\pi f(z)}{\sin \pi z}\\ &= 2\pi \mathrm{i} \left(\sum_{k=-n}^{n} \frac{\pi f(k)}{\pi \cos \pi k} + \sum_{j=1}^{m} \operatorname*{Res}_{z=z_j} \frac{\pi f(z)}{\sin \pi z} \right)\\ &= 2\pi \mathrm{i} \left(\sum_{k=-n}^{n} (-1)^k f(k) + \sum_{j=1}^{m} \operatorname*{Res}_{z=z_j} \frac{\pi f(z)}{\sin \pi z} \right).\end{aligned} \tag{14.2}$$

Now, in order to get (14.1) we need to investigate the curve integral on the left- hand side of (14.2). This integral can be represented as the sum of the following four integrals:

$$I_1 = \int_{-n-1/2}^{n+1/2} \frac{\pi f(x - \mathrm{i}R) \mathrm{d}x}{\sin \pi (x - \mathrm{i}R)},$$

https://doi.org/10.1515/9783111632278-015

$$I_2 = \int_{n+1/2}^{-n-1/2} \frac{\pi f(x+\mathrm{i}R)\mathrm{d}x}{\sin\pi(x+\mathrm{i}R)},$$

$$I_3 = \mathrm{i}\int_{-R}^{R} \frac{\pi f(n+1/2+\mathrm{i}y)\mathrm{d}y}{\sin\pi(n+1/2+\mathrm{i}y)},$$

$$I_4 = \mathrm{i}\int_{R}^{-R} \frac{\pi f(-n-1/2+\mathrm{i}y)\mathrm{d}y}{\sin\pi(-n-1/2+\mathrm{i}y)}.$$

Since

$$|\sin\pi(x\pm\mathrm{i}R)| = \left|\frac{\mathrm{e}^{\mathrm{i}\pi x}\mathrm{e}^{\mp\pi R}-\mathrm{e}^{-\mathrm{i}\pi x}\mathrm{e}^{\pm\pi R}}{2\mathrm{i}}\right| \ge \frac{\mathrm{e}^{\pi R}-\mathrm{e}^{-\pi R}}{2} \ge \frac{1}{4}\mathrm{e}^{\pi R}, \quad R>0$$

then for I_1 and I_2 we have the following estimate:

$$|I_1|,|I_2| \le \frac{4\pi}{\mathrm{e}^{\pi R}}\int_{-n-1/2}^{n+1/2} |f(x\mp\mathrm{i}R)|\mathrm{d}x \le \frac{4\pi}{\mathrm{e}^{\pi R}}\max_{x\in[-n-1/2,n+1/2]}|f(x\mp\mathrm{i}R)|(2n+1).$$

If we choose $R \ge n$ and take into account that $f(z)\to 0$ as $|z|\to+\infty$ (actually we need here only boundedness of f), then when $R\ge n\to\infty$ the right-hand side of the latter inequality tends to zero. Next, since

$$\sin(\pm\pi(n+1/2+\mathrm{i}y)) = \pm(-1)^n\cos(\mathrm{i}\pi y) = \pm(-1)^n\cosh(\pi y)$$

then we have the following estimates for I_3 and I_4:

$$\begin{aligned}
|I_3|,|I_4| &\le \pi\int_{-R}^{R}\frac{|f(n+1/2\pm\mathrm{i}y)|\mathrm{d}y}{\cosh(\pi y)}\\
&\le \pi\max_{y\in[-R,R]}|f(n+1/2\pm\mathrm{i}y)|\int_{-R}^{R}\frac{\mathrm{d}y}{\cosh(\pi y)}\\
&\le \pi\max_{y\in[-R,R]}|f(n+1/2+\mathrm{i}y)|\int_{-\infty}^{\infty}\frac{\mathrm{d}y}{\cosh(\pi y)} \to 0, \quad n\to\infty
\end{aligned}$$

due to the fact that $f(z)\to 0$ as $|z|\to+\infty$ and

$$\int_{-\infty}^{\infty}\frac{\mathrm{d}y}{\cosh(\pi y)} = 2.$$

If we let now $R\ge n\to\infty$ in (14.2), we obtain that

$$0 = 2\pi i\left(\sum_{k=-\infty}^{\infty}(-1)^k f(k) + \sum_{j=1}^{m}\operatorname*{Res}_{z=z_j}\frac{\pi f(z)}{\sin \pi z}\right).$$

It implies (14.1) and, therefore, the theorem is completely proved. □

Theorem 14.2. *Let $f(z)$ be analytic in $\mathbb{C}$ except for the finite number of points $\{z_j\}_{j=1}^m$ with $\operatorname{Im} z_j \neq 0$. Let us assume in addition that $zf(z) \to 0$ as $|z| \to \infty$. Then*

$$\sum_{k=-\infty}^{\infty} f(k) = -\sum_{j=1}^{m}\operatorname*{Res}_{z=z_j}(\pi\cot(\pi z)f(z)). \tag{14.3}$$

Proof. Literally the same as for Theorem 14.1. The only difference is

$$\operatorname*{Res}_{z=k}\pi\cot(\pi z)f(z) = \frac{\pi\cos(\pi k)f(k)}{(\sin \pi z)'|_{z=k}} = f(k).$$ □

Remark. In Theorems 14.1 and Theorem 14.2 some singular points $\{z_j\}_{j=1}^m$ of $f(z)$ may be located on the real line such that they are not equal to some $n \in \mathbb{Z}$.

Example 14.3. Let us show that for real $a \neq 0$ we have

$$\sum_{k=-\infty}^{\infty}\frac{1}{k^2 + a^2} = \frac{\pi}{a}\coth(\pi a).$$

Indeed, let

$$f(z) = \frac{1}{z^2 + a^2}, \quad z \in \mathbb{C}.$$

This function has two singular points $z_1 = ia$ and $z_2 = -ia$. Then Theorem 14.2 gives that

$$\begin{aligned}\sum_{k=-\infty}^{\infty}\frac{1}{k^2 + a^2} &= -\left(\operatorname*{Res}_{z=ia}\frac{\pi\cot(\pi z)}{z^2 + a^2} + \operatorname*{Res}_{z=-ia}\frac{\pi\cot(\pi z)}{z^2 + a^2}\right)\\ &= -\left(\frac{\pi\cot(\pi ia)}{2ia} + \frac{\pi\cot(-\pi ia)}{-2ia}\right) = -\frac{\pi\cot(\pi ia)}{ia} = \frac{\pi}{a}\coth(\pi a).\end{aligned}$$

Example 14.4. Let us show that

$$\sum_{k=1}^{\infty}\frac{1}{k^2} = \frac{\pi^2}{6}.$$

Indeed, let $a = \varepsilon > 0$ and small. Then Example 14.3 implies that

$$\sum_{k=-\infty}^{\infty}\frac{1}{k^2 + \varepsilon^2} = \frac{1}{\varepsilon^2} + 2\sum_{k=1}^{\infty}\frac{1}{k^2 + \varepsilon^2} = \frac{\pi}{\varepsilon}\coth(\pi\varepsilon).$$

So,

$$2\sum_{k=1}^{\infty}\frac{1}{k^2+\varepsilon^2}=\frac{\pi}{\varepsilon}\coth(\pi\varepsilon)-\frac{1}{\varepsilon^2}=\frac{\varepsilon\pi(e^{2\varepsilon\pi}+1)-e^{2\varepsilon\pi}+1}{\varepsilon^2(e^{2\varepsilon\pi}-1)}.$$

Using Taylor's expansion for e^{ξ} near zero, we can easily obtain that the limit of the right-hand side of the latter equality is equal to $\pi^2/3$. Thus,

$$2\sum_{k=1}^{\infty}\frac{1}{k^2}=\frac{\pi^2}{3}.$$

Example 14.5. Let us show that for any $a \notin \mathbb{Z}$ we have

$$\sum_{k=-\infty}^{\infty}\frac{1}{(k+a)^2}=\frac{\pi^2}{\sin^2(\pi a)}.$$

Indeed, let

$$f(z)=\frac{1}{(z+a)^2},\quad z\in\mathbb{C}.$$

This function has one singular point $z=-a$, which is a pole of order 2. Then Theorem 14.2 and the remark after it give that

$$\sum_{k=-\infty}^{\infty}\frac{1}{(k+a)^2}=-\operatorname*{Res}_{z=-a}\frac{\pi\cot(\pi z)}{(z+a)^2}=-\pi\big(\cot(\pi z)\big)'_{z=-a}=\frac{\pi^2}{\sin^2(\pi a)}.$$

Problem 14.6. Prove that

$$\sum_{k=-\infty}^{\infty}\frac{1}{(k+a)^3}=\frac{\pi^3\cot(\pi a)}{\sin^2(\pi a)}.$$

Problem 14.7. Prove that

$$\sum_{k=-\infty}^{\infty}\frac{1}{(k+a)^4}=\frac{\pi^4}{3}\frac{1+2\cos^2(\pi a)}{\sin^4(\pi a)}.$$

Problem 14.8. Show that

$$\sum_{k=1}^{\infty}\frac{(-1)^{k+1}}{k^2}=\frac{\pi^2}{12}.$$

Problem 14.9. Show (using Theorem 14.2) that

$$\sum_{k=1}^{\infty}\frac{1}{k^4+a^4}=\frac{\pi}{2\sqrt{2}a^3}\frac{\sinh(\pi 2\sqrt{2})+\sin(\pi 2\sqrt{2})}{\cosh(\pi 2\sqrt{2})-\cos(\pi 2\sqrt{2})}-\frac{1}{2a^4}.$$

Problem 14.10. Using the result of the previous problem, find the sum

$$\sum_{k=1}^{\infty} \frac{k^2}{k^4 + a^4}.$$

Problem 14.11. Show that

$$\sum_{k=0}^{\infty} \frac{1}{(2k+1)^4} = \frac{\pi^4}{96}.$$

Problem 14.12. Using the result of Problem 14.9, prove that

$$\sum_{k=1}^{\infty} \frac{1}{k^4} = \frac{\pi^4}{90}.$$

15 Entire functions

This chapter is devoted to the introduction of entire functions.

As it was shown (see Corollary 8.4 and the results of Chapter 9), every entire function $f(z)$ can be represented via power series (its Taylor's expansion)

$$f(z) = \sum_{j=0}^{\infty} c_j z^j, \quad z \in \mathbb{C},$$

that converges everywhere in $\mathbb{C}$, hence uniformly on compact sets. The radius of convergence is infinite, which implies that

$$\lim_{j\to\infty} |c_j|^{\frac{1}{j}} = 0 \quad \text{or equivalently} \quad \lim_{j\to\infty} \frac{\log |c_j|}{j} = -\infty.$$

Conversely, any power series satisfying this criterion will represent an entire function. It can be also mentioned that all entire functions (besides polynomials) have an essential singularity at $z = \infty$ (see Theorem 8.2). In addition, if the entire function (its real or imaginary part) is known in the neighborhood of, e. g., $z = 0$, then it is known for the whole complex plane.

From what has been considered earlier, it follows that an entire function in any bounded domain of the complex plane may have only finitely many zeros. Consequently, all zeros of an entire function may accumulate only at infinity and they can be arranged in order of increasing absolute values.

The simplest entire functions are polynomials. The polynomial $f(z)$ having zeros at the points $z_1, z_2, \ldots, z_n$, which are not equal to 0, can be uniquely represented as

$$f(z) = f(0)\left(1 - \frac{z}{z_1}\right)\left(1 - \frac{z}{z_2}\right)\cdots\left(1 - \frac{z}{z_n}\right).$$

The zeros of arbitrary entire function (if they exist) play an equally (as for polynomials) important role in the general case. But there can be infinitely many of them $z_1, z_2, \ldots, z_n, \ldots$ ($\neq 0$), and the product (unlike polynomials)

$$\prod_{n=1}^{\infty}\left(1 - \frac{z}{z_n}\right)$$

may diverge. As a result, the entire function cannot always be composed into such simple factors as $(1 - \frac{z}{z_n})$ and it is necessary to consider different kinds of factors.

Definition 15.1. If $p \in \mathbb{N}$, then the expression

$$E(u,p) := (1-u)\mathrm{e}^{u+u^2/2+\cdots+u^p/p}, \quad E(u,0) := 1-u \tag{15.1}$$

is said to be *primary factor*.

https://doi.org/10.1515/9783111632278-016

It can be easily seen that for $|u| < 1$ we have

$$\left|\log E(u,p)\right| \le |u|^{p+1}(1 + |u| + \cdots) = \frac{|u|^{p+1}}{1 - |u|}. \tag{15.2}$$

Moreover, it can be proved that for $|u| < 1$, it holds that

$$\left|E(u,p) - 1\right| \le |u|^{p+1}.$$

These estimates will define actually the convergence of the product of primary factors. And we can formulate and prove the first result in this direction, which is a well-known Weierstrass' theorem.

Theorem 15.2 (Weierstrass). *Let $\{z_n\}_{n=1}^{\infty}$ be a sequence of complex numbers such that $\lim_{n\to\infty} z_n = \infty$. Then there exists an entire function having the zeros only at these points $z_1, z_2, \ldots$.*

Proof. Let the points $z_1, z_2, \ldots$ be enumerated so that $0 < |z_1| \le |z_2| \le \ldots$. Then for any fixed z the series

$$\sum_{n=1}^{\infty}\left(\frac{|z|}{|z_n|}\right)^{p_n} < \infty$$

converges if the sequence p_n is chosen appropriately. Indeed, since $\lim_{n\to\infty} z_n = \infty$ we can choose n_0 so large that for all $n \ge n_0$ we will have $|z_n| > 2|z|$. So, e. g., if $p_n = n$, then

$$\left(\frac{|z|}{|z_n|}\right)^{n} < \frac{1}{2^n}$$

and, therefore, the series converges. Let us now define the function $f(z)$ as follows:

$$f(z) := \prod_{n=1}^{\infty} E\left(\frac{z}{z_n}, n - 1\right).$$

Then we prove that this function satisfies all required properties. Indeed, if $|z| < \frac{|z_n|}{2}$ then we have that

$$\left|\log E\left(\frac{z}{z_n}, n - 1\right)\right| \le \frac{\frac{|z|^n}{|z_n|^n}}{1 - \frac{|z|}{|z_n|}} < \frac{1}{2^{n-1}}.$$

This estimate yields that for $|z| \le R$ the series

$$\sum_{|z_n|>2R} \log E\left(\frac{z}{z_n}, n - 1\right)$$

converges uniformly and, therefore, the product

$$\prod_{|z_n|>2R} E\left(\frac{z}{z_n}, n-1\right)$$

converges uniformly for $|z| \le R$, too. This means that $f(z)$ is analytic for $|z| \le R$ and its only zeros in $\{z : |z| \le R\}$ are the zeros of the function

$$f_R(z) := \prod_{|z_n|\le 2R} E\left(\frac{z}{z_n}, n-1\right),$$

i. e., the points $z_1, z_2, \ldots$. Since R can be chosen arbitrary large, then theorem is proved. □

Corollary 15.3 (Weierstrass' factorization theorem). *Let function $f(z)$ be entire and $f(0) \neq 0$. Then it can be represented as*

$$f(z) = f(0)P(z)e^{g(z)},$$

where $P(z)$ is some product of primary factors (if $f(z)$ has zeros) and $g(z)$ is some entire function.

Proof. Let function $P(z)$ be some product of primary factors. Let us define a function $\varphi(z)$ as follows:

$$\varphi(z) := \frac{f'(z)}{f(z)} - \frac{P'(z)}{P(z)}.$$

It is not difficult to check that $\varphi(z)$ is an entire function since the poles of order 1 of $\frac{f'(z)}{f(z)}$ are eliminated by the corresponding poles of order 1 of $\frac{P'(z)}{P(z)}$. Consequently, if we define $g(z)$ as

$$g(z) := \int_0^z \varphi(\zeta)d\zeta = \log f(z) - \log f(0) - \log P(z) + \log P(0), \quad P(0) = 1,$$

then $g(z)$ is entire function and

$$e^{g(z)} = \frac{f(z)}{P(z)f(0)}.$$

This completes the proof. □

Remark. It should be mentioned here that the latter representation is not unique. In addition, if $z = 0$ is a zero for $f(z)$ of order $m \in \mathbb{N}$, then $f(0)$ must be replaced by z^m, i. e., the corresponding representation will be

$$f(z) = z^m P(z)e^{g(z)}.$$

The Weierstrass' factorization theorem is too general and not so specified in view of applications since p_n tends to infinity and there is not enough information about entire function $g(z)$. There is however a subclass of all entire functions when this representation will be quite specific.

Definition 15.4. An entire function $f(z)$ is said to be of finite order if there is a number $A > 0$ such that

$$|f(z)| \leq C_A e^{|z|^A}, \quad C_A > 0, z \in \mathbb{C}.$$

The infimum of such numbers A or

$$\rho := \inf A$$

is called the *order of entire function* or just order in short.

Remark. If ρ is the order of entire function $f(z)$, then for any $\epsilon > 0$ (arbitrarily small) there is a number A_ϵ such that

$$|f(z)| \leq A_\epsilon e^{|z|^{\rho+\epsilon}}.$$

Moreover, this number ρ can be calculated as

$$\rho = \overline{\lim_{R\to\infty}} \frac{\log\log(\sup_{|z|\leq R} |f(z)|)}{\log R}.$$

Problem 15.5. Prove the latter remark.

Example 15.6. It is clear that the order of any polynomial $P_n(z)$ is equal to zero. It follows from the fact that for any $\epsilon > 0$ we have

$$\lim_{|z|\to\infty} \frac{|z|^n}{e^{|z|^\epsilon}} = 0.$$

It is also clear that the entire functions e^z, $\cos z$, $\sin z$, $\sinh z$, $\cosh z$ have order 1 but the entire functions $e^{\sqrt{z}}$, $\cos\sqrt{z}$, $\sin\sqrt{z}$, $\cosh\sqrt{z}$, $\sinh\sqrt{z}$ have order 1/2. This fact follows from the elementary estimate:

$$|e^z| \leq e^{|z|}, \quad z \in \mathbb{C}.$$

Problem 15.7. Prove that the order of functions e^{z^k}, $k \in \mathbb{N}$, and e^{e^z} are equal to k and ∞, respectively.

Definition 15.8. Let function $f(z)$ be of order $\rho, 0 < \rho < \infty$. The *type* $\sigma > 0$ of $f(z)$ is defined as

$$\sigma := \overline{\lim_{R\to\infty}} \frac{\log(\sup_{|z|\leq R} |f(z)|)}{R^\rho}.$$

It is easy to check that the types of functions $e^{az^k}, k \in \mathbb{N}, \cos az, \sin az, a \in \mathbb{C}$ are equal to $|a|$. These new concepts—order and type—can be characterized by the following statements.

Proposition 15.9. *The function*

$$f(z) := \sum_{n=0}^{\infty} a_n z^n$$

is an entire function of order ρ and type σ if and only if

$$\frac{1}{\rho} = \varliminf_{n\to\infty} \frac{\log(1/|a_n|)}{n\log n}, \qquad \sigma = \frac{\overline{\lim}_{n\to\infty} n|a_n|^{\rho/n}}{e\rho}.$$

Proof. Denote by μ, $0 \le \mu \le \infty$, the number

$$\mu := \varliminf_{n\to\infty} \frac{\log(1/|a_n|)}{n\log n}.$$

If $\mu \neq \infty$, then for any $\epsilon > 0$ there exists $n_0 \in \mathbb{N}$ such that for all $n \ge n_0$ it holds that

$$\log \frac{1}{|a_n|} > (\mu - \epsilon) n \log n,$$

or equivalently,

$$|a_n| < n^{-n(\mu-\epsilon)}.$$

If, in addition, $\mu > 0$ then it follows that the series $\sum_{n=0}^{\infty} a_n z^n$ converges for all z since

$$\left|\sum_{n=n_0}^{\infty} a_n z^n\right| \le \sum_{n=n_0}^{\infty} n^{-n(\mu-\epsilon)}|z|^n = \sum_{n=n_0}^{\infty}\left(\frac{|z|}{n^{\mu-\epsilon}}\right)^n < \infty.$$

Hence, function $f(z)$ is entire. Next, for $|z| > 1$ we have

$$\begin{aligned}
|f(z)| &\le A|z|^{n_0} + \sum_{n=n_0+1}^{\infty} n^{-n(\mu-\epsilon)}|z|^n \\
&\le A|z|^{n_0} + \sum_{n\le(2|z|)^{1/(\mu-\epsilon)}} n^{-n(\mu-\epsilon)}|z|^n + \sum_{n>(2|z|)^{1/(\mu-\epsilon)}}^{\infty} n^{-n(\mu-\epsilon)}|z|^n \\
&\le A|z|^{n_0} + e^{(2|z|)^{1/(\mu-\epsilon)}\log|z|}\sum_{n=n_0}^{\infty} n^{-n(\mu-\epsilon)} + \sum_{n=1}^{\infty}\frac{1}{2^n} \\
&\le A|z|^{n_0} + Ce^{(2|z|)^{1/(\mu-\epsilon)}\log|z|} + 1 \le Ke^{(2|z|)^{1/(\mu-\epsilon)}\log|z|} \\
&\le K_1 e^{|z|^{1/(\mu-\epsilon_1)}},
\end{aligned}$$

where $\epsilon_1 > 0$ is arbitrarily small. This means that $\rho \le 1/\mu$. For the case $\mu = \infty$, we can easily obtain from here that $\rho = 0$.

On the other hand, for any $\epsilon > 0$ (arbitrarily small), there exists a sequence $k_n \to +\infty$ such that

$$\log \frac{1}{|a_{k_n}|} < (\mu + \epsilon) k_n \log k_n.$$

It leads us to the inequality

$$|a_{k_n}||z|^{k_n} > \left(|z| k_n^{-(\mu+\epsilon)}\right)^{k_n}.$$

For $|z| = (2k_n)^{\mu+\epsilon}$, we obtain from here that

$$|a_{k_n}||z|^{k_n} > 2^{(\mu+\epsilon)k_n} = e^{(1/2 \log 2)(\mu+\epsilon)|z|^{1/(\mu+\epsilon)}}$$

since $k_n = \frac{|z|^{1/(\mu+\epsilon)}}{2}$. Now the Cauchy's inequality (9.6) or

$$M_r \ge |a_n| r^n, \quad r = |z|,$$

implies that for $|z|$ large enough

$$M_r \ge e^{A|z|^{1/(\mu+\epsilon)}}, \quad r = |z|.$$

Hence, $\rho \ge \frac{1}{\mu+\epsilon}$, and due to arbitrariness of $\epsilon > 0$ we get that $\rho \ge \frac{1}{\mu}$. Thus, $\rho = \frac{1}{\mu}$ (the case $\mu = 0$ is included here as well). This proves the first statement of the proposition. A similar proof gives the second statement of the proposition. □

Problem 15.10. Prove that the order and type of functions:
1. $f(z) = \sum_{n=0}^{\infty} \frac{z^n}{2^{n^2}}$ are equal to 0.
2. $f(z) = \sum_{n=2}^{\infty} \frac{z^n}{(n \log n)^n}$ are equal to 1 and 0, respectively.
3. $f(z) = \sum_{n=0}^{\infty} \frac{z^n}{(n!)^a}$, $a > 0$ are equal to $\frac{1}{a}$ and a, respectively.
4. $f(z) = \sum_{n=1}^{\infty} (\frac{e\rho\sigma}{n} z^n)^n$ are equal to ρ and σ, respectively.
5. $f(z) = \sum_{n=1}^{\infty} \frac{z^n}{n^{an}}$ are equal to $\frac{1}{a}$ and $\frac{a}{e}$, respectively.

Further, we may assume for convenience but without loss of generality that $f(0) \ne 0$.

There is a deep connection between the number of zeros of an entire function in the disk $\{z : |z| \le R\}$, denoted as $n(R)$, and the function $f(z)$ itself. More precisely, if $f(z)$ is an entire function of order ρ, then for any $\epsilon > 0$ there is a constant $A_\epsilon > 0$ such that

$$n(R) \le A_\epsilon R^{\rho+\epsilon},$$

i. e., roughly speaking, the number of zeros is as large as $R > 0$. Indeed, due to the Jensen's formula (see Theorem 7.4)

$$\int_0^R \frac{n(t)}{t}\mathrm{d}t = \frac{1}{2\pi}\int_0^{2\pi} \log|f(Re^{i\theta})|\mathrm{d}\theta - \log|f(0)|$$

and this formula holds for all $R > 0$. If function $f(z)$ is of order $\rho < \infty$, then the latter formula implies that

$$\int_0^R \frac{n(t)}{t}\mathrm{d}t \le A_\epsilon R^{\rho+\epsilon}.$$

Next, since function $n(R)$ is not decreasing, then

$$\int_R^{2R} \frac{n(t)}{t}\mathrm{d}t \ge n(R)\int_R^{2R} \frac{\mathrm{d}t}{t} = n(R)\log 2.$$

Consequently, we obtain

$$n(R) \le \frac{1}{\log 2}\int_R^{2R} \frac{n(t)}{t}\mathrm{d}t \le \frac{1}{\log 2}\int_0^{2R} \frac{n(t)}{t}\mathrm{d}t \le A'_\epsilon R^{\rho+\epsilon},$$

which is what is needed to be shown. Moreover, this estimate shows that the series

$$\sum_{n=1}^{\infty} \frac{1}{|z_n|^\alpha} < \infty$$

converges for any $\alpha > \rho$, where $z_1, z_2, \dots$ are the zeros of entire function $f(z)$. To observe this fact, we note that for $\rho < \beta < \alpha$ one can see that $n(R) \le AR^\beta$ and then setting $R = |z_n|$ we obtain that $n < A|z_n|^\beta$, and consequently, $|z_n|^{-\alpha} \le An^{-\alpha/\beta}$. This completes the proof of this fact. The latter property justifies the following definition.

Definition 15.11. Let $f(z)$ be an entire function of order ρ with the zeros $z_1, z_2, \dots$. Denote

$$\rho_1 := \inf \alpha,$$

where infimum is taken over all $\alpha > \rho \ge 0$ for which the series

$$\sum_{n=1}^{\infty} \frac{1}{|z_n|^\alpha} < \infty$$

converges. Then this value ρ_1 is said to be the *zero convergence index.*

It is clear (due to the latter definition) that $\rho_1 \le \rho$ and $\rho_1 = 0$ in the case when given entire function has finitely many zeros. Thus, if $\rho_1 > 0$ then the number of zeros of given entire function is infinite. The importance of this notation leads to the *canonical product* and the famous Hadamard's theorem.

Proposition 15.12. *Let $f(z)$ be an entire function of order ρ with the zeros $z_1, z_2, \ldots$. Then there is an integer number p, which is independent on n such that the product*

$$P(z) := \prod_{n=1}^{\infty} E\left(\frac{z}{z_n}, p\right) = \prod_{n=1}^{\infty}\left(1 - \frac{z}{z_n}\right) e^{z/z_n + \cdots + \frac{1}{p} z^p / z_n^p}$$

converges for all $z \in \mathbb{C}$. This product with the smallest integer p for which this product converges is called canonical, and this integer p is called canonical product order.

Proof. Since function $f(z)$ is of order ρ, then the series

$$\sum_{n=1}^{\infty} \frac{1}{|z_n|^{p+1}} < \infty$$

converges for $p + 1 > \rho_1$. Thus, for any fixed $z \in \mathbb{C}$, the series

$$\sum_{n=1}^{\infty} \frac{|z|^{p+1}}{|z_n|^{p+1}} < \infty$$

converges also. It implies due to Weierstrass' theorem (see Theorem 15.2) that the product

$$\prod_{n=1}^{\infty} E\left(\frac{z}{z_n}, p\right)$$

converges, too. If ρ_1 is not integer, then for the canonical product p can be chosen as $p = [\rho_1]$. If ρ_1 is integer then p can be chosen as $p = \rho_1 - 1$ in the case when the series $\sum_{n=1}^{\infty} \frac{1}{|z_n|^{\rho_1}}$ converges and $p = \rho_1$ when this series diverges. Hence, the proposition is proved. □

Theorem 15.13 (Hadamard). *Let $f(z)$ be an entire function of order ρ with the zeros $z_1, z_2, \ldots$, and with $f(0) \neq 0$. Then the following representation holds:*

$$f(z) = P(z)e^{g(z)}, \quad z \in \mathbb{C},$$

where $P(z)$ is the canonical product and $g(z)$ is a polynomial of degree at most ρ.

Proof. Using the Weierstrass' factorization theorem (see Corollary 15.3), we can take instead of $P(z)$ there the canonical product constructed with respect to the zeros of $f(z)$. It remains to show that $g(z)$ is a polynomial of degree not bigger than ρ. Let $\nu = [\rho]$. Then p from the canonical product satisfies $p \leq \nu$. Differentiating now the logarithmic derivative of $f(z)$ to order ν, we obtain

$$\left(\frac{f'(z)}{f(z)}\right)^{(\nu)} = g^{(\nu+1)}(z) - \nu! \sum_{n=1}^{\infty} \frac{1}{(z_n - z)^{\nu+1}}, \quad z \neq z_n.$$

Consider now the function

$$F_R(z) := \frac{f(z)}{f(0)} \prod_{|z_n| \le R} \left(1 - \frac{z}{z_n}\right)^{-1}.$$

Then for $|z| = 2R$ and $|z_n| \le R$, we will have that $|1 - \frac{z}{z_n}| \ge 1$, and consequently,

$$|F_R(z)| \le \frac{|f(z)|}{|f(0)|} \le A_\epsilon e^{(2R)^{\rho+\epsilon}}.$$

Using the maximum modulus principle of analytic function, we can extend this inequality for all $|z| \le 2R$. Considering function $h_R(z) := \log F_R(z)$ with $h_R(0) = 0$, we may conclude that $h_R(z)$ is analytic and for $|z| \le R$ it holds that

$$\operatorname{Re} h_R(z) \le A'_\epsilon R^{\rho+\epsilon}.$$

Therefore the Cauchy formula yields for $|z| = r < R$ that

$$|h_R^{(\nu+1)}(z)| \le A'_\epsilon R \frac{2^{\nu+2}(\nu+1)!}{(R-r)^{\nu+2}} R^{\rho+\epsilon}.$$

That is why for $|z| = R/2$ we obtain

$$|h_R^{(\nu+1)}(z)| \le A''_\epsilon R^{\rho+\epsilon-\nu-1}.$$

This implies that for $|z| = R/2$ we have

$$\begin{aligned} g^{(\nu+1)}(z) &= h^{(\nu+1)}(z) + \nu! \sum_{|z_n|>R} \frac{1}{(z_n - z)^{\nu+1}} \\ &= O(R^{\rho+\epsilon-\nu-1}) + O\left(\sum_{|z_n|>R} \frac{1}{|z_n|^{\nu+1}}\right). \end{aligned}$$

This estimate is also valid for all $|z| < R/2$. Since $\nu + 1 > \rho$ and $\epsilon > 0$ is arbitrary (small), then letting $R \to \infty$ we obtain that $g^{(\nu+1)} \equiv 0$ for all $z \in \mathbb{C}$. This proves the theorem. □

Remark. Under the conditions of the latter theorem, if $z = 0$ is a zero of $f(z)$ of order $m \in \mathbb{N}$ then the Hadamard representation has the form

$$f(z) = z^m P(z) e^{g(z)}.$$

Example 15.14.

1. Consider the entire function $f(z) = \sin z$ of order 1. The zeros of this function are $z_n = n\pi$, $n \in \mathbb{Z}$. Then due to the Hadamard's theorem

$$\begin{aligned}\sin z &= z e^{az+b} \prod_{n=-\infty, n\neq 0}^{\infty} \left(1 - \frac{z}{n\pi}\right) e^{z/n\pi} \\ &= z e^{az+b} \prod_{n=1}^{\infty} \left(1 - \frac{z}{n\pi}\right) e^{z/n\pi} \left(1 + \frac{z}{n\pi}\right) e^{-z/n\pi} \\ &= z e^{az+b} \prod_{n=1}^{\infty} \left(1 - \frac{z^2}{(n\pi)^2}\right).\end{aligned}$$

   But since $\lim_{z\to 0} \frac{\sin z}{z} = 1$, then necessarily we have that $e^b = 1$. Moreover, since function $\frac{\sin z}{z}$ is even, then we must have that

$$e^{az} \prod_{n=1}^{\infty} \left(1 - \frac{z^2}{(n\pi)^2}\right) = e^{-az} \prod_{n=1}^{\infty} \left(1 - \frac{z^2}{(n\pi)^2}\right).$$

   This implies that $a = 0$, and we obtain finally that

$$\sin z = z \prod_{n=1}^{\infty} \left(1 - \frac{z^2}{(n\pi)^2}\right).$$

2. Similarly, we easily obtain that

$$\cos z = \prod_{n=1}^{\infty} \left(1 - \frac{z^2}{(\pi/2 + n\pi)^2}\right).$$

3. The entire functions $\sinh z$ and $\cosh z$ of order 1 have the zeros $z_n = in\pi$, $n \in \mathbb{Z}$, and $z_n = i(\pi/2 + n\pi)$, $n \in \mathbb{Z}$, respectively. The application of the Hadamard's theorem leads to the representations

$$\sinh z = z \prod_{n=1}^{\infty} \left(1 + \frac{z^2}{(n\pi)^2}\right), \quad \cosh z = \prod_{n=1}^{\infty} \left(1 + \frac{z^2}{(\pi/2 + n\pi)^2}\right).$$

The Hadamard's theorem can be effectively applied for the investigation of the number of zeros for concrete entire functions.

Example 15.15. Suppose $\lambda \neq 0$ and $Q(z) \neq 0$ is a polynomial. Then the entire function $f(z) = e^{\lambda z} - Q(z)$ has infinitely many zeros. The order of $f(z)$ is equal to 1. Assuming now on the contrary that $f(z)$ has finitely many zeros we conclude (using Hadamard's theorem) that

$$e^{\lambda z} - Q(z) = e^{az} P(z),$$

where $P(z)$ is a polynomial, too. It is clear (from the behavior of $f(z)$ at the infinity) that $a = \lambda$, and hence,

$$e^{\lambda z} = \frac{Q(z)}{1 - P(z)},$$

i. e., $e^{\lambda z}$ is rational. This contradiction proves the statement.

Problem 15.16.
1. Prove that each of the equations $\sin z = z^2$, $\log z = z^3$, $\tan z = az + b$ have infinitely many solutions.
2. Find all zeros of function $e^{e^z} - 1$ and show that the zero convergence index is infinite.

Problem 15.17. Let an entire function $f(z)$ have the form

$$f(z) := \prod_{n=1}^{\infty}\left(1 + \frac{z}{r_n}\right), \quad r_n > 0,$$

and assume it is of order ρ, $0 < \rho < 1$. Show that for $\sigma, \rho < \sigma < 1$ we have

$$\int_0^{\infty} \frac{\log f(x)}{x^{\sigma+1}} dx = \frac{\pi}{\sigma \sin \pi\sigma} \sum_{n=1}^{\infty} \frac{1}{r_n^{\sigma}}, \quad \int_0^{\infty} \frac{\log |f(-x)|}{x^{\sigma+1}} dx = \frac{\pi}{\sigma \tan \pi\sigma} \sum_{n=1}^{\infty} \frac{1}{r_n^{\sigma}}.$$

Hint: Use the integrals from Problem 13.47.

If $f(z)$ is an entire function of order ρ, then its derivative $f'(z)$ is also an entire function having the same order ρ. Indeed, let us denote by $M_r' = \max_{|z|=r} |f'(z)|$. Then

$$f(z) = \int_0^z f'(\zeta) d\zeta + f(0),$$

where the integral is taken over the line connecting 0 and z. Consequently, using the maximum modulus principle, we have

$$\begin{aligned} M_r = \max_{|z|=r} |f(z)| &\le \left| \int_0^z |f'(\zeta)| |d\zeta| \right| + |f(0)| \\ &\le M_r' |z| + |f(0)| = M_r' r + |f(0)|, \quad r = |z|. \end{aligned}$$

On the other hand, using the Cauchy formula, we have

$$f'(z) = \frac{1}{2\pi i} \int_{|\zeta - z| = r} \frac{f(\zeta)}{(\zeta - z)^2} d\zeta$$

and this implies the estimate

$$M'_r \le \frac{1}{2\pi} \int\limits_{|\zeta - z| = r} \frac{|f(\zeta)|}{|\zeta - z|^2} |d\zeta| \le \frac{M_{2r}}{r}.$$

Combining the latter estimates, we obtain that

$$\frac{M_r - |f(0)|}{r} \le M'_r \le \frac{M_{2r}}{r}.$$

This double inequality proves the needed statement (see the proof of Proposition 15.9). More deeper statement is contained in the following theorem, which is the generalization of the corresponding fact for polynomials.

Theorem 15.18 (Laguerre). *Let $f(z)$ be an entire function of order $\rho < 2$. Assume that $f(z)$ is real-valued for real z and has only real zeros. Then the zeros of $f'(z)$ are also real and they are separated from each other by the zeros of $f(z)$.*

Proof. Since $\rho < 2$, then using the Hadamard's theorem we have the representation

$$f(z) = Cz^m e^{az} \prod_{n=1}^{\infty} \left(1 - \frac{z}{z_n}\right) e^{z/z_n},$$

where $z_1, z_2, \dots, z_n, \dots$ are real zeros of $f(z)$ and $C, a, m \in \mathbb{N}_0$ some real numbers (m is the order of the zero $z = 0$). It follows that

$$\frac{f'(z)}{f(z)} = \frac{m}{z} + a + \sum_{n=1}^{\infty} \left(\frac{1}{z - z_n} + \frac{1}{z_n} \right)$$

and

$$\operatorname{Im}\left(\frac{f'(z)}{f(z)}\right) = -\frac{my}{x^2 + y^2} - \sum_{n=1}^{\infty} \frac{y}{(x - z_n)^2 + y^2}.$$

Since the right-hand side is equal to zero if and only if $y = 0$, then $f'(z)$ may have only real zeros. Further, since

$$\left(\frac{f'(z)}{f(z)}\right)' = -\frac{m}{z^2} - \sum_{n=1}^{\infty} \frac{1}{(z - z_n)^2}$$

we can observe that this derivative for real z is real and negative for $z \ne z_n$. This means that $\frac{f'(z)}{f(z)}$ is decreasing for real z. Moreover, for $z_n < z < z_{n+1}$ this function has only one zero (even of order 1 due to the monotonicity) since it changes the sign there. This proves the theorem. □

Remark. The proof of this theorem evidently shows that the series

$$\sum_{n=1}^{\infty} \frac{1}{|z_n|^{\alpha}}, \quad \sum_{n=1}^{\infty} \frac{1}{|z'_n|^{\alpha}},$$

where z'_n denotes the zeros of $f'(z)$ converge or diverge simultaneously.

Remark. The following examples:

1. $f(z) = ze^{z^2}, f'(z) = (2z^2 + 1)e^{z^2}$,
2. $f(z) = (z^2 - 4)e^{z^2/3}, f'(z) = \frac{2z}{3}(z^2 - 1)e^{z^2/3}$

show that for $\rho = 2$ the Laguerre's theorem does not hold; in the first case, the zeros of $f'(z)$ are not real, and in the second case the zeros of $f(z)$ and $f'(z)$ are not separated by each other.

The zero convergence index ρ_1 (see Definition 15.11) is actually equal to the order of canonical product of entire function constructed by the zeros $z_1, z_2, \dots z_n, \dots$. We know from above that $\rho_1 \le \rho$, where ρ here is an order of the canonical product $P(z)$. It remains to check that $\rho \le \rho_1$. In order to show this consider $\log P(z)$ and estimate it as follows:

$$|\log P(z)| \le \left|\sum_{|z_n|\le k|z|} \log E\left(\frac{z}{z_n}, p\right)\right| + \left|\sum_{|z_n|>k|z|} \log E\left(\frac{z}{z_n}, p\right)\right| =: I_1 + I_2,$$

where $E(\frac{z}{z_n}, p)$ is defined in (15.1) and constant $k > 1$. From inequality (15.2), it follows that

$$I_2 \le \sum_{|z_n|>k|z|} \frac{(\frac{|z|}{|z_n|})^{p+1}}{1 - |z|/|z_n|} \le C|z|^{p+1} \sum_{n=1}^{\infty} \frac{1}{|z_n|^{p+1}} \le C|z|^{\rho_1},$$

if $p + 1 = \rho_1$, or

$$I_2 \le C_\epsilon |z|^{\rho_1+\epsilon},$$

if $p + 1 > \rho_1 + \epsilon$ (see the definition of ρ_1 in the proof of Proposition 15.12). The term I_1 can be estimated as follows:

$$\begin{aligned} I_1 &\le \sum_{|z_n|\le k|z|} \left|\log E\left(\frac{z}{z_n}, p\right)\right| \le \sum_{|z_n|\le k|z|} \left(\left|\log\left|E\left(\frac{z}{z_n}, p\right)\right|\right| + 2\pi\right) \\ &\le C \sum_{|z_n|\le k|z|} \left(\left(\frac{|z|}{|z_n|}\right)^p + 2\pi\right) \le C|z|^p \sum_{|z_n|\le k|z|} \left(\frac{1}{|z_n|^p} + \frac{2\pi}{|z|^p}\right) \\ &\le C|z|^p \sum_{|z_n|\le k|z|} \frac{1}{|z_n|^p} \le C|z|^p \sum_{|z_n|\le k|z|} \frac{|z_n|^{\rho_1+\epsilon-p}}{|z_n|^{\rho_1+\epsilon}} \le C|z|^{\rho_1+\epsilon}. \end{aligned}$$

Combining the estimates for I_2 and I_1, we obtain finally that $\rho \le \rho_1$. This proves the statement. It follows from these considerations that an entire function of not integer order must have infinitely many zeros.

Example 15.19. Let $f(z)$ be an entire function of order $\rho < 1$. Assume that all its zeros are real and negative. Using the Hadamard's theorem, we obtain for $f(z)$ the following representation:

$$f(z) = f(0)\prod_{n=1}^{\infty}\left(1+\frac{z}{z_n}\right), \quad f(0) \neq 0, z_n > 0,$$

where $-z_n < 0$ are the zeros of $f(z)$. We may assume without loss of generality that $f(0) = 1$. Consequently, for $z \in \mathbb{R}$ we have that

$$\begin{aligned}\log f(z) &= \sum_{n=1}^{\infty}\log\left(1+\frac{z}{z_n}\right) = \sum_{n=1}^{\infty} n\left(\log\left(1+\frac{z}{z_n}\right) - \log\left(1+\frac{z}{z_{n+1}}\right)\right) \\ &= \sum_{n=1}^{\infty} n \int_{z_n}^{z_{n+1}} \frac{z\mathrm{d}t}{t(z+t)} = z\int_0^{\infty}\frac{n(t)\mathrm{d}t}{t(z+t)},\end{aligned}$$

where $n(t)$ denotes (as above) the number of zeros of $f(z)$ not bigger than t. This representation leads to the equality

$$f(z) = \exp\left(z\int_0^{\infty}\frac{n(t)\mathrm{d}t}{t(z+t)}\right),$$

which gives the estimates of $f(z)$ in this concrete case.

The distribution of the zeros of entire function and the canonical product also allow us to estimate the function

$$m_r := \inf_{|z|=r}|f(z)|.$$

Proposition 15.20. *Let $f(z)$ be an entire function of order ρ. Then for any $\epsilon > 0$ (arbitrarily small) there exists $R_0 > 0$ such that for all $|z| \geq R_0$ it holds that*

$$m_r \geq \mathrm{e}^{-r^{\rho+\epsilon}}, \quad r = |z|.$$

Proof. Using the canonical product for $f(z)$ (see Proposition 15.12), we have that

$$\begin{aligned}\log|P(z)| &= \log\prod_{n=1}^{\infty}\left|E\left(\frac{z}{z_n},p\right)\right| = \sum_{n=1}^{\infty}\log\left|1-\frac{z}{z_n}\right| + \sum_{n=1}^{\infty}\left|\frac{z}{z_n}+\cdots+\frac{1}{p}\left(\frac{z}{z_n}\right)^p\right| \\ &\geq \sum_{|z_n|\leq k|z|}\log\left|1-\frac{z}{z_n}\right| - \sum_{|z_n|>k|z|}\left|\log\left|1-\frac{z}{z_n}\right|\right| \\ &\quad - \sum_{|z_n|\leq k|z|}\left(\frac{|z|}{|z_n|}+\cdots+\frac{1}{p}\left(\frac{|z|}{|z_n|}\right)^p\right) - \sum_{|z_n|>k|z|}\left(\frac{|z|}{|z_n|}+\cdots+\frac{1}{p}\left(\frac{|z|}{|z_n|}\right)^p\right) \\ &\geq \sum_{|z_n|\leq k|z|}\log\left|1-\frac{z}{z_n}\right| - C_1\sum_{|z_n|>k|z|}\left(\frac{|z|}{|z_n|}\right)^{p+1} - C_2\sum_{|z_n|\leq k|z|}\left(\frac{|z|}{|z_n|}\right)^p \\ &\quad - C_3\sum_{|z_n|>k|z|}\left(\frac{|z|}{|z_n|}\right)^{p+1}\end{aligned}$$

if we choose constant $k > 1$. Thus, for any $\epsilon > 0$ we have that

$$\log|P(z)| \geq \sum_{|z_n|\leq k|z|} \log\left|1 - \frac{z}{z_n}\right| - C_\epsilon |z|^{\rho+\epsilon}.$$

Next, consider $h > \rho$ such that

$$|z - z_n| < |z_n|^{-h}.$$

It means that the sequence of radii $\{|z_n|^{-h}\}_{n=1}^{\infty}$ for these disks tend to zero as $n \to \infty$. Moreover, since $h > \rho$ then the series

$$\sum_{n=1}^{\infty} \frac{1}{|z_n|^h} < \infty$$

converges. Due to this property of these radii, there is R_0 so large that for all $|z| \geq R_0$ the union of all these disks belongs to the disk $\{z : |z| \leq R_0\}$. If now $|z| \geq R_0$, then for $|z| \geq \frac{1}{k}|z_n|$ we have that

$$\left|1 - \frac{z}{z_n}\right| > |z_n|^{-1-h} \geq (k|z|)^{-1-h},$$

and consequently,

$$\begin{aligned}\log|P(z)| &\geq \sum_{|z_n|\leq k|z|} \log\left|1 - \frac{z}{z_n}\right| - C_\epsilon |z|^{\rho+\epsilon} \\ &> -(1+h)\log(k|z|) \sum_{|z_n|\leq k|z|} 1 - C_\epsilon |z|^{\rho+\epsilon} > -C'_\epsilon |z|^{\rho+\epsilon} > -|z|^{\rho+\epsilon'}\end{aligned}$$

if $\epsilon' > \epsilon$ and $|z|$ is sufficiently large. Further, due to the Hadamard's representation

$$f(z) = P(z)e^{g(z)},$$

where $g(z)$ is a polynomial of degree not bigger than ρ, and thus we have that

$$|e^{g(z)}| \geq e^{-K|z|^\rho}, \quad K > 0, |z| \leq R_0.$$

Combining the estimates for $P(z)$ and $e^{g(z)}$, we obtain the needed estimate for m_r. Hence, the proposition is proved. □

The latter estimate of m_r does not show possible behavior of this function at the infinity with respect to r. However, there are some situations when it is possible to characterize more precisely this behavior at the infinity. The following proposition holds.

Proposition 15.21. *Let $f(z)$ be a nonconstant entire function of order $\rho < 1/2$. Then there exists infinitely large sequence $\{r_j\}_{j=1}^{\infty}$ such that*

$$\lim_{j\to\infty} m_{r_j} = +\infty.$$

Proof. First, we note that there is no half-line $\arg z =$ constant such that $f(z)$ is bounded on it. Indeed, a plane cut along such half-line represents an angle of magnitude 2π. Since $2\pi < \pi/\rho$ for $\rho < 1/2$, then $f(z)$ being bounded on this half-line will be bounded everywhere (see Theorem 7.17). Thus, the Liouville's theorem implies that $f(z) \equiv$ constant. Next, the Hadamard's theorem yields the representation

$$f(z) = cz^m \prod_{n=1}^{\infty}\left(1 - \frac{z}{z_n}\right).$$

Let us define now a new function

$$\varphi(z) := cz^m \prod_{n=1}^{\infty}\left(1 + \frac{z}{|z_n|}\right).$$

Since for each $n \in \mathbb{N}$, we have

$$\left|1 - \frac{z}{z_n}\right| \ge \left|1 - \frac{|z|}{|z_n|}\right|$$

then

$$m_r = \min_{|z|=r}|f(z)| \ge |\varphi(-r)|.$$

But the function $\varphi(-r)$ is not bounded since $\varphi(z)$ is entire function of the same order ρ as $f(z)$. This proves the proposition. □

Remark. The entire functions may grow up as fast as any increasing function. More precisely, for any positive and increasing function $g(t), t \in \mathbb{R}_+$, there exists an entire function $f(z)$ such that $f(x) > g(|x|)$ for all real x. Such a function $f(z)$ may have the form

$$f(z) = C + \sum_{n=1}^{\infty}\left(\frac{z}{n}\right)^{k_n}$$

for some constant C and a strictly increasing sequence of even $k_n \in \mathbb{N}$. This sequence k_n can be chosen as follows:

$$\left(1 + \frac{1}{n}\right)^{k_n} \ge g(n+2).$$

The constant C is chosen as $C = g(2)$.

Problem 15.22.

1. Show that for $\alpha > 1$ the function

$$F_\alpha(z) := \int_0^\infty e^{-t^\alpha} \cos(zt)dt$$

   is entire of order $\frac{\alpha}{\alpha-1}$.
2. Let $|q| < 1$ and $k > 1$. Prove that the function

$$f(z) := \sum_{n=-\infty}^{\infty} q^{n^k} e^{inz}$$

   is entire and find its order.
3. Let $\sigma > 1$. Prove that the function

$$f_\sigma(z) := \prod_{n=1}^{\infty}\left(1 + \frac{z}{n^\sigma}\right)$$

   is entire of order $1/\sigma$.
4. Show that the functions

$$f(z) := \prod_{n=1}^{\infty}\left(1 + \frac{z}{e^n}\right), \quad f(z) := \prod_{n=1}^{\infty}\left(1 + \frac{e^z}{e^{n^\alpha}}\right), \quad \alpha > 0,$$

   are entire of order 0 and $1 + 1/\alpha$, respectively.
5. Assume that $f(z)$ is an entire function with the zeros $z_1, z_2, \ldots, z_n, \ldots$ such that the series

$$\sum_{n=1}^{\infty} \frac{1}{|z_n|} < \infty$$

   converges. Show that for any $\epsilon > 0$ there exists $C_\epsilon > 0$ so that

$$e^{-\epsilon|z|} < |f(z)| \le C_\epsilon e^{\epsilon|z|}$$

   for $|z|$ large enough.
6. Show that if the function

$$f(z) := \sum_{n=0}^{\infty} a_n z^n$$

   is an entire of order $\rho < 1$ then

$$|f^{(n)}(0)| \le Ce^{An}, \quad C > 0, A > 0, n = 0, 1, 2, \ldots.$$

7. Assume that $f(z)$ is an entire function of order 0, i. e., for any $\epsilon > 0$ there is $C_\epsilon > 0$ such that $|f(z)| \leq C_\epsilon e^{\epsilon|z|}$. Prove that

$$|f^{(n)}(0)| \leq C'_\epsilon e^{\epsilon n}, \quad n = 0, 1, 2, \ldots .$$

8. Show that the function

$$f(z) := \sum_{n=0}^{\infty} \frac{z^n}{(n+a)^s n!}, \qquad a > 0,$$

is entire of order $\rho = 1$ for any $s \in \mathbb{C}$.

9. Show that the function

$$f(z) := \sum_{n=0}^{\infty} \frac{\cosh \sqrt{n}}{n!} z^n,$$

is entire of order $\rho = 1$. Show also that all its zeros are real and negative.

10. Prove that if the function

$$f(z) := \sum_{n=0}^{\infty} a_n z^n$$

is entire of order ρ then the order of an entire function

$$F(z) := \sum_{n=0}^{\infty} |a_n|^p z^n, \quad p > 0,$$

is ρ/p.

Exercises

1. Find the primitives of the following functions:
   a) $f(z) = \sin z \cos z$, b) $f(z) = \cos^2 z$, c) $f(z) = ze^{2z}$,
   d) $f(z) = z^2 \sin z$, e) $f(z) = z \sin z^2$, f) $f(z) = e^z \sin z$.
2. Let f be analytic in the whole $\mathbb{C}$ such that

$$|f(z)| \le \left|\frac{z+1}{z-1}\right|$$

   for all $z \in \mathbb{C}$. Prove that f is constant function.
3. Let f be analytic in the disk $\{z : |z| < R\}$. Assume that f is nonconstant. Let us define the function

$$g(r) := \max_{|z| \le r} |f(z)|, \quad 0 < r < R.$$

   Prove that $g(r_1) < g(r_2)$ whenever $0 < r_1 < r_2 < R$.
4. Let $f(z) = \cos z$, $z \in \mathbb{C}$. Find $\max_{|z|\le 1} |f(z)|$.
5. Define all values of $z \in \mathbb{C}$ where the following functions are analytic:

$$\int_0^\infty e^{-zt^2} dt, \quad \int_0^\infty \frac{\sin t}{t^z} dt, \quad \int_0^\infty \frac{\cos t}{t^z} dt$$

   and define their derivatives.
6. Investigate the convergence of the function sequence f_n, $n = 1, 2, \ldots$ in the set $E \subset \mathbb{C}$ when
   a) $f_n(z) = \frac{nz}{z+n}$, $E = \{z : |z| < 1\}$,
   b) $f_n(z) = \frac{nz}{nz+1}$, $E = \{z : |z| > 1\}$.
   Is the convergence uniform in E?
7. Find the radius of convergence and disk of convergence for the following series:
   a) $\sum_{k=0}^\infty \frac{1}{2^k+1} z^k$, b) $\sum_{k=1}^\infty \frac{1}{k^2}(z-1)^k$, c) $\sum_{k=0}^\infty k^2 z^k$, d) $\sum_{k=0}^\infty \frac{k^3}{3^k} z^k$.
8. Find the radius of convergence for the series

$$\sum_{k=0}^\infty \left(\frac{1}{1-i/2}\right)^{k+1} (z - i/2)^k.$$

   Find also the sum of the series.
9. Find the function $f(z) = \sum_{k=0}^\infty kz^k$ for $|z| < 1$.
10. Find the Taylor series for $f(z) = \sin z$ around the point $z = \pi/4$.
11. Find the Taylor series for $f(z) = (z-1)^{-2}$ around the point $z = 2$.
12. Find the order of the zero $z = 0$ of $f(z) = e^z - 1 - \sin z$.

https://doi.org/10.1515/9783111632278-017

13. Problem 9.22. Apply this problem to prove that if f is an analytic function in the unit disk such that

$$f\left(\frac{n}{2n+1}\right) = f\left(\frac{n}{2n+1}\mathrm{i}\right), \quad n = 2, 3, \ldots$$

then $f^{(10)}(0) = 0$.

14. Find the Laurent series for f at $z_0 = 0$ and investigate the type of singular point 0 and evaluate the residue, when
a) $f(z) = \frac{1-\cos z}{z}$, b) $f(z) = \frac{e^{z^2}}{z^3}$.

15. Find the Laurent series for $f(z) = \frac{1}{z(z+1)(z+2)}$ at $z_0 = 0$.

16. Evaluate the integral $\int_0^{2\pi} \frac{1}{a+\cos t}\mathrm{d}t$, $a > 1$.

17. Evaluate the integral $\int_0^{\infty} \frac{1}{x^6+1}\mathrm{d}x$.

18. Evaluate the integrals

$$\int_{-\infty}^{\infty} \frac{x \sin bx}{x^2 + a^2}\mathrm{d}x, \quad \int_{-\infty}^{\infty} \frac{\cos bx}{x^2 + a^2}\mathrm{d}x, \quad a, b > 0.$$

19. Evaluate the integral

$$\text{p.v.} \int_{-\infty}^{\infty} \frac{x \sin x}{x^2 - \pi^2}\mathrm{d}x.$$

20. Evaluate the integrals

$$\int_{-\infty}^{\infty} \frac{\sin x}{x - \omega}\mathrm{d}x, \quad \int_{-\infty}^{\infty} \frac{\cos x}{x - \omega}\mathrm{d}x,$$

where $\operatorname{Im} \omega \neq 0$.

21. Show that for $a \geq 0$ and $b \in \mathbb{R}$ we have

$$\int_0^{\infty} \mathrm{e}^{a \cos bx} \sin(a \sin bx)\frac{\mathrm{d}x}{x} = \frac{\operatorname{sgn} b}{2}\pi(\mathrm{e}^a - 1).$$

22. Prove that for $\lambda > 0$ it holds that

$$\int_{-\infty}^{\infty} \mathrm{e}^{-\lambda x^2} \cos(2\lambda a x)\mathrm{d}x = \sqrt{\frac{\pi}{\lambda}}\mathrm{e}^{-\lambda a^2}.$$

Hint: Integrate the function $f(z) = \mathrm{e}^{-\lambda z^2}$ over the rectangle with the vertices at the points $-R, R, R + \mathrm{i}a, -R + \mathrm{i}a$ and let $R \to \infty$.

23. Show that for $a \geq 0, b \geq 0$

$$\int_0^\infty \frac{\cos 2ax - \cos 2bx}{x^2} \mathrm{d}x = \pi(b-a).$$

24. Prove that for $b > 0$ and $0 < a < 3$

$$\int_0^\infty \frac{x^{a-1} \sin(\pi a/2 - bx)}{x^2 + r^2} \mathrm{d}x = \frac{\pi}{2} r^{a-2} \mathrm{e}^{-br}.$$

25. Evaluate the series

$$\sum_{k=1}^{\infty} \frac{(-1)^k}{k^4}.$$

26. Evaluate the series

$$\sum_{k=-\infty}^{\infty} \frac{1}{k + \mathrm{i}a}, \qquad a \in \mathbb{R} \setminus \{0\}.$$

27. Based on Example 14.5 show that

$$\sum_{k=1}^{\infty} \frac{1}{(2k-1)^2} = \frac{\pi^2}{8}, \quad \sum_{k=1}^{\infty} \frac{1}{k^2} = \frac{\pi^2}{6}.$$

28. Prove that:
   a)

$$\sum_{k=-\infty}^{\infty} \frac{1}{(z - k\pi)^2} = \frac{1}{\sin^2 z}, \qquad k \in \mathbb{Z}, z \neq k\pi.$$

   b)

$$\sum_{k=1}^{\infty} \frac{1}{z^2 - k^2\pi^2} = \frac{z \cot z - 1}{2z^2}, \qquad z \neq k\pi, k \in \mathbb{N}.$$

29. Show that

$$\frac{z}{\mathrm{e}^z - 1} = \sum_{k=0}^{\infty} \frac{B_k}{k!} z^k,$$

where $B_k = (\frac{z}{\mathrm{e}^z - 1})^{(k)}(0)$ are called *Bernoulli's numbers*.

30. Prove that

$$\sum_{k=1}^{\infty} \frac{1}{k^{2m}} = (-1)^{m-1} \frac{B_{2m}}{(2m)!} 2^{2m-1} \pi^{2m}, \quad m \in \mathbb{N},$$

where B_{2m} are Bernoulli's numbers (see previous exercise).

31. Prove that for any $a, b \in \mathbb{C}$, and for all complex z,

$$e^{az} - e^{bz} = (a-b)ze^{(a+b)z/2} \prod_{k=1}^{\infty} \left(1 + \frac{(a-b)^2 z^2}{4k^2 \pi^2}\right).$$

Hint: Use Example 15.14.

32. Show that if the series $\sum_{k=1}^{\infty} |p_k|^2$ converges then the product $\prod_{k=1}^{\infty} (1-p_k)e^{p_k}$ converges also. Hint: Use the fact that $p_k \to 0$ implies $(1-p_k)e^{p_k} = 1 + O(p_k^2)$.

33. Let function $g(t)$ of the real variable $t > 0$ be defined as

$$g(t) := \int_0^{\infty} \frac{e^{-(x-i)^2 t}}{e^{i2\pi(x-i)} - 1} dx - \int_0^{\infty} \frac{e^{-(x+i)^2 t}}{e^{i2\pi(x+i)} - 1} dx.$$

Show that

a)

$$g(t) = \sum_{n=-\infty}^{\infty} e^{-n^2 \pi t}.$$

b)

$$g(t) = \frac{1}{\sqrt{t}} g\left(\frac{1}{t}\right).$$

c)

$$g(t) = o(1) \quad \text{as } t \to \infty, \quad g(t) = o(t^{-1/2}) \quad \text{as } t \to 0.$$

Hint: Consider the integral

$$\int \frac{e^{-z^2 \pi t}}{e^{i2\pi z} - 1} dz$$

over the rectangle with the vertices at the points $\pm(N + 1/2) \pm i$, $N \in \mathbb{N}$ and let $N \to \infty$.

Part III

16 Conformal mappings

We return now to the geometric properties of nonzero derivative. Let f be analytic in a domain D and let $z_0 \in D$ be an arbitrary point. If $f'(z_0) \neq 0$, then this is equivalent to (see Cauchy–Riemann conditions)

$$|f'(z_0)|^2 = \left(\frac{\partial u}{\partial x}\right)^2 + \left(\frac{\partial v}{\partial x}\right)^2 = \frac{\partial u}{\partial x}\frac{\partial v}{\partial y} - \frac{\partial u}{\partial y}\frac{\partial v}{\partial x} > 0.$$

It means that the Jacobian of the transformation from (x, y) to (u, v) is non-zero at (x_0, y_0), and thus in the neighborhood of $(u_0, v_0) = (u(x_0, y_0), v(x_0, y_0))$ there exists an inverse function $z = x + iy = f^{-1}(w)$, $w = u + iv$ such that $z = f^{-1}(w)$ is analytic at $w_0 = u_0 + iv_0$ and

$$(f^{-1}(w))'(w_0) = \frac{1}{f'(z_0)}.$$

This fact can be interpreted as follows: in the neighborhood of z_0 the function $w = f(z)$ is univalent and analytic. The same is true in the neighborhood of the point $w_0 = f(z_0)$ for the inverse function $z = f^{-1}(w)$. So, this property is local.

Another geometric property of analytic function with nonzero derivative is the following. Let f be analytic in the domain D and $f'(z_0) \neq 0$ for $z_0 \in D$. Consider two arbitrary curves γ_1 and γ_2 on the z-plane, which intersect at the point z_0. Assume that the angle between γ_1 and γ_2 at z_0 is $\varphi_2 - \varphi_1$, and the angle between Γ_1 and Γ_2 at $w_0 = f(z_0)$ in the w-plane is equal to $\phi_2 - \phi_1$, where Γ_j is the image of γ_j under the mapping f for $j = 1, 2$; see Figure 16.1. If $z_1 = z_0 + \Delta z_1 \in \gamma_1$ and $z_2 = z_0 + \Delta z_2 \in \gamma_2$, then $f(z_1) = f(z_0 + \Delta z_1) = f(z_0) + \Delta f_1 = w_0 + \Delta w_1 \in \Gamma_1$ and $f(z_2) = f(z_0 + \Delta z_2) = f(z_0) + \Delta f_2 = w_0 + \Delta w_2 \in \Gamma_2$.

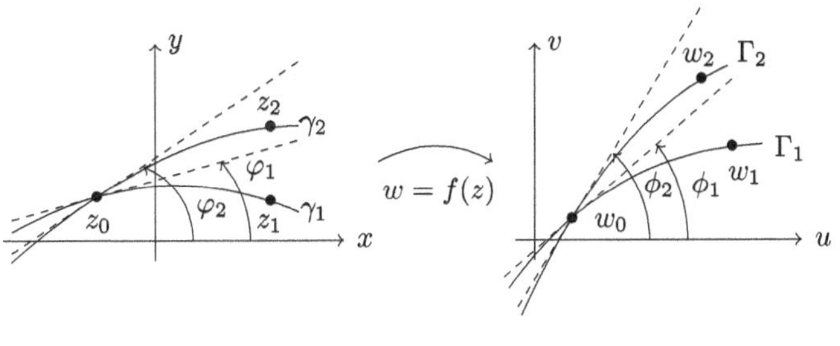

Figure 16.1: Mapping preserving angles.

Moreover,

$$\lim_{\Delta z_1 \to 0} \arg \frac{\Delta w_1}{\Delta z_1} = \lim_{\Delta z_1 \to 0} [\arg \Delta w_1 - \arg \Delta z_1] = \phi_1 - \varphi_1$$

and

https://doi.org/10.1515/9783111632278-018

$$\lim_{\Delta z_2 \to 0} \arg \frac{\Delta w_2}{\Delta z_2} = \lim_{\Delta z_2 \to 0} [\arg \Delta w_2 - \arg \Delta z_2] = \phi_2 - \varphi_2.$$

By the existence of $f'(z_0) \neq 0$ and due to the independence of this derivative with respect to direction, we obtain

$$\phi_1 - \varphi_1 = \phi_2 - \varphi_2 = \arg f'(z_0). \tag{16.1}$$

So, we may conclude that the transformation $w = f(z)$ preserves the angles with respect to orientation and magnitude. In addition, since $f'(z_0) \neq 0$ then

$$|\Delta w| = k|\Delta z| + o(|\Delta z|), \quad k = |f'(z_0)|, \tag{16.2}$$

i. e., there is the *factor of stretching* in all directions.

It is also proved earlier (see Problem 12.17) that if f is analytic in the domain D and univalent there, then $f'(z) \neq 0$ for all $z \in D$.

These properties justify the following definition.

Definition 16.1. The mapping $f : D \to \mathbb{C}$ is called *conformal* at $z_0 \in D$ if it preserves the angles and the factor of stretching at this point. If f is conformal at each point in D, then f is called conformal in D.

There is a very deep connection between analytic functions and conformal mappings.

Theorem 16.2. *The mapping $f : D \to \mathbb{C}$ is conformal in D if and only if f is analytic and univalent in D.*

Proof. Let f be analytic and univalent in the domain D. Then applying Problem 12.17 we conclude that $f'(z) \neq 0$ everywhere in D. Hence, by (16.1) and (16.2), f is conformal at each point $z \in D$ and, therefore, it is conformal in D.

Conversely, let z_0 be an arbitrary point in D and let $w_0 = f(z_0)$. By the conditions of this theorem, we have

$$\arg(w_2 - w_0) - \arg(w_1 - w_0) = \alpha + o(\max(|w_1 - w_0|, |w_2 - w_0|))$$

and

$$\arg(z_2 - z_0) - \arg(z_1 - z_0) = \alpha + o(\max(|z_1 - z_0|, |z_2 - z_0|)),$$

where $\alpha = \varphi_2 - \varphi_1 = \phi_2 - \phi_1$; see Figure 16.1. Moreover,

$$\frac{|w_2 - w_0|}{|z_2 - z_0|} = k + o(1), \quad \frac{|w_1 - w_0|}{|z_1 - z_0|} = k + o(1)$$

as $|z_2 - z_0|, |z_1 - z_0| \to 0$. These equalities imply that

$$\frac{w_2 - w_0}{z_2 - z_0} = ke^{i\varphi} + o(1), \quad \frac{w_1 - w_0}{z_1 - z_0} = ke^{i\varphi} + o(1),$$

where (since α is the same in both equalities)

$$\arg \frac{w_2 - w_0}{z_2 - z_0} = \varphi + o(1), \quad \arg \frac{w_1 - w_0}{z_1 - z_0} = \varphi + o(1)$$

as $|z_2 - z_0|, |z_1 - z_0| \to 0$. Since γ_1 and γ_2 are arbitrary then z_2 and z_1 are arbitrary, too. Hence, we may conclude that there exists

$$\lim_{z \to z_0} \frac{f(z) - f(z_0)}{z - z_0} = ke^{i\varphi} = f'(z_0),$$

and $f'(z_0) \neq 0$ (or $k \neq 0$), i. e., f is analytic and univalent in D. □

Remark. Theorem 16.2 says that univalent functions and only they realize conformal mappings.

The next important property of conformal mappings is contained in the following theorem.

Theorem 16.3 (Boundary correspondence principle). *Let D be a simply connected domain with the boundary ∂D, which is a closed curve γ. Let also $f \in H(D) \cap C(\overline{D})$. Assume that f maps γ to the closed curve $\Gamma := f(\gamma)$ bijectively with the same direction of orientation as for γ. Then $f : D \to \operatorname{int}\Gamma$ is surjective and conformal.*

Proof. Due to Theorem 16.2, it suffices to show that f is univalent in D and f maps D onto $\operatorname{int}\Gamma$. Let us consider two different points $w_1 \in \operatorname{int}\Gamma$ and $w_2 \in \mathbb{C} \setminus \overline{\operatorname{int}\Gamma}$ and two different functions

$$F_1(z) = f(z) - w_1, \quad F_2(z) = f(z) - w_2, \quad z \in D.$$

If z goes over γ, then $w = f(z)$ goes over Γ and the direction of orientation over these curves are the same. Thus, using the principle of argument (see Theorem 12.4), we obtain that

$$\frac{1}{2\pi} \operatorname*{Var}_{\gamma} \operatorname{Arg} F_1(z) = N(F_1) = 1, \quad \frac{1}{2\pi} \operatorname*{Var}_{\gamma} \operatorname{Arg} F_2(z) = N(F_2) = 0,$$

where $N(F_1)$ and $N(F_2)$ denote the number of zeros of F_1 and F_2, respectively. It means that for any $w_1 \in \operatorname{int}\Gamma$ there is only one point $z_1 \in D$ such that $w_1 = f(z_1)$ and for any $w_2 \in \mathbb{C} \setminus \overline{\operatorname{int}\Gamma}$ there are no points $z \in D$ such that $w_2 = f(z)$, i. e., f maps D onto $\operatorname{int}\Gamma$ and it is univalent in D. □

There is one more important property of conformal mappings: Schwarz reflection principle (or Schwarz symmetry principle).

Definition 16.4. Let $D \subset \mathbb{C}$ be a domain. The set

$$J(D) = \{\zeta \in \mathbb{C} : \zeta = \overline{z}, z \in D\} \tag{16.3}$$

is called the *conjugate domain.*

This definition implies that $J(D)$ is a domain and that if $f(z)$ is analytic in D then $g(z) := \overline{f(\overline{z})}$ is analytic in $J(D)$. Indeed, since $f(z)$ is analytic in D then for each $z_0 \in D$ the Taylor's expansion holds, i. e.,

$$f(z) = \sum_{j=0}^{\infty} a_j (z - z_0)^j, \quad a_j = \frac{f^{(j)}(z_0)}{j!}$$

for $|z - z_0| < R$ with $R = \operatorname{dist}(z_0, \partial D)$. Thus, if $\zeta, \zeta_0 \in J(D)$, then $\overline{\zeta}, \overline{\zeta_0} \in D$ and

$$f(\overline{\zeta}) = \sum_{j=0}^{\infty} a_j (\overline{\zeta} - \overline{\zeta_0})^j$$

or

$$g(\zeta) = \overline{f(\overline{\zeta})} = \sum_{j=0}^{\infty} \overline{a_j} (\zeta - \zeta_0)^j,$$

i. e., $g(\zeta)$ is analytic, too, with $\zeta_0 = \overline{z_0}$ from above.

Theorem 16.5 (Schwarz reflection principle). *Let D be a domain in the upper half of the complex plane whose boundary includes an interval $I := (a, b)$ of the real axis. Let $f \in H(D) \cap C(\overline{D})$. Suppose that $f(x + \mathrm{i}0)$ is real for all $x \in I$ and define the function*

$$F(z) = \begin{cases} f(z), & z \in D \cup I, \\ \overline{f(\overline{z})}, & z \in J(D). \end{cases} \tag{16.4}$$

Then F is analytic on $D \cup I \cup J(D)$.

Proof. Since $f(z) \in H(D)$ and $\overline{f(\overline{z})} \in H(J(D))$, then it remains to show that $F(z)$ is analytic at each point $x_0 \in I$. First, we check that $F(z)$ is continuous everywhere in $D \cup I \cup J(D)$. Continuity of F in $D \cup I$ follows from the conditions of the theorem. The definition (16.4) of F and the real-valuedness of $f(x + \mathrm{i}0)$ imply that

$$F(x - \mathrm{i}0) = \overline{f(\overline{x - \mathrm{i}0})} = f(x + \mathrm{i}0) = F(x + \mathrm{i}0).$$

This proves that F is continuous. Next, we introduce the closed curves (see Figure 16.2)

$$\Gamma^+ := \{\zeta : |\zeta - x_0| = \delta, \operatorname{Im} \zeta > 0\} \cup [x_1, x_2]$$

and

$$\Gamma^- := \{\zeta : |\zeta - x_0| = \delta, \operatorname{Im} \zeta < 0\} \cup [x_1, x_2].$$

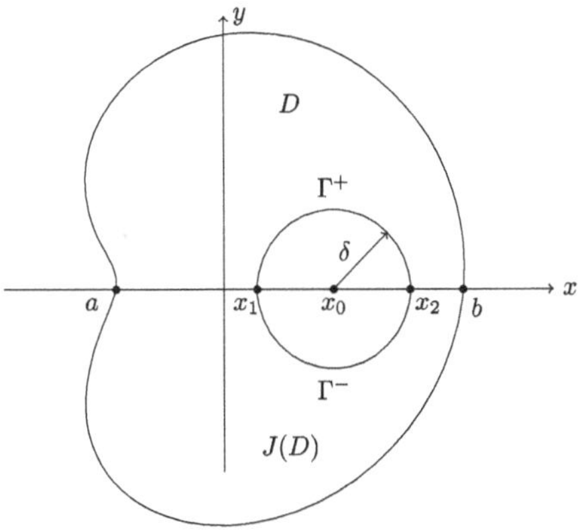

Figure 16.2: Curves $\Gamma^{\pm}$.

By Theorem 5.8, we obtain

$$\frac{1}{2\pi i} \int\limits_{|\zeta - x_0| = \delta} \frac{F(\zeta) d\zeta}{\zeta - x_0} = \frac{1}{2\pi i} \text{p.v.} \int\limits_{\Gamma^+} \frac{F(\zeta) d\zeta}{\zeta - x_0} + \frac{1}{2\pi i} \text{p.v.} \int\limits_{\Gamma^-} \frac{F(\zeta) d\zeta}{\zeta - x_0}$$
$$= \frac{1}{2} F(x_0 + i0) + \frac{1}{2} F(x_0 - i0) = F(x_0).$$

Hence, the Cauchy integral formula yields that F is analytic also at x_0. □

Problem 16.6. Suppose that $f(z) = \sum_{j=0}^{\infty} a_j z^j$ and this series converges for $|z| < r$ and f is real for $x \in (-r, r)$. Show that all a_j are real and $f(z) = \overline{f(\overline{z})}$ for all $|z| < r$.

There is a key question at this point: Is there, in fact, a conformal mapping from a given domain D to some other domain, e. g., unit disk? The theoretical answer is the celebrated *Riemann mapping theorem*, which we give without a proof.

Theorem 16.7 (Riemann). *If D is any simply-connected domain, not equal to the whole complex plane $\mathbb{C}$, then there exists a conformal mapping of D onto $\{w : |w| < 1\}$. This mapping is uniquely determined by the value $f(z_0)$ and $\arg f'(z_0)$ at one arbitrary point $z_0 \in D$, e. g., by the values $f(z_0) = 0$ and $f'(z_0) > 0$.*

Remark. The assumption that the domain D is not equal to the entire complex plane $\mathbb{C}$ is essential. Indeed, if we assume that there exists a conformal mapping $f(z)$ of the complex plane $\mathbb{C}$ onto the unit disk $\{w : |w| < 1\}$, then $f(z)$ is bounded entire function. Hence, due to Liouville theorem $f \equiv$ constant and $f'(z) \equiv 0$. The same is true if $D = \mathbb{C} \setminus \{z_0\}$ with some fixed point $z_0 \in \mathbb{C}$ since z_0 is a removable singularity for $f(z)$ therefore again $f(z) \equiv$ constant and $f'(z) \equiv 0$. That is why the equivalent formulation of the Riemann mapping theorem includes the assumption that the boundary of $D \subset \mathbb{C}$ has more than two points.

Example 16.8. Let $w = f(z) = z^n$, $n \in \mathbb{N}$, $n > 1$ be a power function. Then this function carries out one-to-one mapping of the domain of its univalence, i. e., the sector $\phi < \arg z < \phi + \frac{2\pi}{n}$ (ϕ is fixed), onto the extended w-plane cut along the ray $\arg z = n\phi$. Its derivative $f'(z) = nz^{n-1}$ is nonzero and is bounded everywhere within the given sector with the exception of $z = 0$ and $z = \infty$. Thus, this function establishes the conformal mapping.

The general power function $w = f(z) = z^a$, $a \in \mathbb{R}_+$ maps the sector $\frac{2\pi k}{a} < \arg z < \frac{2\pi(k+1)}{a}$, $k \in \mathbb{Z}$ of its many branches (infinitely many for irrational a and finitely many for rational a) onto the extended w-plane (the ray $\arg z = \frac{2\pi k}{a}$ is mapped onto the positive real semiaxis). Moreover, this function maps the given sector conformally onto the cut w-plane.

Example 16.9. Let us show that the logarithm function $w = f(z) = \log z$ maps conformally the sector $\phi_1 < \arg z < \phi_2$ onto infinite strip $\phi_1 < \operatorname{Im} w < \phi_2$. Indeed, if $z = e^{i \arg z}$, then the value of the main branch of w is equal to

$$f(z) = \log r + i \arg z, \quad \operatorname{Re} w = \log r, \quad \operatorname{Im} w = \arg z.$$

When r changes from 0 to $+\infty$, then $\operatorname{Re} w = \log r$ changes from $-\infty$ to $+\infty$. Hence, the given sector is transformed onto the infinite strip conformally if we use the main branch of $\operatorname{Log} z$.

Example 16.10. Let $f(z) = e^{i\frac{\pi}{a}z}$, $a > 0$. Then f maps $\{z : 0 < \operatorname{Re} z < a\}$ onto $\{w : \operatorname{Im} w > 0\}$ conformally. Indeed, if $z = x + iy$, $0 < x < a$ then

$$e^{i\frac{\pi}{a}(x+iy)} = e^{-\frac{\pi}{a}y} e^{i\frac{\pi}{a}x} = e^{-\frac{\pi}{a}y}\left(\cos\frac{\pi}{a}x + i\sin\frac{\pi}{a}x\right)$$

so that $|f(z)| = e^{-\frac{\pi}{a}y} \in (0, +\infty)$ if $y \in \mathbb{R}$ and $\arg f(z) = \frac{\pi}{a}x \in (0, \pi)$ if $0 < x < \pi$. Since in addition $f'(z) = i\frac{\pi}{a}e^{i\frac{\pi}{a}z} \neq 0$ for all z and f is one-to-one transformation, then f is conformal.

Example 16.11. Consider a linear-fractional transformation

$$w = f(z) = \frac{az + b}{cz + d}, \quad ad - bc \neq 0, c \neq 0.$$

We call it a *nondegenerate* (or *regular*) linear-fractional transformation. This transformation is well-defined and analytic everywhere on $\overline{\mathbb{C}} \setminus \{-d/c\}$. Its derivative is equal to

$$f'(z) = \frac{ad - bc}{(cz + d)^2}, \quad z \neq -d/c, \quad f'(\infty) = \frac{bc - ad}{c^2}$$

and it is not equal to zero everywhere on $\overline{\mathbb{C}} \setminus \{-d/c\}$. We know that f maps $\overline{\mathbb{C}}$ onto $\overline{\mathbb{C}}$ bijectively (see Example 2.7). So, f is conformal. Let us represent it in the form

$$f(z) = \begin{cases} \lambda \frac{\alpha+z}{\beta+z}, & a \neq 0, \\ \lambda \frac{1}{\beta+z}, & a = 0, \end{cases} \tag{16.5}$$

where $\lambda = a/c$, $\alpha = b/a$, and $\beta = d/c$ if $a \neq 0$ and $\lambda = b$, $\beta = d/c$ if $a = 0$. The following theorem holds.

Theorem 16.12. *If $z_1 \neq z_2$, $z_2 \neq z_3$, $z_1 \neq z_3$, and $w_1 \neq w_2$, $w_2 \neq w_3$, $w_1 \neq w_3$ then the correspondence*

$$z_j \to w_j, \quad j = 1, 2, 3$$

defines uniquely a nondegenerate linear-fractional transformation (16.5), *where $a \neq 0$. Moreover,*

$$\lambda = \frac{Aw_2 - Bw_1}{A - B}, \quad \alpha = \frac{Bw_1z_2 - Aw_2z_1}{Aw_2 - Bw_1}, \quad \beta = \frac{Bz_2 - Az_1}{A - B}, \tag{16.6}$$

where

$$A = \frac{w_1 - w_3}{w_2 - w_3}, \quad B = \frac{z_1 - z_3}{z_2 - z_3}.$$

Proof. Using (16.5) for $a \neq 0$, we have

$$w_1 - w_3 = \lambda \frac{(z_1 - z_3)(\alpha - \beta)}{(\beta + z_1)(\beta + z_3)}, \quad w_2 - w_3 = \lambda \frac{(z_2 - z_3)(\alpha - \beta)}{(\beta + z_2)(\beta + z_3)}.$$

Here, $\beta \neq \alpha$ since $ad \neq bc$. These equalities imply that

$$\frac{w_1 - w_3}{w_2 - w_3} = \frac{z_1 - z_3}{z_2 - z_3} \frac{\beta + z_2}{\beta + z_1}$$

or

$$\beta = \frac{z_2 - z_1 \frac{w_1 - w_3}{w_2 - w_3} \frac{z_2 - z_3}{z_1 - z_3}}{\frac{w_1 - w_3}{w_2 - w_3} \frac{z_2 - z_3}{z_1 - z_3} - 1} = \frac{z_2 - \frac{A}{B} z_1}{\frac{A}{B} - 1} = \frac{Bz_2 - Az_1}{A - B}.$$

It proves (16.6) for β. Next,

$$w_1 = \lambda \frac{\alpha + z_1}{\beta + z_1}, \quad w_2 = \lambda \frac{\alpha + z_2}{\beta + z_2}$$

imply that

$$\frac{w_1}{w_2} = \frac{\alpha + z_1}{\beta + z_1}\frac{\beta + z_2}{\alpha + z_2}$$

or

$$\alpha = \frac{w_2 z_1(\beta + z_2) - w_1 z_2(\beta + z_1)}{w_1(\beta + z_1) - w_2(\beta + z_2)} = \frac{Bw_1 z_2 - Aw_2 z_1}{Aw_2 - Bw_1} = \frac{Aw_2 z_1 - Bw_1 z_2}{Bw_1 - Aw_2}.$$

This proves (16.6) for α. Finally,

$$\lambda = \frac{w_1(\beta + z_1)}{\alpha + z_1} = \frac{w_1(\frac{Bz_2 - Az_1}{A-B} + z_1)}{\frac{Aw_2 z_1 - Bw_1 z_2}{Bw_1 - Aw_2} + z_1} = \frac{w_1\frac{B(z_2 - z_1)}{A-B}}{\frac{Bw_1 z_1 - Bw_1 z_2}{Bw_1 - Aw_2}} = \frac{Aw_2 - Bw_1}{A - B}$$

proving the claim for λ. The formulae (16.6) show that α, β, and λ are uniquely determined by the correspondence $z_j \to w_j$, $j = 1, 2, 3$ if z_j and w_j are mutually distinct points. □

Corollary 16.13. *If we denote*

$$w := f(z) = \lambda\frac{\alpha + z}{\beta + z}, \quad a \neq 0,$$

then Theorem 16.12 *says that*

$$\frac{w_1 - w_3}{w_2 - w_3} : \frac{w_1 - w}{w_2 - w} = \frac{z_1 - z_3}{z_2 - z_3} : \frac{z_1 - z}{z_2 - z}. \tag{16.7}$$

Proof. As it is proved in Theorem 16.12,

$$\frac{w_1 - w_3}{w_2 - w_3} = \frac{z_1 - z_3}{z_2 - z_3}\frac{\beta + z_2}{\beta + z_1}.$$

Similarly, if $w \neq w_1$, $w \neq w_2$ and $z \neq z_1$, $z \neq z_2$, we obtain

$$\frac{w_1 - w}{w_2 - w} = \frac{z_1 - z}{z_2 - z}\frac{\beta + z_2}{\beta + z_1}.$$

Hence, (16.7) follows straightforwardly from the latter equalities. □

Corollary 16.14. *For the case $a = 0$ ($b \neq 0$ necessarily) instead of three correspondences it is enough to have only two different points $z_1 \neq z_2$ and $w_1 \neq w_2$, respectively, with $w_1 \neq 0$ and $w_2 \neq 0$. In that case (see* (16.5)*),*

$$\beta = \frac{w_1 z_1 - w_2 z_2}{w_2 - w_1}, \quad \lambda = \frac{w_1 w_2(z_1 - z_2)}{w_2 - w_1}. \tag{16.8}$$

Problem 16.15. Show (16.8) for the case $a = 0$, $b \neq 0$, $c \neq 0$ in the non-degenerate linear-fractional transformation.

Example 16.16. Let us find $w = f(z)$, which is a conformal mapping of the unit disk $\{z : |z| < 1\}$ onto the domain $\{w : \operatorname{Im} w > 0\}$. Let z_j and w_j be as in the Figure 16.3. By

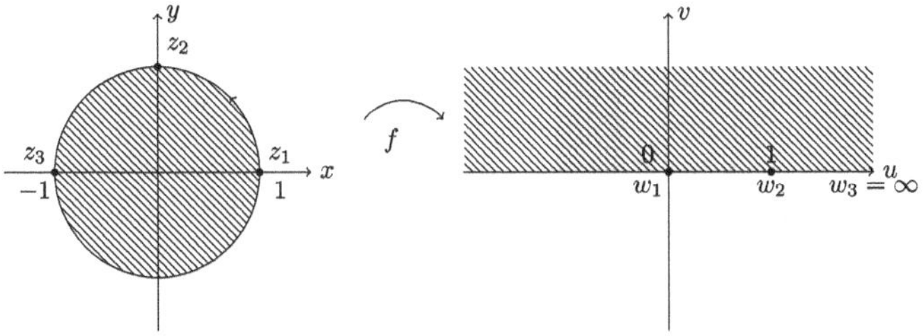

Figure 16.3: Domains in Example 16.16.

Theorem 16.12 and Theorem 16.3, we have

$$\frac{0-\infty}{1-\infty} : \frac{0-w}{1-w} = \frac{1+1}{i+1} : \frac{1-z}{i-z}.$$

So

$$\frac{w-1}{w} = \frac{2}{i+1}\frac{z-i}{z-1}$$

or

$$w = e^{-i\frac{\pi}{2}}\frac{z-1}{z+1}.$$

Problem 16.17. Using Example 16.16, show that

$$w = \frac{z+e^{-i\pi/2}}{e^{-i\pi/2}-z}$$

maps conformally the domain $\{z : \operatorname{Im} z > 0\}$ onto the unit disk $\{w : |w| < 1\}$. Find all linear-fractional transformations, which map the domain $\{z : \operatorname{Im} z > 0\}$ onto the unit disk $\{w : |w| < 1\}$.

Problem 16.18.

1. Investigate the conformal properties of the function $w = \tan z$. Hint: Represent $\tan z$ in the form

$$w = \frac{1}{i}\frac{e^{2iz}-1}{e^{2iz}+1}.$$

2. Prove that the function $w = \tan^2 z$ maps conformally infinite strip $\{z : 0 < \operatorname{Re} < \frac{\pi}{2}\}$ onto the unit disk $\{w : |w| < 1\}$ with the cut along the real axis from $w = -1$ to $w = 0$ (twice in different directions).

3. Prove that the function

$$w = \left(\frac{1+z}{1-z}\right)^2$$

maps conformally the upper half of the disk $\{z : |z| < 1, \operatorname{Im} z > 0\}$ onto the upper half-plane $\{w : \operatorname{Im} w > 0\}$.

Problem 16.19. Show that

$$w = f(z) = \mathrm{e}^{\mathrm{i}\alpha} \frac{z - z_0}{z\overline{z_0} - 1}$$

maps conformally the unit disk $\{z : |z| < 1\}$ onto the unit disk $\{w : |w| < 1\}$ such that an arbitrary point $z_0, |z_0| < 1$ is transferred to $w_0 = 0$ and α is an arbitrary real parameter. Show that if $\arg f'(z_0)$ is prescribed then α is uniquely determined.

Example 16.20. Assume that the function $w = f(z)$ is univalent and analytic, and maps unit disk $\{z : |z| \le 1\}$ onto unit disk $\{w : |w| \le 1\}$ such that $z_0 = 0$ transfers to $w_0 = 0$ and $f'(0) = 1$. Then necessarily $f(z) \equiv z$. Indeed, since $|f(z)| \le 1$ for $|z| = 1$ and $f(0) = 0$, then the Schwarz's lemma (see Corollary 7.11) says that

$$|w| = |f(z)| \le |z|, \quad |z| \le 1.$$

Applying this lemma to the inverse function $z = f^{-1}(w)$, we obtain that

$$|z| = |f^{-1}(w)| \le |w|, \quad |w| \le 1.$$

It means that actually

$$|f(z)| = |z|, \quad |z| \le 1,$$

and thus

$$f(z) = az, \quad |a| = 1.$$

But since $f'(0) = 1$, we obtain that $a = 1$. It might be noted here that we did not assume from the beginning that $f(z)$ is a linear-fractional transformation. However, it is not so difficult to show that any univalent and analytic function, which maps the unit disk onto the unit disk is necessarily a linear-fractional transformation.

Problem 16.21. Prove the latter assertion.

Problem 16.22. Show that a nondegenerate linear-fractional transformation maps lines and circles on the extended complex plane onto lines or circles.

Problem 16.23. Find the conditions on $0 < r_1 < r_2$ and $0 < R_1 < R_2$, which guarantee the existence of the conformal mapping of the annulus $\{z : r_1 < |z| < r_2\}$ onto the annulus $\{w : R_1 < |w| < R_2\}$.

Example 16.24. Consider a nonconcentric ring (annulus), i. e., the set which is formed by two circles $\{z : |z - a_1| = R_1\}$ and $\{z : |z - a_2| = R_2\}$ such that $0 < R_2 < R_1$ and the first circle is located inside of the second one. We assume without loss of generality that a_1 and a_2 are real; see Figure 16.4.

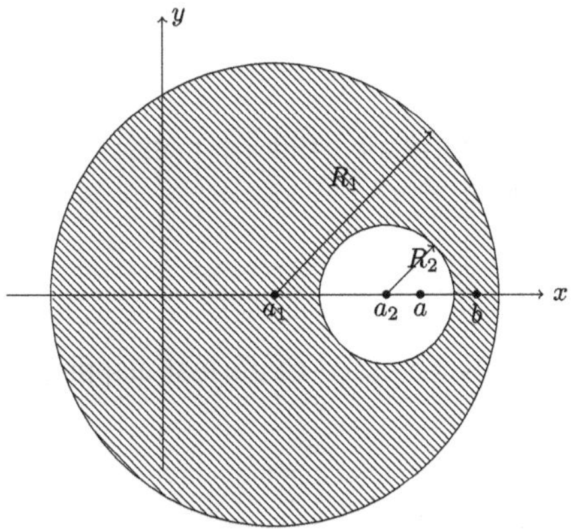

Figure 16.4: Two nonconcentric circles in Example 16.24.

The task is to map conformally this annulus onto the domain $\{w : \operatorname{Im} w > 0\}$. Let now a and b be two real numbers such that they are symmetric with respect to the first and second circle at the same time, i. e., they satisfy the equations

$$(a - a_1)(b - a_1) = R_1^2, \quad (a - a_2)(b - a_2) = R_2^2. \tag{16.9}$$

Solving these equations, we can easily obtain a and b uniquely ($a < b$). Then the map

$$w_1 = \frac{z - a}{z - b}$$

transfers conformally the given nonconcentric ring to the concentric one centered at 0. Indeed, if $z - a_1 = R_1 e^{i\varphi}$, then

$$\begin{aligned} w_1 &= \frac{z - a}{z - b} = \frac{(z - a_1) - (a - a_1)}{(z - a_1) - (b - a_1)} = \frac{R_1 e^{i\varphi} - (a - a_1)}{R_1 e^{i\varphi} - \frac{R_1^2}{a - a_1}} \\ &= \frac{a - a_1}{R_1} \frac{R_1 e^{i\varphi} - (a - a_1)}{(a - a_1) - R_1 e^{-i\varphi}} e^{-i\varphi}. \end{aligned}$$

This equality implies that

$$|w_1|\big|_{|z-a_1|=R_1} = \left|\frac{a-a_1}{R_1}\right| = \frac{a-a_1}{R_1} =: r_1.$$

Similarly, we obtain that

$$|w_1|\big|_{|z-a_2|=R_2} = \left|\frac{a-a_2}{R_2}\right| = \frac{a-a_2}{R_2} =: r_2.$$

Let us note that for $0 < R_2 < R_1$ it follows that $r_2 < r_1$ since $b > a$. The next step is: We consider

$$w_2 = \log w_1$$

with the main branch of logarithm. Under this transformation, this symmetric (or concentric) annulus is transferred conformally to the set

$$\{w_2 : \log r_2 < \operatorname{Re} w_2 < \log r_1\}.$$

Using now Example 16.10, we may conclude that the required conformal mapping is given by

$$w = e^{i\frac{\pi}{\log(r_1/r_2)}}\left(\log\frac{z-a}{z-b} - \log r_2\right),$$

where a and b are from (16.9).

Example 16.25. Let us find the conformal mapping of the crescent shape (lune) formed by two arcs of two different circles shown in Figure 16.5.

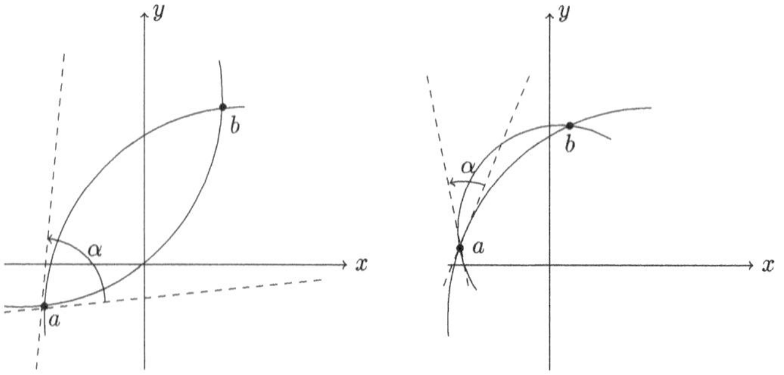

Figure 16.5: Mapping a lune.

We consider first

$$w_1 = \frac{z-a}{z-b},$$

where a and b are the two intersecting points of these circles. Then this conformal mapping transfers this lune to the angle of span α (this angle is the same as for the lune due to conformality), with the vertex in the origin (Figure 16.6).

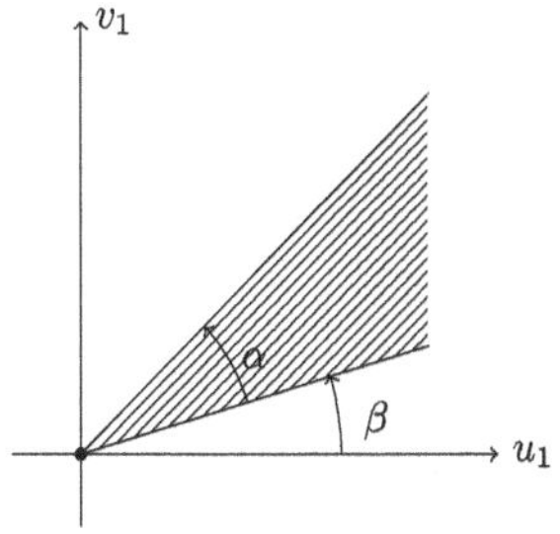

Figure 16.6: Sector given by angles α and β.

Indeed, if $z = \rho_0 e^{i\varphi}$, $\varphi_0 \le \varphi \le \varphi_0'$ for the part of the first circle in the boundary of the lune and $z = \rho_1 e^{i\varphi}$, $\varphi_1 \le \varphi \le \varphi_1'$ for the second circle, then

$$\left.\frac{z-a}{z-b}\right|_{z=\rho_0 e^{i\varphi}} = \frac{\rho_0 e^{i\varphi} - \rho_0 e^{i\varphi_0}}{\rho_0 e^{i\varphi} - \rho_0 e^{i\varphi_0'}} = \frac{e^{i\varphi} - e^{i\varphi_0}}{e^{i\varphi} - e^{i\varphi_0'}}$$
$$= e^{i(\varphi_0-\varphi_0')} \frac{e^{i(\varphi-\varphi_0)/2} - e^{-i(\varphi-\varphi_0)/2}}{e^{i(\varphi-\varphi_0')/2} - e^{-i(\varphi-\varphi_0')/2}} = e^{i(\varphi_0-\varphi_0')} \frac{\sin(\varphi-\varphi_0)/2}{\sin(\varphi-\varphi_0')/2}.$$

Similarly,

$$\left.\frac{z-a}{z-b}\right|_{z=\rho_1 e^{i\varphi}} = e^{i(\varphi_1-\varphi_1')} \frac{\sin(\varphi-\varphi_1)/2}{\sin(\varphi-\varphi_1')/2}.$$

These formulae show that the arcs are mapped to the rays starting from the origin because

$$\frac{v_1}{u_1} = \tan(\varphi_0 - \varphi_0'), \quad \frac{v_1}{u_1} = \tan(\varphi_1 - \varphi_1'),$$

respectively, for these two arcs. Next,

$$w_2 = e^{-i\beta} w_1$$

maps conformally the sector $\{w_1 : \beta < \arg w_1 < \alpha + \beta\}$ to the sector $\{w_2 : 0 < \arg w_2 < \alpha\}$, see Figure 16.7.

Finally,

$$w = w_2^{\pi/\alpha} = \left(e^{-i\beta}\frac{z-a}{z-b}\right)^{\pi/\alpha}$$

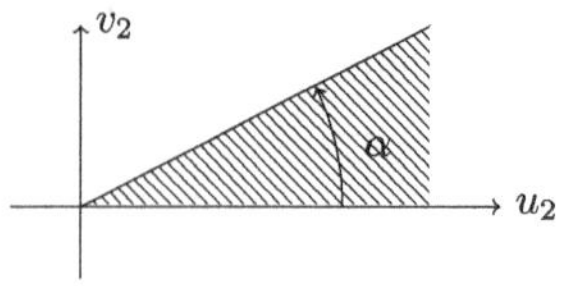

Figure 16.7: Sector given by angle α.

maps conformally the latter sector onto the domain $\{w : \operatorname{Im} w > 0\}$. Indeed,

$$w = w_2^{\pi/\alpha} = e^{\frac{\pi}{\alpha}(\log|w_2| + i\arg w_2)} = e^{i\frac{\pi}{\alpha}\arg w_2} e^{\frac{\pi}{\alpha}\log|w_2|}.$$

This is equivalent that $\arg w = \frac{\pi}{\alpha}\arg w_2 \in (0, \pi)$, and $\operatorname{Re} w \in (-\infty, \infty)$, $\operatorname{Im} w > 0$. Here, we have used the boundary correspondence principle.

Example 16.26. Consider the Zhukovski function (see Example 2.13) $w = \frac{1}{2}(z + \frac{1}{z})$. Then any circle $\{z : |z| = r_0\}, r_0 \neq 1$ is transformed to the value $w = u + iv$ such that

$$u(r_0, \phi) = \frac{1}{2}\left(r_0 + \frac{1}{r_0}\right)\cos\phi, \quad v(r_0, \phi) = \frac{1}{2}\left(r_0 - \frac{1}{r_0}\right)\sin\phi.$$

Eliminating the angle ϕ, we obtain

$$\frac{u^2}{1/4(r_0 + \frac{1}{r_0})^2} + \frac{v^2}{1/4(r_0 - \frac{1}{r_0})^2} = 1.$$

Thus, the Zhukovski function maps the family of the concentric circles onto the family of confocal ellipses, and this map is conformal.

Let us now find the image of the rays $\{z : \arg z = \phi\}$ under the mapping defined by the Zhukovski function. To do this, eliminate from the previous relations for u and v the parameter r_0 and put $\phi = \phi_0$. Then

$$\frac{u^2}{\cos^2\phi_0} - \frac{v^2}{\sin^2\phi_0} = 1, \quad \phi_0 \neq 0, \frac{\pi}{2}, \pi, \frac{3\pi}{2}.$$

This relation shows that the Zhukovski function transforms segments of these rays into branches of the hyperbola. To sum up, this function defines a transformation of the orthogonal system of polar coordinates in the z-plane into an orthogonal curvilinear system of coordinates in the w-plane whose coordinate lines are the confocal families of ellipses and hyperbolas.

Problem 16.27. Find the curve in w-plane, which is the image of the line $\{z : \operatorname{Re} z = 1\}$ in z-plane under the mapping defined by the Zhukovski function.

Problem 16.28. Show that the Zhukovski function maps conformally:

1. $\{z : |z| < 1\}$ onto $\mathbb{C} \setminus [-1, 1]$,

2. $\{z : |z| < 1, \operatorname{Im} z < 0\}$ onto $\{w : \operatorname{Im} w > 0\}$ and
3. $\{z : |z| > 1\}$ onto $\mathbb{C} \setminus [-1, 1]$.

Example 16.29. Let function $w = f(z)$ be defined using the Cauchy type integral

$$f(z) := \int_0^z \frac{d\zeta}{\sqrt{1-\zeta^2}},$$

where the integration goes from 0 to z (independently on the curve) if $\zeta \neq \pm 1$ do not belong to this curve. This function (in general) is multivalued due to the square root. If we consider the first quadrant of the z-plane, then this function is univalent there and let us consider the image of this part of z-plane under the mapping of $f(z)$. If $\zeta = \mathrm{i}s$, $z = \mathrm{i}y$ with $s, y \in \mathbb{R}_+$, then we have that

$$w = \mathrm{i} \int_0^y \frac{ds}{\sqrt{1+s^2}},$$

is pure imaginary, and for $y \to +\infty$ we have that $\operatorname{Im} w \to +\infty$. If $z = x + \mathrm{i}0$, $x \in (0,1)$, then w is also real and changes from 0 to

$$\int_0^1 \frac{dt}{\sqrt{1-t^2}} = \frac{\pi}{2}.$$

Next, if $z = x + \mathrm{i}0$ and $x \to +\infty$, then

$$w = \frac{\pi}{2} + \mathrm{i} \int_1^x \frac{ds}{\sqrt{s^2-1}} = u + \mathrm{i}v$$

so that $v \to +\infty$. Using now the boundary correspondence principle (see Theorem 16.3), we may conclude that this function $f(z)$ maps conformally the first quadrant of z-plane onto the infinite strip $\{w : 0 < \operatorname{Re} w < \frac{\pi}{2}, \operatorname{Im} w > 0\}$.

Schwarz–Christoffel integral

In the complex w-plane, let us consider a polygon (n-gon) with vertices at the points $A_1, A_2, \ldots, A_n$ and with interior angles at these vertices $\alpha_1\pi, \alpha_2\pi, \ldots, \alpha_n\pi$, respectively. Obviously,

$$\sum_{j=1}^{n} \alpha_j = n - 2.$$

We assume that $0 < \alpha_j < 2$ (for $0 < \alpha_j < 1$ this polygon is convex). We want to construct a conformal mapping of the upper half-plane onto the interior of such polygon. This problem is solved by means of the so-called *Schwarz–Christoffel integral*, i. e., by the integral of the form

$$f(z) := C\int_{z_0}^{z}(\zeta-a_1)^{\alpha_1-1}(\zeta-a_2)^{\alpha_2-1}\cdots(\zeta-a_n)^{\alpha_n-1}\mathrm{d}\zeta + C_1,$$

where $z, z_0 \in \{z : \operatorname{Im} z > 0\}$, C, C_1 are given complex constants and $a_1, a_2, \ldots, a_n$ are real numbers arranged in the increasing order. In the integrand, we choose those branches of functions $(\zeta - a_j)^{\alpha_j-1}$, which are equal to $(x-a_j)^{\alpha_j-1}$ with real variable $x > a_j$. In that case, this function $f(z)$ is a single-valued analytic function in the upper half-plane. We will show that if a_j are chosen appropriately then $w = f(z)$ defines a conformal mapping of the upper half-plane $\{z : \operatorname{Im} z > 0\}$ onto the interior of this polygon.

First, we show that $f(z)$ is bounded for all $z, \operatorname{Im} z \geq 0$. Since $\alpha_j > 0$, the integral that defines $f(z)$ is bounded in the neighborhood of the singularities a_j. Moreover, for $|\zeta| \to \infty$ we can write the integrand in the form

$$\frac{1}{\zeta^2}\left(1-\frac{a_1}{\zeta}\right)^{\alpha_1-1}\left(1-\frac{a_2}{\zeta}\right)^{\alpha_2-1}\cdots\left(1-\frac{a_n}{\zeta}\right)^{\alpha_n-1}.$$

From this expression, it follows that the integral is convergent as $z \to \infty$, $\operatorname{Im} z \geq 0$. Thus, function $w = f(z)$ defines a mapping of this half-plane onto some bounded domain D. Next, we have

$$f'(z) = C(z-a_1)^{\alpha_1-1}(z-a_2)^{\alpha_2-1}\cdots(z-a_n)^{\alpha_n-1}.$$

It means that $f'(z) \neq 0$ everywhere in the upper half-plane except at the points $z = a_j$, $j = 1, 2, \ldots, n$. If $z = x + \mathrm{i}0$ is real and varies on every one of the intervals $a_j < x < a_{j+1}$, $j = 1, 2, \ldots, n-1$ of the real axis, then the argument of the derivative does not change because

$$\arg(x-a_j)^{\alpha_j-1} = \begin{cases} 0, & x > a_j, \\ \pi(\alpha_j - 1), & x < a_j. \end{cases}$$

In view of the geometric meaning of the argument of derivative (see, e. g., (16.1)), this equality shows that the segments $a_j < x < a_{j+1}$ are mapped by $f(z)$ onto rectilinear segments. The points a_j are transformed by $f(z)$ into points A_j—the ends of the corresponding straight-line segments A_jA_{j+1} into which the function $f(z)$ maps $[a_j, a_{j+1}]$ of the real axis. Moreover, when the point z traverses the entire real axis in the positive direction, the point $w = f(z)$ corresponding to it makes a complete circuit of the closed polygonal line $A_1A_2\cdots A_n$. Now let us determine the size of the angles between adjacent segments of the polygonal line obtained. Considering the variation of $\arg f(z)$ as z passes through the points a_j along the real axis in the positive direction (during which the singular point a_j is taken around an arc of infinitesimal radius in the upper half-plane) the argument changes its value $\pi(1-\alpha_j)$. It means that the angle between the vectors $\overline{A_{j-1}A_j}$ and $\overline{A_jA_{j+1}}$ is equal to $\pi(1-\alpha_j), \alpha_j < 1$. It corresponds to the angle $\pi\alpha_j$ of this polygon. Taking into account the behavior of the integrand as $\zeta \to \infty$, we may conclude finally that if z goes through the real axis, function $w = f(z)$ describes a closed curve, which is a polygon with the angles $\alpha_1\pi, \alpha_2\pi, \dots, \alpha_n\pi$, and the upper half-plane transforms to the interior of this polygon. Moreover, this map is conformal due to the fact that $f'(z) \neq 0$. These considerations can be extended to the case when one (or several) of the points a_j is infinite.

Consider some examples.

Example 16.30. Let us find a function that conformally maps the upper half-plane $\{z : \operatorname{Im} z > 0\}$ onto the sector $\{w : 0 < \arg w < \alpha\pi, 0 < \alpha < 2\}$. Since the given sector is a polygon with vertices $A_1(w = 0)$ and $A_2(w = \infty)$, the Schwarz–Christoffel integral takes the form

$$w = f(z) = C\int_{z_0}^{z} \zeta^{\alpha-1} d\zeta + C_1.$$

Putting $z_0 = 0$, we find that $C_1 = 0$ and, therefore,

$$w = \frac{C}{\alpha} z^{\alpha}.$$

Requiring now, e. g., that in addition to two points $z = 0$ and $z = \infty$ there occurs a supplementary correspondence of boundary points as follows:

$$z = 1 \to w = 1,$$

we determine that $C = \alpha$ and, finally, $w = z^{\alpha}$.

Example 16.31. Let us find a function that conformally maps the upper half-plane $\{z : \operatorname{Im} z > 0\}$ onto the rectangle with vertices $A_1(w = a)$, $A_2(w = a + ib)$, $A_3(w = -a + ib)$ and $A_4(w = -a)$. Consider now only one part of the interior of this rectangle, which is located in the first quadrant of the complex w-plane, i. e., rectangle with the vertices $O(w = 0)$, $O'(w = ib)$ and A_1, A_2 from above. Therefore, we can establish the following correspondence of the points (boundary points):

$$z = 1 \to w = a, \quad z = 0 \to w = 0, \quad z = \frac{1}{k} \to w = a + ib,$$

where the parameter $0 < k < 1$ is to be determined. Then on the basis of the Schwarz reflection principle (see Theorem 16.5) the function, which defines a conformal mapping of the upper half-plane onto a given rectangle may be represented as follows:

$$w = f(z) = C \int_0^z (\zeta - 1)^{-1/2} \left(\zeta - \frac{1}{k}\right)^{-1/2} (\zeta + 1)^{-1/2} \left(\zeta + \frac{1}{k}\right)^{-1/2} d\zeta$$

$$= C' \int_0^z \frac{d\zeta}{\sqrt{(1-\zeta^2)(1-k^2\zeta^2)}}.$$

It remains to determine the constants C' and k'. First, we have that (see the correspondence of the boundary points)

$$a = C' \int_0^1 \frac{d\zeta}{\sqrt{(1-\zeta^2)(1-k^2\zeta^2)}}, \quad \zeta \in \mathbb{R}.$$

Second, we have similarly that

$$a + ib = C' \left(\int_0^1 \frac{d\zeta}{\sqrt{(1-\zeta^2)(1-k^2\zeta^2)}} + i \int_1^{1/k} \frac{d\zeta}{\sqrt{(\zeta^2-1)(1-k^2\zeta^2)}} \right), \quad \zeta \in \mathbb{R}.$$

Solving this transcendental system with respect to C' and k, we obtain the needed result. The integrals from above are called *elliptic integrals of the first kind*.

Problem 16.32.

1. Show that the function

$$w = f(z) = C \int_0^z (1+\zeta)^{-5/6} \zeta^{-1/2} (1-\zeta)^{-2/3} d\zeta,$$

   where

$$\frac{1}{C} = \int_0^1 (1+\zeta)^{-5/6} \zeta^{-1/2} (1-\zeta)^{-2/3} d\zeta, \quad \zeta \in \mathbb{R},$$

   maps conformally the upper half-plane onto the triangle with the vertices $w = i\sqrt{3}$, $w = 0$, $w = 1$.

2. Prove that the function

$$w = f(z) = \int_0^z \frac{d\zeta}{(1-\zeta^2)^{2/3}}$$

   maps conformally the upper half-plane onto a triangle and find the triangle.

3. Prove that the function

$$w = f(z) = \int_0^z \frac{d\zeta}{\sqrt{1-\zeta^4}}$$

maps conformally the unit disk onto a quadrant and find the quadrant.

Problem 16.33.

1. Let function $f(z)$ be analytic on the closed unit disk $\{z : |z| \leq 1\}$. Prove that if

$$|f(z)| \leq M, \quad |z| = 1, \quad \text{and} \quad f(a) = 0, \quad |a| < 1,$$

then

$$|f(z)| \leq M \frac{|z-a|}{|\bar{a}z - 1|}, \quad |z| \leq 1.$$

2. Let function $f(z)$ be analytic on the closed unit disk $\{z : |z| \leq 1\}$. Prove that if

$$|f(z)| \leq M, \quad |z| = 1, \quad \text{and} \quad f(0) = a, \quad 0 < a < M,$$

then

$$|f(z)| \leq M \frac{M|z| + a}{a|z| + M}, \quad |z| \leq 1.$$

There is an application of conformal mappings also in the theory of partial differential equations.

Let D be a simply-connected and bounded domain on the complex plane $\mathbb{C}$.

Definition 16.34. A function $G(z, \zeta)$ is said to be *Green's function* for the Laplace operator Δ in the domain D if the following conditions are satisfied:

1. $G(z, \zeta) = \frac{1}{2\pi} \log|z - \zeta| + g(z, \zeta)$ for $z, \zeta \in D$,
2. $\Delta_z g(z, \zeta) = 0$ for $z, \zeta \in D$,
3. $g(z, \zeta) = -\frac{1}{2\pi} \log|z - \zeta|$ for $z \in \partial D, \zeta \in D$.

Remark. This definition implies (in particular) that $G(z, \zeta) = 0$ for $z \in \partial D$ and $\zeta \in D$.

With the Green's function in hand, the solution of the *inhomogeneous boundary value problem*

$$\begin{cases} \Delta u(z) = F(z), & z \in D, \\ u(z) = u_0(z), & z \in \partial D \end{cases}$$

is given by the superposition principle as

$$u(z) = \int_D G(z,\zeta)F(\zeta)d\xi d\eta + \int_{\partial D} \partial_{\nu_\zeta} G(z,\zeta)d\sigma(\xi,\eta),$$

where $z = x + iy$, $\zeta = \xi + i\eta$, and ∂_ν is the outward normal derivative with respect to ζ on the boundary ∂D.

Using the principles of conformal mappings, we may construct the Green's function for arbitrary simply-connected bounded domain D. Indeed, let ζ be an arbitrary fixed point from D. Let $h(z,\zeta)$ be a function, which maps conformally D onto the unit disk $\{w : |w| < 1\}$ such that $h(\zeta,\zeta) = 0$. This function exists due to Riemann mapping theorem (see Theorem 16.7). Moreover, $h'_z(z,\zeta) \neq 0$ for all $z \in D$ (see Problem 12.17 and Theorem 16.2). Hence, $h(z,\zeta)$ has a zero of order 1 at $z = \zeta$. This fact allows us to represent $h(z,\zeta)$ in the form

$$h(z,\zeta) = (z-\zeta)\psi(z,\zeta), \quad \psi(\zeta,\zeta) \neq 0.$$

It implies that

$$\frac{1}{2\pi}\log\left|h(z,\zeta)\right| = \frac{1}{2\pi}\log|z-\zeta| + g(z,\zeta),$$

where $g(z,\zeta) = \frac{1}{2\pi}\log|\psi(z,\zeta)|$. We prove that

$$G(z,\zeta) := \frac{1}{2\pi}\log\left|h(z,\zeta)\right|$$

is the Green's function for Δ in D. Indeed, since $h(z,\zeta) \in H(D)$ (ζ is a parameter) and $h'_z(z,\zeta) \neq 0$ for all $z \in D$, then $\psi(z,\zeta) \neq 0$ for all $z \in D$ and analytic there. Thus, $g(z,\zeta) = \frac{1}{2\pi}\log|\psi(z,\zeta)|$ is harmonic in D (see Problem 7.2). Next, since $|h(z,\zeta)| = 1$ for all $z \in \partial D$ and for all $\zeta \in D$ (see Theorem 16.7) then

$$g(z,\zeta) = -\frac{1}{2\pi}\log|z-\zeta|, \quad z \in \partial D, \zeta \in D.$$

This proves that $G(z,\zeta)$ is the needed Green's function.

Problem 16.35. Show that the Green's function for the unit disk is given by

$$G(z,\zeta) = \frac{1}{2\pi}\log\left|\frac{z-\zeta}{z\bar{\zeta}-1}\right|.$$

Hint: Use the fact that a nondegenerate linear-fractional transformation

$$w = \frac{z-\zeta}{z\bar{\zeta}-1}, \quad |\zeta|, |z| < 1$$

maps conformally unit disk onto itself such that $w = 0$ for $z = \zeta$.

Problem 16.36. Using Problem 16.35 show that the Green's function for simply-connected bounded domain D can be written as

$$G(z,\zeta) = \frac{1}{2\pi}\log\left|\frac{g(\zeta) - g(z)}{g(z)\overline{g(\zeta)} - 1}\right|,$$

where g maps conformally D onto the unit disk.

17 Laplace transform

Let f be a function (possibly complex-valued) of one real variable t. We denote by $\mathcal{F}^+$ the class of functions (and write $f \in \mathcal{F}^+$), which satisfy the conditions:

1. $f(t) \equiv 0, t < 0$,
2. $f(t)$ is continuous for $t \geq 0$,
3. there exist $M > 0$ and $a > 0$ such that $|f(t)| \leq M e^{at}$ for any $t \geq 0$.

The value $s := \inf a$ is called the *growth index* of f.

Problem 17.1. Show that if the growth index of $f \in \mathcal{F}^+$ is equal to $s \geq 0$ then the growth index of $t^\mu f(t)$ for any $\mu \geq 0$ is also equal to s. In particular, the growth index of t^μ for any $\mu > 0$ is equal to zero.

Definition 17.2. Let f be a function from the class $\mathcal{F}^+$. The *Laplace transform* of f, denoted by $\mathcal{L}(f)(p)$ is defined by

$$\mathcal{L}(f)(p) := \int_0^\infty e^{-pt} f(t) dt, \quad p \in \mathbb{C}. \tag{17.1}$$

Theorem 17.3 (Existence). *Suppose $f \in \mathcal{F}^+$ with growth index $s \geq 0$. Then the Laplace transform $\mathcal{L}(f)(p)$ is well-defined analytic function in the domain $\{p \in \mathbb{C} : \operatorname{Re} p > s\}$. Moreover,*

$$\lim_{\operatorname{Re} p \to +\infty} \mathcal{L}(f)(p) = 0 \tag{17.2}$$

uniformly with respect to $\operatorname{Im} p \in \mathbb{R}$.

Proof. Let $p = x + iy$ and $f \in \mathcal{F}^+$ with growth index $s \geq 0$. Then for any $\varepsilon > 0$ there is $M_\varepsilon > 0$ such that $|f(t)| \leq M_\varepsilon e^{(s+\varepsilon)t}$, $t \geq 0$. It implies for any fixed $x = \operatorname{Re} p > s$ that

$$\begin{aligned} |\mathcal{L}(f)(p)| &\leq \left| \int_0^\infty e^{-(x+iy)t} f(t) dt \right| \leq \int_0^\infty e^{-xt} |f(t)| dt \\ &\leq M_\varepsilon \int_0^\infty e^{-(x-s-\varepsilon)t} dt = \frac{M_\varepsilon}{x - s - \varepsilon} \end{aligned} \tag{17.3}$$

if ε is chosen such that $0 < \varepsilon < x - s$. This proves well-posedness of (17.1) for $\operatorname{Re} p > s$. In addition, (17.3) shows that the integral in (17.1) converges uniformly for all $x = \operatorname{Re} p \geq s_0 > s$. Let us prove now that $\mathcal{L}(f)(p)$ is analytic in the domain $\{p \in \mathbb{C} : \operatorname{Re} p > s\}$. If p_0 and Δp are chosen so that $\operatorname{Re} p_0, \operatorname{Re}(p_0 + \Delta p) > s$, then

https://doi.org/10.1515/9783111632278-019

$$\frac{\mathcal{L}(f)(p_0+\Delta p)-\mathcal{L}(f)(p_0)}{\Delta p}=\int_0^\infty \mathrm{e}^{-p_0t}f(t)\frac{\mathrm{e}^{-t\Delta p}-1}{\Delta p}\,\mathrm{d}t.$$

But it is known that

$$\lim_{\Delta p\to 0}\frac{\mathrm{e}^{-t\Delta p}-1}{\Delta p}=-t.$$

Due to this fact, Problem 17.1 and the fact that the integral in (17.1) converges uniformly for $\operatorname{Re} p \ge s_0 > s$, we may consider the limit $\Delta p \to 0$ under the integral sign. Hence, we obtain the existence of the limit

$$\begin{aligned}\lim_{\Delta p\to 0}\frac{\mathcal{L}(f)(p_0+\Delta p)-\mathcal{L}(f)(p_0)}{\Delta p}&=\int_0^\infty \mathrm{e}^{-p_0t}f(t)\lim_{\Delta p\to 0}\frac{\mathrm{e}^{-t\Delta p}-1}{\Delta p}\,\mathrm{d}t\\&=-\int_0^\infty \mathrm{e}^{-p_0t}tf(t)\,\mathrm{d}t=-\mathcal{L}(tf)(p_0).\end{aligned}$$

The latter formula proves the analyticity of $\mathcal{L}(f)(p)$ for all $\operatorname{Re} p > s$ and also the equality

$$\mathcal{L}(tf)(p)=-(\mathcal{L}(f))'(p). \tag{17.4}$$

Finally, (17.2) follows from (17.3) straightforwardly. □

Corollary 17.4. *Formula* (17.4) *can be generalized as*

$$\mathcal{L}(t^nf)(p)=(-1)^n(\mathcal{L}(f))^{(n)}(p),\quad n=1,2,\dots. \tag{17.5}$$

Proof. It follows from (17.4) by induction using the fact that any analytic function is infinitely many times differentiable. □

Example 17.5. Let us show that

$$\mathcal{L}(t^n)(p)=\frac{n!}{p^{n+1}},\quad \operatorname{Re} p>0 \tag{17.6}$$

for any $n = 0,1,2,\dots$. Indeed, Problem 17.1 gives that for each $n = 0,1,2,\dots$ the growth index of $t^n \in \mathcal{F}^+$ is equal to zero. Formula (17.5) yields

$$\begin{aligned}\mathcal{L}(t^n)(p)&=(-1)^n(\mathcal{L}(1))^{(n)}(p)=(-1)^n\frac{\mathrm{d}^n}{\mathrm{d}p^n}\int_0^\infty \mathrm{e}^{-pt}\,\mathrm{d}t=(-1)^n\frac{\mathrm{d}^n}{\mathrm{d}p^n}\frac{1}{p}\\&=(-1)^n\frac{(-1)^n n!}{p^{n+1}}=\frac{n!}{p^{n+1}}\end{aligned}$$

for $\operatorname{Re} p > 0$ and $n = 0,1,2,\dots$.

Example 17.6. Let $f \in \mathcal{F}^+$ and $f(t) = e^{at}$, $t \geq 0$ with $\operatorname{Re} a \geq 0$. Then, by definition,

$$\mathcal{L}(e^{at})(p) = \int_0^\infty e^{-(p-a)t} dt = \frac{1}{p-a}, \quad \operatorname{Re} p > \operatorname{Re} a \tag{17.7}$$

is well-defined in the domain $\{p : \operatorname{Re} p > \operatorname{Re} a\}$. In particular, for real ω we have

$$\mathcal{L}(e^{i\omega t}) = \frac{1}{p - i\omega}, \quad \mathcal{L}(\sin \omega t) = \frac{\omega}{p^2 + \omega^2}, \quad \mathcal{L}(\cos \omega t) = \frac{p}{p^2 + \omega^2} \tag{17.8}$$

for $\operatorname{Re} p > 0$.

Remark. For $\operatorname{Re} a < 0$, we have $|e^{at}| \leq 1$ for $t \geq 0$ and, therefore, the growth index is $s = 0$. In that case, (17.7) holds for $\operatorname{Re} p > 0$ (even for $\operatorname{Re} p \geq 0$).

Problem 17.7.

1. Show that if $f \in \mathcal{F}^+$ is periodic with period $T > 0$ then

$$\mathcal{L}(f)(p) = \frac{1}{1 - e^{-pT}} \int_0^T e^{-pt} f(t) dt, \quad \operatorname{Re} p > 0.$$

2. Show that if $a > 0$ then

$$\mathcal{L}(\sinh(at)) = \frac{a}{p^2 - a^2}, \quad \mathcal{L}(\cosh(at)) = \frac{p}{p^2 - a^2}$$

   for $\operatorname{Re} p > a$.

3. Show that if $a > 0$ then

$$\mathcal{L}\left(\frac{\sinh(at)}{t}\right) = \frac{1}{2} \log \frac{p+a}{p-a}$$

   for $\operatorname{Re} p > a$.

4. Show that if $f, g \in \mathcal{F}^+$ and $tf(t) = g'(t)$ then

$$\mathcal{L}(f)(p) = \int_p^\infty z\mathcal{L}(g)(z) dz,$$

   where the integral on the right-hand side is a primitive (with minus sign) for the analytic function $z\mathcal{L}(g)(z)$. In particular,

$$\mathcal{L}(tf)(p) = \mathcal{L}(g')(p) = -(\mathcal{L}(f))'(p) = p\mathcal{L}(g)(p).$$

5. Show that if $f, g \in \mathcal{F}^+$, and $f(t) = \int_t^\infty g(\tau) d\tau$ then

$$\mathcal{L}(f)(p) = -\frac{1}{p}\mathcal{L}(g)(p), \quad \operatorname{Re} p > 0.$$

Definition 17.8. Let $f_1, f_2 \in \mathcal{F}^+$. The *convolution* $g := f_1 * f_2 = f_2 * f_1$ of f_1 and f_2 is defined by

$$g(t) = \int_0^t f_1(\tau)f_2(t-\tau)\mathrm{d}\tau = \int_0^t f_2(\tau)f_1(t-\tau)\mathrm{d}\tau. \tag{17.9}$$

Remark. The growth index of $g = f_1 * f_2$ is $\max(s_1, s_2)$, where s_1 and s_2 are the growth indices of f_1 and f_2, respectively.

We collect some properties of the Laplace transform in class $\mathcal{F}^+$ in the following theorem.

Theorem 17.9.
1. *Suppose* $f_k \in \mathcal{F}^+$ *with growth indices* $s_k \geq 0$ *for* $k = 1, 2, \dots, m$. *Then* $f(t) := \sum_{k=1}^m c_k f_k(t)$, $c_k \in \mathbb{C}$ *belongs to the class* $\mathcal{F}^+$ *with the growth index* $s = \max(s_1, \dots, s_m)$ *and*

$$\mathcal{L}(f)(p) = \sum_{k=1}^m c_k \mathcal{L}(f_k)(p), \quad \operatorname{Re} p > s.$$

2. *Let* f_1 *and* f_2 *have growth indices* s_1 *and* s_2, *respectively. Then* $g = f_1 * f_2 \in \mathcal{F}^+$ *with the growth index* $s = \max(s_1, s_2)$ *and*

$$\mathcal{L}(g)(p) = \mathcal{L}(f_1 * f_2)(p) = \mathcal{L}(f_1)(p)\mathcal{L}(f_2)(p), \quad \operatorname{Re} p > s. \tag{17.10}$$

3. *Let* $f \in \mathcal{F}^+$ *with the growth index* s *and let* $f \in C^{(n)}[0, \infty)$. *Then* $\mathcal{L}(f^{(n)})(p)$ *exists for* $\operatorname{Re} p > s$ *and*

$$\mathcal{L}(f^{(n)})(p) = p^n\left[\mathcal{L}(f)(p) - \frac{f(0)}{p} - \dots - \frac{f^{(n-1)}(0)}{p^n}\right]. \tag{17.11}$$

4. *If* $f \in \mathcal{F}^+$ *with the growth index* $s \geq 0$ *and* $\lambda \in \mathbb{C}$, *then*

$$\mathcal{L}(\mathrm{e}^{-\lambda t}f)(p) = \mathcal{L}(f)(p+\lambda), \quad \operatorname{Re} p > \max(0, s - \operatorname{Re}\lambda). \tag{17.12}$$

Proof. 1. Follows from the linearity of integral and from the fact that for two functions f_1 and f_2 with growth indices s_1 and s_2 the growth index of the sum $f_1 + f_2$ is $\max(s_1, s_2)$.

2. By the definition of convolution, we have for $\varepsilon > 0$ small enough that

$$\begin{aligned}|g(t)| &\leq M_\varepsilon^{(1)} M_\varepsilon^{(2)} \int_0^t \mathrm{e}^{(s_1+\varepsilon)\tau}\mathrm{e}^{(s_2+\varepsilon)(t-\tau)}\mathrm{d}\tau \\ &= M_\varepsilon^{(1)} M_\varepsilon^{(2)} \mathrm{e}^{(s_2+\varepsilon)t} \int_0^t \mathrm{e}^{(s_1-s_2)\tau}\mathrm{d}\tau\end{aligned}$$

$$
\begin{aligned}
&= M_\varepsilon^{(1)} M_\varepsilon^{(2)} \mathrm{e}^{(s_2+\varepsilon)t} \frac{\mathrm{e}^{(s_1-s_2)t} - 1}{s_1 - s_2} \\
&= M_\varepsilon^{(1)} M_\varepsilon^{(2)} (\mathrm{e}^{(s_1+\varepsilon)t} - \mathrm{e}^{(s_2+\varepsilon)t}) \frac{1}{s_1 - s_2} \\
&\le M_\varepsilon^{(1)} M_\varepsilon^{(2)} \frac{\mathrm{e}^{(s_1+\varepsilon)t} + \mathrm{e}^{(s_2+\varepsilon)t}}{|s_1 - s_2|}
\end{aligned}
$$

for $s_1 \neq s_2$. This shows that the growth index s for g is equal to $\max(s_1, s_2)$. Next, for $\operatorname{Re} p > s$ we have that

$$
\begin{aligned}
\mathcal{L}(g)(p) &= \int_0^\infty \mathrm{e}^{-pt} \left(\int_0^t f_1(\tau) f_2(t-\tau) \mathrm{d}\tau \right) \mathrm{d}t \\
&= \int_0^\infty f_1(\tau) \int_\tau^\infty \mathrm{e}^{-pt} f_2(t-\tau) \mathrm{d}t \mathrm{d}\tau \\
&= \int_0^\infty f_1(\tau) \int_0^\infty \mathrm{e}^{-p(\xi+\tau)} f_2(\xi) \mathrm{d}\xi \mathrm{d}\tau \\
&= \int_0^\infty \mathrm{e}^{-p\tau} f_1(\tau) \int_0^\infty \mathrm{e}^{-p\xi} f_2(\xi) \mathrm{d}\xi \mathrm{d}\tau = \mathcal{L}(f_1)(p) \mathcal{L}(f_2)(p).
\end{aligned}
$$

We have used here Fubini's theorem and the fact that $\operatorname{Re} p > s = \max(s_1, s_2)$. For the case $s_1 = s_2$, the proof is similar.

3. We proceed by induction with respect to n. For $n = 1$, we assume that $f \in \mathcal{F}^+$ with growth index s and $f' \in C[0, \infty)$. Then for $\operatorname{Re} p > s$ we obtain formally by integration by parts that

$$
\mathcal{L}(f')(p) = \int_0^\infty \mathrm{e}^{-pt} f'(t) \mathrm{d}t = \mathrm{e}^{-pt} f(t)\big|_0^\infty + p \int_0^\infty \mathrm{e}^{-pt} f(t) \mathrm{d}t = -f(0) + p\mathcal{L}(f)(p).
$$

The right-hand side exists and is finite due to the fact that $f \in \mathcal{F}^+$ with growth index $s \ge 0$ and $\operatorname{Re} p > s$. This proves (17.11) for $n = 1$. Let us assume that (17.11) holds for any $n \ge 1$. Then by induction hypothesis we may write

$$
\begin{aligned}
\mathcal{L}(f^{(n+1)})(p) &= \mathcal{L}((f^{(n)})')(p) = -f^{(n)}(0) + p\mathcal{L}(f^{(n)})(p) \\
&= -f^{(n)}(0) + p\left(p^n \left[\mathcal{L}(f)(p) - \frac{f(0)}{p} - \dots - \frac{f^{(n-1)}(0)}{p^n} \right] \right) \\
&= p^{n+1} \left[\mathcal{L}(f)(p) - \frac{f(0)}{p} - \dots - \frac{f^{(n)}(0)}{p^{n+1}} \right].
\end{aligned}
$$

This proves (17.11) by induction.

4. If $f \in \mathcal{F}^+$ with growth index s, then for any $\varepsilon > 0$ there is $M_\varepsilon > 0$ such that

$$\begin{aligned}|\mathrm{e}^{-\lambda t} f(t)| &= \mathrm{e}^{-t \operatorname{Re}\lambda} |f(t)| \\ &\leq M_\varepsilon \mathrm{e}^{(s+\varepsilon)t - t \operatorname{Re}\lambda} \leq M_\varepsilon \begin{cases} \mathrm{e}^{(s+\varepsilon-\operatorname{Re}\lambda)t}, & s > \operatorname{Re}\lambda, \\ \mathrm{e}^{\varepsilon t}, & s \leq \operatorname{Re}\lambda. \end{cases}\end{aligned}$$

This means that the growth index for $\mathrm{e}^{-\lambda t} f(t)$ is equal to $s_\lambda := \max(0, s - \operatorname{Re}\lambda)$. Next,

$$\mathcal{L}(\mathrm{e}^{-\lambda t} f(t))(p) = \int_0^\infty \mathrm{e}^{-(p+\lambda)t} f(t)\mathrm{d}t = \mathcal{L}(f(t))(p + \lambda)$$

for $\operatorname{Re} p > s_\lambda$. □

The next result shows how we can recover the original function $f \in \mathcal{F}^+$ if its Laplace transform is known.

Theorem 17.10 (Mellin's formula). *Let $\mathcal{L}(f)(p)$ be the Laplace transform of $f \in \mathcal{F}^+$ with growth index $s \geq 0$. Then*

$$\begin{aligned}f(t) &= \lim_{A\to+\infty} \frac{1}{2\pi\mathrm{i}} \int_{\operatorname{Re} p - \mathrm{i}A}^{\operatorname{Re} p + \mathrm{i}A} \mathrm{e}^{pt} \mathcal{L}(f)(p)\mathrm{d}p \\ &= \frac{1}{2\pi\mathrm{i}} \int_{\operatorname{Re} p - \mathrm{i}\infty}^{\operatorname{Re} p + \mathrm{i}\infty} \mathrm{e}^{pt} \mathcal{L}(f)(p)\mathrm{d}p =: \mathcal{L}^{-1}(\mathcal{L}(f))(t), \qquad (17.13)\end{aligned}$$

where the integration is carried out over the line for fixed $\operatorname{Re} p$ *such that* $\operatorname{Re} p > s$ *and where* $\mathcal{L}^{-1}$ *denotes the* inverse Laplace transform.

Proof. Let us define

$$\varphi(t) = \mathrm{e}^{-xt} f(t), \quad x > s.$$

Since $x > s$, then for any $0 < \varepsilon < x - s$ we have

$$|\varphi(t)| \leq M_\varepsilon \mathrm{e}^{-(x-s-\varepsilon)t}.$$

It means that φ tends to zero as $t \to +\infty$ exponentially and $\varphi(t) \equiv 0$ for $t < 0$. Using now the Fourier inversion formula,

$$\varphi(t) = \frac{1}{2\pi} \int_{-\infty}^{\infty} \int_{-\infty}^{\infty} \varphi(\eta) \mathrm{e}^{\mathrm{i}\xi(t-\eta)} \mathrm{d}\eta \mathrm{d}\xi$$

we obtain

$$e^{-xt}f(t) = \frac{1}{2\pi}\int_{-\infty}^{\infty}\int_{0}^{\infty} e^{-x\eta}f(\eta)e^{i\xi(t-\eta)}d\eta d\xi$$
$$= \frac{1}{2\pi}\int_{-\infty}^{\infty}\int_{0}^{\infty} e^{-(x+i\xi)\eta}f(\eta)e^{i\xi t}d\eta d\xi = \frac{1}{2\pi}\int_{-\infty}^{\infty} \mathcal{L}(f)(x+i\xi)e^{i\xi t}d\xi.$$

So,

$$f(t) = \frac{1}{2\pi}\int_{-\infty}^{\infty} \mathcal{L}(f)(x+i\xi)e^{(x+i\xi)t}d\xi = \frac{1}{2\pi i}\int_{x-i\infty}^{x+i\infty} \mathcal{L}(f)(x+i\xi)e^{(x+i\xi)t}d(i\xi),$$

where the integral is understood in the sense of principal value at infinity (as in Fourier inversion formula). This proves (17.13). □

Remark. Formula (17.13) shows that the result of inversion is actually independent on $\operatorname{Re} p$ if $\operatorname{Re} p > s$.

Example 17.11. Let us evaluate the inverse Laplace transform of the function

$$\frac{1}{p^3(p^2+1)}, \quad \operatorname{Re} p > 0.$$

Using (17.10) and Examples 17.5 and 17.6, we have

$$\frac{1}{p^3(p^2+1)} = \frac{1}{p^3}\frac{1}{p^2+1} = \mathcal{L}\left(\frac{t^2}{2}\right)\mathcal{L}(\sin t) = \mathcal{L}\left(\frac{t^2}{2} * \sin t\right).$$

Therefore,

$$\mathcal{L}^{-1}\left(\frac{1}{p^3(p^2+1)}\right) = \int_0^t \frac{\tau^2}{2}\sin(t-\tau)d\tau = \frac{t^2}{2} + \cos t - 1.$$

Example 17.12. Let us evaluate the inverse Laplace transform of the function

$$\frac{p}{(p+a)(p+b)}, \quad a, b \in \mathbb{C}.$$

Let us first assume that $a \neq b$. Then the Mellin's formula reads as

$$f(t) = \frac{1}{2\pi i}\int_{\operatorname{Re} p-i\infty}^{\operatorname{Re} p+i\infty} e^{pt}\frac{p}{(p+a)(p+b)}dp.$$

Using now Jordan's lemma in the left half-plane (see Corollary 13.19), where $\operatorname{Re} p > -\operatorname{Re} a$, $\operatorname{Re} p > -\operatorname{Re} b$ and $\operatorname{Re} p > 0$, we obtain (Figure 17.1)

$$f(t) = \operatorname*{Res}_{p=-a} \frac{pe^{pt}}{(p+a)(p+b)} + \operatorname*{Res}_{p=-b} \frac{pe^{pt}}{(p+a)(p+b)} = \frac{be^{-bt} - ae^{-at}}{b-a}.$$

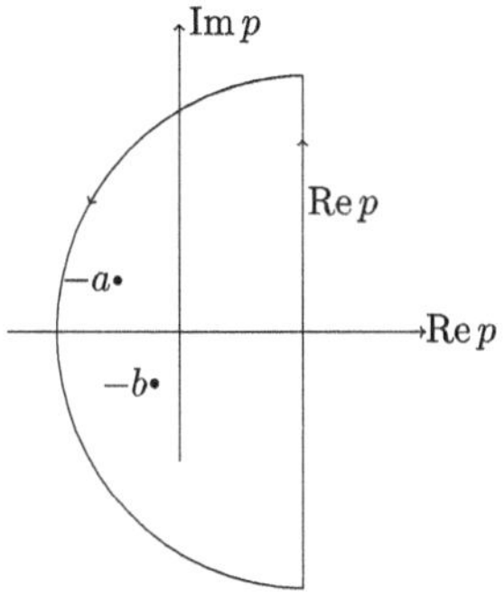

Figure 17.1: Enclosing the poles inside closed curve of integration.

For the second case $a = b$, we may proceed by the same manner or use limiting process $b \to a$ in the latter formula to obtain that

$$f(t) = e^{-at} - ate^{-at}.$$

In particular, for $a = b = 0$ we have that $f(t) \equiv 1$.

Problem 17.13. Show that for $c > 0$ the following is true:

1.

$$\frac{1}{2\pi i} \int_{c-i\infty}^{c+i\infty} \frac{a^p}{p} dp = \begin{cases} 1, & a \geq 1, \\ 0, & 0 < a < 1. \end{cases}$$

2.

$$\frac{1}{2\pi i} \int_{c-i\infty}^{c+i\infty} \frac{a^p}{p^2} dp = \begin{cases} \log a, & a \geq 1, \\ 0, & 0 < a < 1. \end{cases}$$

3.

$$\frac{1}{2\pi i} \int_{c-i\infty}^{c+i\infty} \frac{a^{-p}}{\sin \pi p} dp = \frac{1}{\pi(1+a)}, \quad c < 1, 0 < a \leq 1.$$

Problem 17.14. Using Mellin's formula find the inverse Laplace transforms of the following functions:

1. $F(p) = \frac{1}{p^4-1}$, $\operatorname{Re} p > 1$.

2. $F(p) = \frac{p}{(p-1)^2}$, $\operatorname{Re} p > 1$.
3. $F(p) = \frac{e^{-ap} - e^{-bp}}{p}$, $0 \le a < b$, $\operatorname{Re} p > 0$.
4. $F(p) = \frac{e^{-ap} - e^{-bp}}{1+p^2}$, $0 \le a < b$, $\operatorname{Re} p > 0$.
5. $F(p) = \log \frac{p+b}{p+a}$, $a \neq b$, $\operatorname{Re} p > \max(0, -\operatorname{Re} a, -\operatorname{Re} b)$.
6. $F(p) = p \log \frac{p^2-a^2}{p^2}$, $a > 0$, $\operatorname{Re} p > 0$.
7. $F(p) = \frac{n!}{(p-a)^{n+1}}$, $\operatorname{Re} p > \operatorname{Re} a$.
8. $F(p) = \frac{p^2-\omega^2}{(p^2+\omega^2)}$, $\operatorname{Re} p > |\operatorname{Im} \omega|$.
9. $F(p) = \frac{2p\omega}{(p^2+\omega^2)}$, $\operatorname{Re} p > |\operatorname{Im} \omega|$.
10. $F(p) = \frac{\pi}{2} - \arctan \frac{p}{\omega}$, $\operatorname{Re} p > |\operatorname{Im} \omega|$.
11. $F(p) = \frac{\omega}{(p^2+\omega^2)} \coth \frac{p\pi}{2\omega}$, $\operatorname{Re} p > |\operatorname{Im} \omega|$.

Problem 17.15. Show that

$$\mathcal{L}^{-1}(FG) = \mathcal{L}^{-1}(F) * \mathcal{L}^{-1}(G),$$

where F and G satisfy all conditions of Theorem 17.10.

The next theorem characterizes (gives certain sufficient conditions) the set of analytic functions that are Laplace transforms of some function from the class $\mathcal{F}^+$.

Theorem 17.16. *Let $F(p)$ be a function of complex variable p, which satisfies the conditions:*
1. *$F(p)$ is analytic for $\operatorname{Re} p > s \ge 0$,*
2. *$\lim_{|p|\to+\infty} F(p) = 0$ uniformly in $\arg p$ with $\operatorname{Re} p > s$,*
3. *for any $x > s$, we have*

$$\int_{-\infty}^{\infty} |F(x + iy)| dy \le M,$$

where M does not depend on $x > s$.

Then for any fixed $\operatorname{Re} p > s$ there exists the limit

$$\lim_{A\to+\infty} \frac{1}{2\pi i} \int_{\operatorname{Re} p - iA}^{\operatorname{Re} p + iA} e^{pt} F(p) dp =: f(t)$$

such that $F(p) = \mathcal{L}(f)(p)$.

Proof. We first prove that the Mellin's transform of $F(p)$ (the limit from above) is well-defined and independent on $\operatorname{Re} p > s$, i. e., this limit depends only on single variable t. Indeed, for $p = x + iy$ and any $A > 0$, we have the estimate

$$\left|\frac{1}{2\pi \mathrm{i}} \int_{\operatorname{Re} p-\mathrm{i}A}^{\operatorname{Re} p+\mathrm{i}A} \mathrm{e}^{pt} F(p)\mathrm{d}p\right| \le \frac{1}{2\pi} \int_{-\infty}^{\infty} |\mathrm{e}^{(x+\mathrm{i}y)t} F(x+\mathrm{i}y)|\mathrm{d}y \le \frac{M}{2\pi}\mathrm{e}^{xt}. \tag{17.14}$$

This implies (by the Cauchy criterion) the existence of the limit above. Next, we will show that this integral is independent on x and defines the function $f(t)$ of only the variable t. In the domain $\operatorname{Re} p > s$, consider a closed curve Γ consisting of segments of straight lines $[x_1 - \mathrm{i}A, x_1 + \mathrm{i}A]$ and $[x_2 - \mathrm{i}A, x_2 + \mathrm{i}A]$ parallel to the imaginary axis, and of the straight lines connecting them $[x_1 - \mathrm{i}A, x_2 - \mathrm{i}A]$, $[x_1 + \mathrm{i}A, x_2 + \mathrm{i}A]$, which are parallel to the real axis. Here, $x_1 > s$, $x_2 > s$ are arbitrary and fixed; see Figure 17.2.

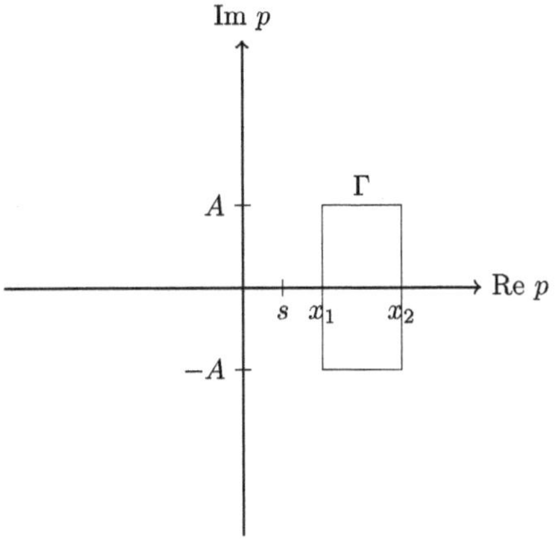

Figure 17.2: The curve of integration in the proof of Theorem 17.16.

Due to hypothesis of the theorem, the function $\mathrm{e}^{pt}F(p)$ is analytic in the domain $\operatorname{Re} p > s$. That is why by the Cauchy theorem we have

$$0 = \int_{\Gamma} \mathrm{e}^{pt} F(p)\mathrm{d}p = \int_{x_1-\mathrm{i}A}^{x_1+\mathrm{i}A} \mathrm{e}^{pt} F(p)\mathrm{d}p - \int_{x_2-\mathrm{i}A}^{x_2+\mathrm{i}A} \mathrm{e}^{pt} F(p)\mathrm{d}p.$$

Since $A > 0$ is arbitrary, then this yields that the integral

$$\int_{\operatorname{Re} p-\mathrm{i}\infty}^{\operatorname{Re} p+\mathrm{i}\infty} \mathrm{e}^{pt} F(p)\mathrm{d}p$$

is independent on $\operatorname{Re} p > s$. Thus, the integral from the formulation of this theorem is a function f of only the single variable t. Furthermore (due to the independence on $\operatorname{Re} p > s$), it follows from the evaluation of the integral in (17.14) that the function f is a function with the growth index s. Now we show that $f(t) \equiv 0$ for $t < 0$. Assuming that

$t < 0$, consider in the domain $\operatorname{Re} p > s$ a closed curve Γ_R consisting of the straight-line segment $[x - \mathrm{i}R, x + \mathrm{i}R]$ and of the arc of the semicircle $|p - x| = R$ completing it. Applying, first the Cauchy theorem for Γ_R, and then the Jordan's lemma (see Theorem 13.17), we can easily obtain that

$$0 = \int_{\Gamma_R} \mathrm{e}^{pt} F(p)\mathrm{d}p = \int_{x-\mathrm{i}R}^{x+\mathrm{i}R} \mathrm{e}^{pt} F(p)\mathrm{d}p - \int_{|p-x|=R} \mathrm{e}^{pt} F(p)\mathrm{d}p \to 2\pi \mathrm{i} f(t)$$

as $R \to \infty$. This proves that $f \in \mathcal{F}^+$. Construct now the Laplace transform of $f(t)$ and consider its value for some arbitrary point p_0 with $\operatorname{Re} p_0 > s$:

$$\int_0^\infty \mathrm{e}^{-p_0 t} f(t)\mathrm{d}t = \frac{1}{2\pi \mathrm{i}} \int_0^\infty \mathrm{e}^{-p_0 t} \left(\int_{\operatorname{Re} p - \mathrm{i}\infty}^{\operatorname{Re} p + \mathrm{i}\infty} \mathrm{e}^{pt} F(p)\mathrm{d}p \right) \mathrm{d}t.$$

Since the inner integral is independent on $\operatorname{Re} p$, we can choose a value $s < \operatorname{Re} p < \operatorname{Re} p_0$ and change then the order of integration (due to Condition 3 of theorem). Hence, we obtain

$$\begin{aligned} \int_0^\infty \mathrm{e}^{-p_0 t} f(t)\mathrm{d}t &= \frac{1}{2\pi \mathrm{i}} \int_{\operatorname{Re} p - \mathrm{i}\infty}^{\operatorname{Re} p + \mathrm{i}\infty} F(p) \left(\int_0^\infty \mathrm{e}^{-(p_0 - p)t} \mathrm{d}t \right) \mathrm{d}p \\ &= \frac{1}{2\pi \mathrm{i}} \int_{\operatorname{Re} p - \mathrm{i}\infty}^{\operatorname{Re} p + \mathrm{i}\infty} \frac{F(p)}{p_0 - p} \mathrm{d}p. \end{aligned}$$

The latter integral can be computed with the aid of residue theorem. Indeed, using the fact (see Condition 2 of the theorem) that $\frac{F(p)}{p_0 - p} = o(\frac{1}{p})$ as $p \to \infty$, we obtain (see the same closed curve Γ_R from above), letting $R \to \infty$, that

$$\begin{aligned} F(p_0) &= -\operatorname*{Res}_{p=p_0} \frac{F(p)}{p_0 - p} = \frac{1}{2\pi \mathrm{i}} \int_{\Gamma_R} \frac{F(p)}{p_0 - p} \mathrm{d}p \\ &= \frac{1}{2\pi \mathrm{i}} \int_{\operatorname{Re} p - \mathrm{i}R}^{\operatorname{Re} p + \mathrm{i}R} \frac{F(p)}{p_0 - p} \mathrm{d}p + \frac{1}{2\pi \mathrm{i}} \int_{|p-x|=R} \frac{F(p)}{p_0 - p} \mathrm{d}p \\ &\to \frac{1}{2\pi \mathrm{i}} \int_{\operatorname{Re} p - \mathrm{i}\infty}^{\operatorname{Re} p + \mathrm{i}\infty} \frac{F(p)}{p_0 - p} \mathrm{d}p. \end{aligned}$$

Thus, due to arbitrariness of p_0 we have $F(p) = \mathcal{L}(f)(p)$. This completes the proof. □

Example 17.17. Let us find (using Theorem 17.16) the original function $f(t)$ if its Laplace transform $F(p)$ is analytic function in the domain $\operatorname{Re} p > s \geq 0$ and such that $z = \infty$ is a regular point. Thus, first we have that a Laurent's expansion of $F(p)$ has the form

$$F(p) = \sum_{j=0}^{\infty} \frac{c_j}{p^j},$$

and second, we assume (in order to apply Theorem 17.16) that $c_0 = 0$. Applying now the Mellin's formula (17.13) and (equivalently) Theorem 17.16, we have

$$\frac{1}{2\pi\mathrm{i}} \int_{\operatorname{Re} p-\mathrm{i}A}^{\operatorname{Re} p+\mathrm{i}A} \mathrm{e}^{pt} \sum_{j=1}^{\infty} \frac{c_j}{p^j} \mathrm{d}p = \sum_{j=1}^{\infty} c_j \frac{1}{2\pi\mathrm{i}} \int_{\operatorname{Re} p-\mathrm{i}A}^{\operatorname{Re} p+\mathrm{i}A} \frac{\mathrm{e}^{pt}}{p^j} \mathrm{d}p \to \sum_{j=1}^{\infty} c_j \mathcal{L}^{-1}\left(\frac{1}{p^j}\right),$$

as $A \to \infty$ for $t > 0$. Since $\mathcal{L}^{-1}(\frac{1}{p^j}) = \frac{t^{j-1}}{(j-1)!}$, $j = 1, 2, \ldots$ (see (17.6)), then we have finally that the original function f is equal to

$$f(t) = \sum_{j=1}^{\infty} c_j \frac{t^{j-1}}{(j-1)!} = \sum_{j=0}^{\infty} c_{j+1} \frac{t^j}{j!}, \quad t > 0.$$

Euler's gamma and beta functions

Let $p = x + \mathrm{i}y$ with $x = \operatorname{Re} p > 0$.

Definition 17.18. *Euler's gamma function* Γ is defined as

$$\Gamma(p) := \int_0^{\infty} t^{p-1} \mathrm{e}^{-t} \mathrm{d}t = \int_0^{\infty} t^{x-1} \mathrm{e}^{\mathrm{i}y \log t} \mathrm{e}^{-t} \mathrm{d}t, \quad p \in \mathbb{C}, x > 0, \tag{17.15}$$

in particular, $\Gamma(1) = 1$.

Remark. It can be written (formally) as

$$\Gamma(p) = \mathcal{L}(t^{p-1})(1), \quad \operatorname{Re} p > 0.$$

Theorem 17.19. *If* $\operatorname{Re} p > 0$, *then integral* (17.15) *converges (even absolutely) and the following properties are satisfied:*

1. $\Gamma(p + n) = (p + n - 1)(p + n - 2) \cdots p\Gamma(p)$, $n = 1, 2, \ldots$, *in particular,* $\Gamma(n + 1) = n!$,
2. $\lim_{p \to -n} (p + n)\Gamma(p) = \frac{(-1)^n}{n!}$, $n = 0, 1, 2, \ldots$,
3. $\Gamma(\frac{1}{2}) = \sqrt{\pi}$, $\Gamma(n + \frac{1}{2}) = \sqrt{\pi} \frac{(2n-1)!!}{2^n}$, $n = 1, 2, \ldots$.

Proof. 1. Consider $\Gamma(p+1)$ with $\operatorname{Re} p > 0$. Then integration by parts in (17.15) yields

$$\Gamma(p+1) = -t^p e^{-t}\big|_0^\infty + \int_0^\infty p t^{p-1} e^{-t} dt = p\Gamma(p)$$

since $\operatorname{Re} p > 0$. Further, using induction with respect to n and integration by parts we obtain equality (1) of this theorem.

2. Using (1) from above, we have that for $n = 0, 1, 2, \ldots$

$$(p+n)\Gamma(p) = \frac{\Gamma(p+n+1)}{(p+n-1)(p+n-2)\cdots p}.$$

Taking the limit as $p \to -n$, we easily obtain (2) of this theorem.

3. Using the new coordinate $r = \sqrt{t}$ in (17.15) and polar coordinates, we have that

$$\left(\Gamma\left(\frac{1}{2}\right)\right)^2 = 4\int_0^\infty\int_0^\infty e^{-r^2-s^2} dr ds = 2\pi\int_0^\infty \rho e^{-\rho^2} d\rho = \pi\int_0^\infty e^{-\xi} d\xi = \pi.$$

This proves the first equality in (3) of this theorem. The second equality in (3) now follows immediately from (1). Thus, the theorem is completely proved. □

Corollary 17.20. *Euler's gamma function $\Gamma(p)$ is well-defined for all $p \in \mathbb{C} \setminus \{0, -1, -2, \ldots\}$. Moreover, it is analytic there and*

$$\Gamma(p) \approx \frac{(-1)^n}{(p+n)n!}, \quad p \to -n,$$

i. e., $\Gamma(p)$ has a pole of order 1 at any point $p = 0, -1, -2, \ldots$ and

$$\operatorname*{Res}_{p=-n} \Gamma(p) = \frac{(-1)^n}{n!}.$$

Proof. The gamma function $\Gamma(p)$ for any $p \in \mathbb{C} \setminus \{0, -1, -2, \ldots\}$ can be defined by

$$\Gamma(p) := \frac{\Gamma(p+n)}{p(p+1)\cdots(p+n-1)}, \quad \operatorname{Re} p > -n, \tag{17.16}$$

where due to arbitrarily chosen $n \in \mathbb{N}_0$ and due to $\operatorname{Re}(p+n) > 0$, the value $\Gamma(p+n)$ is well-defined by the corresponding integral (17.16) with $p+n$ instead of p. Next, for $\operatorname{Re} p > 0$ the formal derivative of $\Gamma(p)$ is equal to

$$\Gamma'(p) = \int_0^\infty t^{p-1}(\log t)e^{-t} dt.$$

Since this integral converges absolutely, then $\Gamma(p)$ is analytic for all $p \in \mathbb{C}$, $\operatorname{Re} p > 0$. Hence, formula (17.16) leads to the analyticity of $\Gamma(p)$ in the required domain. The fact that $\operatorname{Res}_{p=-n} \Gamma(p) = \frac{(-1)^n}{n!}$ is a direct consequence of (2) from Theorem 17.19. □

Example 17.21. Let us show that

$$2^{2p-1}\Gamma(p)\Gamma(p+1/2) = \sqrt{\pi}\Gamma(2p), \quad \operatorname{Re} p > 0. \tag{17.17}$$

By the definition of the Euler's gamma function, the left-hand side of (17.17) can be rewritten as

$$\int_0^\infty\int_0^\infty \mathrm{e}^{-(s+t)}(2\sqrt{st})^{2p-1}t^{-1/2}\mathrm{d}s\mathrm{d}t = 4\int_0^\infty\int_0^\infty \mathrm{e}^{-(\alpha^2+\beta^2)}(2\alpha\beta)^{2p-1}\alpha\mathrm{d}\alpha\mathrm{d}\beta$$

using also the new variables $\alpha := \sqrt{s}$, $\beta := \sqrt{t}$. Due to the symmetricity of the right-hand side with respect to α and β, it can be represented as

$$\begin{aligned}
&2\int_0^\infty\int_0^\infty \mathrm{e}^{-(\alpha^2+\beta^2)}(2\alpha\beta)^{2p-1}(\alpha+\beta)\mathrm{d}\alpha\mathrm{d}\beta \\
&\quad = 4\int_0^\infty \mathrm{d}\beta\int_\beta^\infty \mathrm{e}^{-(\alpha^2+\beta^2)}(2\alpha\beta)^{2p-1}(\alpha+\beta)\mathrm{d}\alpha
\end{aligned}$$

if we consider two symmetric cases $\alpha \le \beta$ and $\beta \le \alpha$. Applying now the new variables (again) $u := \alpha^2 + \beta^2$, $v := 2\alpha\beta$ with the Jacobian of transformation $\frac{1}{4\sqrt{u^2-v^2}}$, we obtain that the latter integral will be equal

$$\int_0^\infty \mathrm{d}v\int_v^\infty \mathrm{e}^{-u}v^{2p-1}\frac{1}{\sqrt{u-v}}\mathrm{d}u = \int_0^\infty v^{2p-1}\mathrm{e}^{-v}\mathrm{d}v\int_0^\infty \frac{\mathrm{e}^{-\eta}}{\sqrt{\eta}}\mathrm{d}\eta = \Gamma(1/2)\Gamma(2p).$$

Combining all together, we obtain (17.17). In particular, if $p = n$, $n \in \mathbb{N}$ then we obtain very useful result

$$\Gamma(n+1/2) = \frac{\sqrt{\pi}(2n)!}{2^{2n-1}n!}.$$

Problem 17.22. Prove that for any $p \in \mathbb{C}$, $\operatorname{Re} p > 0$ it holds that

$$p\Gamma'(p) + \Gamma(p) = \Gamma'(p+1).$$

Problem 17.23. Generalize (17.6) and show that, for $\nu \ge 0$,

$$\mathcal{L}(t^\nu)(p) = \frac{\Gamma(\nu+1)}{p^{\nu+1}}, \quad \operatorname{Re} p > 0,$$

where $p^{\nu+1}$ is the multi-valued analytic function given by

$$p^{\nu+1} = pp^{\nu} = pe^{\nu \log p} = pe^{\nu[\log|p| + \mathrm{i}\operatorname{Arg} p]} = p|p|^{\nu} e^{\mathrm{i}\nu \operatorname{Arg} p}.$$

Problem 17.24. Show that gamma function $\Gamma(p)$ has no zeros on the complex plane $\mathbb{C}$ and, therefore $\frac{1}{\Gamma(p)}$ is an entire function.

Problem 17.25. Let $\psi := \frac{\Gamma'(p)}{\Gamma(p)}$, $p \in \mathbb{C} \setminus \mathbb{Z}$. Show that:
1. $\psi(p+1) - \psi(p) = \frac{1}{p}$,
2. $\psi(p) - \psi(1-p) = -\frac{\pi}{\tan(\pi p)}$.

Definition 17.26. *Euler's beta function* B *is defined as (see, for clarification, Definition 17.18)*

$$B(p,q) := \int_0^1 t^{p-1}(1-t)^{q-1}\mathrm{d}t, \quad p, q \in \mathbb{C}, \operatorname{Re} p, \operatorname{Re} q > 0. \tag{17.18}$$

Theorem 17.27. *If* $\operatorname{Re} p > 0$ *and* $\operatorname{Re} q > 0$, *the integral* (17.18) *converges absolutely and the following properties are fulfilled:*
1. $B(p,q) = \int_0^\infty \frac{t^{q-1}}{(1+t)^{p+q}}\mathrm{d}t$,
2. $B(p,q) = \frac{\Gamma(p)\Gamma(q)}{\Gamma(p+q)}$,
3. $B(p,q) = (b-a)^{1-p-q}\int_a^b (t-a)^{p-1}(b-t)^{q-1}\mathrm{d}t$, $b > a$,
4. $B(p,1-p) = \frac{\pi}{\sin(\pi p)}$ *for all* $p \in \mathbb{C} \setminus \mathbb{Z}$.

Proof. Using Definition 17.26 and changing the variable $t := \frac{1}{x} - 1$, we obtain

$$\begin{aligned} B(p,q) &= \int_0^1 x^{p-1}(1-x)^{q-1}\mathrm{d}x = \int_\infty^0 (1+t)^{1-p}\left(1 - \frac{1}{1+t}\right)^{q-1}\left(-\frac{\mathrm{d}t}{(1+t)^2}\right) \\ &= \int_0^\infty \frac{t^{q-1}}{(1+t)^{p+q}}\mathrm{d}t. \end{aligned}$$

This proves (1).

To prove (2) we use first the identity (see Definition 17.18)

$$\frac{\Gamma(p)}{t^p} = \int_0^\infty x^{p-1}e^{-tx}\mathrm{d}x, \quad t > 0.$$

Similarly, we obtain

$$\frac{\Gamma(p+q)}{(1+t)^p} = \int_0^\infty x^{p+q-1} e^{-(1+t)x} dx.$$

Using (1) and the latter equality, we have

$$\begin{aligned}\Gamma(p+q)B(p,q) &= \Gamma(p+q)\int_0^\infty t^{q-1}\frac{\int_0^\infty x^{p+q-1}e^{-(1+t)x}dx}{\Gamma(p+q)}dt\\ &= \int_0^\infty t^{q-1}\int_0^\infty x^{p+q-1}e^{-x-xt}dxdt\\ &= \int_0^\infty x^{p+q-1}e^{-x}\left(\int_0^\infty t^{q-1}e^{-xt}dt\right)dx.\end{aligned}$$

Changing the variable $v = xt$ in the inner integral, we obtain

$$\Gamma(p+q)B(p,q) = \int_0^\infty x^{p-1}e^{-x}\left(\int_0^\infty v^{q-1}e^{-v}dv\right)dx = \Gamma(p)\Gamma(q).$$

Thus, (2) is proved.

Part (3) in this theorem follows immediately if we introduce a new variable $v := \frac{t-a}{b-a}$ in the corresponding integral.

To prove (4), we first assume that $0 < \operatorname{Re} p < 1$ and use (1). Then we have

$$B(p, 1-p) = \int_0^\infty \frac{t^{-p}}{1+t}dt.$$

Next, consider now any branch of multivalued complex-valued function of complex variable z of the form

$$f(z) = \frac{1}{z^p(1+z)} = \frac{1}{e^{x\log z}e^{iy\log z}(1+z)}, \quad p = x + iy.$$

This function is analytic in the domain $\Omega_{R,\epsilon}$ with a cut along the positive axis and a cut around the origin of radius $\epsilon > 0$ small enough inside of the circle of radius $R > 0$ big enough, except one point $z = -1$. Thus, by the residue theorem

$$\int_{\partial\Omega_{R,\epsilon}} f(z)dz = 2\pi i \operatorname*{Res}_{z=-1} f(z) = 2\pi i(-1)^{-p} = 2\pi i e^{-ip\pi}.$$

At the same time, the integral over the boundary $\partial\Omega_{R,\epsilon}$ is equal to

$$\int_{\partial\Omega_{R,\epsilon}} f(z)\mathrm{d}z = \int_{|z|=R} f(z)\mathrm{d}z + \int_{\epsilon}^{R} f(z)\mathrm{d}z + \int_{|z|=\epsilon} f(z)\mathrm{d}z + \int_{R}^{\epsilon} f(z)\mathrm{d}z$$
$$=: I_1 + I_2 + I_3 + I_4.$$

Integral I_1 tends to 0 as $R \to +\infty$ due to Jordan's lemma (see Theorem 13.17). Integral I_3 can be rewritten as

$$I_3 = \mathrm{i}\epsilon^{1-p} \int_0^{2\pi} \frac{\mathrm{e}^{\mathrm{i}\theta}}{\mathrm{e}^{\mathrm{i}p\theta}(1+\epsilon \mathrm{e}^{\mathrm{i}\theta})} \mathrm{d}\theta \to 0$$

as $\epsilon \to 0$ since $0 < \operatorname{Re} p < 1$. Letting now $R \to +\infty$ and $\epsilon \to 0$, we obtain from (17.18) that

$$2\pi \mathrm{i} \mathrm{e}^{-\mathrm{i}p\pi} = \int_0^{\infty} \frac{t^{-p}}{1+t} \mathrm{d}t - \mathrm{e}^{-\mathrm{i}2p\pi} \int_0^{\infty} \frac{t^{-p}}{1+t} \mathrm{d}t.$$

This equality implies that

$$B(p, 1-p) = \int_0^{\infty} \frac{t^{-p}}{1+t} \mathrm{d}t = \frac{2\pi \mathrm{i}}{\mathrm{e}^{\mathrm{i}\pi p} - \mathrm{e}^{-\mathrm{i}\pi p}} = \frac{\pi}{\sin(p\pi)}, \quad 0 < \operatorname{Re} p < 1.$$

Using (2) of this theorem, we obtain also that for these values p,

$$\Gamma(p)\Gamma(1-p) = \frac{\pi}{\sin(p\pi)}, \quad 0 < \operatorname{Re} p < 1. \tag{17.19}$$

Taking into account now Corollary 17.20 (see (17.16)), we can extend equality (17.19) for all $p \in \mathbb{C} \setminus \mathbb{Z}$. This proves (4). ☐

Example 17.28. Let us show that function $\psi(p) = \frac{\Gamma'(p)}{\Gamma(p)}$ (see Problem 17.25) is actually equal to

$$\psi(p) = -\frac{1}{p} + \sum_{k=1}^{\infty} \left(\frac{1}{k} - \frac{1}{k+p} \right) - \gamma,$$

where γ is the *Euler's constant*, i. e.,

$$\gamma = \lim_{n\to\infty} (1 + 1/2 + \cdots + 1/n - \log n).$$

Indeed, using Parts (2) and (4) of Theorem 17.27, we see that

$$\frac{\Gamma(p-h)\Gamma(h)}{\Gamma(p)} = B(p-h,h) = \int_0^1 ((1-t)^{p-1}-1)\frac{t^h}{(1-t)^h}\frac{dt}{t} + B(h,1-h)$$
$$= \frac{1}{h} + \int_0^1 ((1-t)^{p-1}-1)\frac{dt}{t} + o_h(1).$$

At the same time, the left-hand side of the latter equality (putting the value $\frac{1}{h}$ to the left-hand side) can be represented as

$$\frac{\Gamma(p-h)\Gamma(h)}{\Gamma(p)} - \frac{1}{h} = \Gamma(h+1)\frac{\Gamma(p-h)-\Gamma(p)}{\Gamma(p)h} + \frac{\Gamma(h+1)-1}{h}.$$

Letting now $h \to 0$ in both sides, we obtain

$$\frac{\Gamma'(p)}{\Gamma(p)} = \int_0^1 (1-(1-t)^{p-1})\frac{dt}{t} + \Gamma'(1).$$

Using the representation,

$$\frac{1}{t} = \sum_{n=0}^{\infty}(1-t)^n, \quad 0 < t < 1,$$

we calculate the right-hand side of the latter equality as follows:

$$\frac{\Gamma'(p)}{\Gamma(p)} = \sum_{n=0}^{\infty}\int_0^1 ((1-t)^n - (1-t)^{n+p-1})dt + \Gamma'(1)$$
$$= \sum_{n=0}^{\infty}\left(\frac{1}{n+1} - \frac{1}{n+p}\right) + \Gamma'(1)$$
$$= -\frac{1}{p} + \sum_{n=1}^{\infty}\left(\frac{1}{n} - \frac{1}{n+p}\right) + \sum_{n=1}^{\infty}\left(\frac{1}{n+1} - \frac{1}{n}\right) + 1 + \Gamma'(1).$$

Since $1 + \sum_{n=1}^{\infty}(\frac{1}{n+1} - \frac{1}{n}) = 0$, we obtain that

$$\frac{\Gamma'(p)}{\Gamma(p)} = -\frac{1}{p} + \sum_{n=1}^{\infty}\left(\frac{1}{n} - \frac{1}{n+p}\right) + \Gamma'(1).$$

It remains only to show that $\Gamma'(1) = -\gamma$. Indeed, integrating the latter equality from 1 to p, we obtain

$$\log\Gamma(p) = -\log p + \sum_{n=1}^{\infty}\int\left(\frac{1}{n} - \frac{1}{n+p}\right)dp + \Gamma'(1)(p-1),$$

i. e.,

$$\Gamma(p) = \frac{e^{\Gamma'(1)(p-1)}}{p} e^{\sum_{n=1}^{\infty}(\frac{p-1}{n} - \log(n+p) + \log(n+1))}.$$

It implies, if we put $p = 2$, that

$$1 = \frac{e^{\Gamma'(1)}}{2} e^{\sum_{n=1}^{\infty}(\frac{1}{n} - \log(n+2) + \log(n+1))}.$$

Thus, $\Gamma'(1) = -\lim_{N\to\infty}(\sum_{n=1}^{N}\frac{1}{n} - \log(N+2)) = -\gamma$. This completes the proof.

Problem 17.29. Prove that for $p \to 0$,

$$\Gamma'(p) \approx -\frac{\gamma}{p} - \frac{1}{p^2}.$$

Remark. Analogously, to the gamma function Γ beta function B can be extended for all $p, q \in \mathbb{C} \setminus \mathbb{Z}$ (see (17.16) and (2) of Theorem 17.27).

Problem 17.30. Define the regions of complex variable p where the following integrals represent analytic functions:

1.
$$\int_0^\infty e^{-pt^2} dt,$$

2.
$$\int_0^\infty \frac{\sin t}{t^p} dt, \quad \int_0^\infty \frac{\cos t}{t^p} dt,$$

3.
$$\int_0^\infty \frac{\sin tp}{t} dt.$$

Problem 17.31. Prove that ($\operatorname{Re} p, \operatorname{Re} q > 0$):
1. $B(p,q) = B(p,q+1) + B(p+1,q)$,
2. $B(p+1,q) = B(p,q)\frac{p}{p+q}$,
3. $\lim_{p\to 0} pB(p,q) = \frac{1}{q}$.

Problem 17.32. Prove that

$$\int_{-1}^{1}(1-x^2)^n dx = 2^{n+1}\frac{n!}{(2n+1)!!}, \quad n \in \mathbb{N}.$$

Problem 17.33. Evaluate the integrals

$$\int_0^{\frac{\pi}{2}} (\sin\theta)^7(\cos\theta)^5 \mathrm{d}\theta, \quad \int_{-1}^{1} (1+t)^2(1-t)^3 \mathrm{d}t.$$

Problem 17.34. Evaluate the integral

$$\int_0^\infty \frac{x^{\pi/5-1}}{1+x^{2\pi}} \mathrm{d}x.$$

Problem 17.35. Prove that for $c > 0$ and $a > 0$,
1.

$$\frac{1}{2\pi \mathrm{i}} \int_{c-\mathrm{i}\infty}^{c+\mathrm{i}\infty} \Gamma(p) a^{-p} \mathrm{d}p = \mathrm{e}^{-a},$$

2.

$$\frac{1}{2\pi \mathrm{i}} \int_{c-\mathrm{i}\infty}^{c+\mathrm{i}\infty} \Gamma(p)\Gamma(1-p) a^{-p} \mathrm{d}p = \frac{1}{1+a}, \quad c < 1, a \le 1.$$

We have one more instructive example of the use of Euler's gamma function. Namely, the following result holds:

$$\int_0^\infty \frac{1}{(\cosh t)^{2p}} \mathrm{d}t = 2^{2p-2} \frac{(\Gamma(p))^2}{\Gamma(2p)}, \quad \operatorname{Re} p > 0. \tag{17.20}$$

To show (17.20), let us rewrite the integral in the left-hand side as

$$2^{2p} \int_0^\infty \frac{\mathrm{e}^{2pt}}{(\mathrm{e}^{2t}+1)^{2p}} \mathrm{d}t = 2^{2p-1} \int_1^\infty \eta^{p-1}(1+\eta)^{-2p} \mathrm{d}\eta.$$

In this equality, we have used a new variable $\eta := \mathrm{e}^{2t}$. Introducing one more new variable $v = \frac{2}{1+\eta}$, the latter integral can be transformed to the integral

$$\int_0^1 (2v - v^2)^{p-1} \mathrm{d}v$$

or, with $v = 1 - \sqrt{\xi}$, to the integral

$$\frac{1}{2}\int_0^1 \xi^{-1/2}(1-\xi)^{p-1} \mathrm{d}\xi = \frac{1}{2} B(1/2, p) = \frac{1}{2}\frac{\Gamma(1/2)\Gamma(p)}{\Gamma(p+1/2)}$$

and using (17.17) we finally obtain (17.20). In particular,

$$\int_0^\infty \frac{1}{(\cosh t)^3}\,dt = \frac{\pi}{4}.$$

We consider now applications of Laplace transform to differential equations with constant coefficients and to some class of integral equations. Let us consider the initial value problem (or Cauchy problem) of the form

$$\begin{aligned} &a_0 y^{(n)}(t) + a_1 y^{(n-1)}(t) + \cdots + a_n y(t) = f(t), \quad t > 0, \\ &y(0) = y_0, \quad y'(0) = y_1, \quad \ldots, \quad y^{(n-1)}(0) = y_{n-1}, \end{aligned} \tag{17.21}$$

where a_j, y_j are given complex constants ($a_0 \neq 0$) and f is a given function. The task is to determine $y(t)$. Due to linearity of (17.21), this problem can be represented as the sum of two separate problems: (a) homogeneous equation ($f = 0$) and (b) homogeneous initial conditions ($y_j = 0$). Next, in order to solve problem (a) it suffices to find the fundamental system of solutions, i. e., the system $\{\varphi_j(t)\}_{j=0}^{n-1}$ such that

$$a_0\varphi_j^{(n)}(t) + a_1\varphi_j^{(n-1)}(t) + \cdots + a_n\varphi_j(t) = 0, \quad j = 0, 1, \ldots, n-1$$

with

$$\varphi_j^{(k)}(0) = \begin{cases} 1, & k = j, \\ 0, & k \neq j \end{cases} \tag{17.22}$$

for $k = 1, 2, \ldots, n-1$. In that case, the solution of (a) is given by

$$u(t) = \sum_{j=0}^{n-1} y_j \varphi_j(t), \tag{17.23}$$

where the constants y_j are from (17.21). Since we know that (see (17.11))

$$\mathcal{L}(\varphi_j^{(k)})(p) = p^k\left[F_j(p) - \frac{\varphi_j(0)}{p} - \cdots - \frac{\varphi_j^{(k-1)}(0)}{p^k}\right], \quad F_j = \mathcal{L}(\varphi_j),$$

then (17.23) implies

$$\mathcal{L}(\varphi_j^{(k)})(p) = \begin{cases} p^k F_j(p), & k \leq j, \\ p^k [F_j(p) - \frac{1}{p^{j+1}}], & k > j. \end{cases} \tag{17.24}$$

Using (17.24) and applying the Laplace transform to the homogeneous equation from (17.21), we obtain

$$a_0 p^n \left[F_j(p) - \frac{1}{p^{j+1}}\right] + a_1 p^{n-1}\left[F_j(p) - \frac{1}{p^{j+1}}\right] + \cdots$$
$$+ a_{n-j-1} p^{j+1}\left[F_j(p) - \frac{1}{p^{j+1}}\right] + a_{n-j} p^j F_j(p) + \cdots + a_n F_j(p) = 0.$$

This equation can be rewritten as

$$F_j(p) = \frac{Q_j(p)}{P_n(p)}, \quad j = 0, 1, 2, \ldots, n-1, \tag{17.25}$$

where $P_n(p) = a_0 p^n + a_1 p^{n-1} + \cdots + a_n$ is the characteristic polynomial of the differential operator from (17.21) and

$$Q_j(p) = a_0 p^{n-j-1} + a_1 p^{n-j-2} + \cdots + a_{n-j-1}, \quad j = 0, 1, \ldots, n-1. \tag{17.26}$$

To solve (17.25) with respect to $\mathcal{L}^{-1}(F_j(p))(t)$, we apply Mellin's formula for fixed $\operatorname{Re} p > s$, where $s \geq 0$ is to the right of all singular points of $Q_j(p)/P_n(p)$. We obtain

$$\varphi_j(t) = \mathcal{L}^{-1}(F_j)(t) = \frac{1}{2\pi \mathrm{i}} \int\limits_{\operatorname{Re} p - \mathrm{i}\infty}^{\operatorname{Re} p + \mathrm{i}\infty} \mathrm{e}^{pt} \frac{Q_j(p)}{P_n(p)} \mathrm{d}p.$$

Jordan's lemma in the left half-plane gives

$$\varphi_j(t) = \sum_{l=1}^{m} \operatorname*{Res}_{p=p_l}\left(\mathrm{e}^{pt} \frac{Q_j(p)}{P_n(p)}\right), \tag{17.27}$$

where $p_l, l = 1, 2, \ldots, m$ are the singular points of $Q_j(p)/P_n(p)$. Now the problem (a) is solved by (17.23) and (17.27).

For solving the problem (b), i. e., the problem (17.21) with nonhomogeneous equation ($f \neq 0$) and with homogeneous initial conditions ($y_j = 0$) we use (17.11) and easily obtain

$$P_n(p)\mathcal{L}(v)(p) = \mathcal{L}(f)(p),$$

where P_n is a characteristic polynomial and v is the solution of the problem. Applying Mellin's formula gives

$$v(t) = \frac{1}{2\pi \mathrm{i}} \int\limits_{\operatorname{Re} p - \mathrm{i}\infty}^{\operatorname{Re} p + \mathrm{i}\infty} \mathrm{e}^{pt} \frac{\mathcal{L}(f)(p)}{P_n(p)} \mathrm{d}p, \tag{17.28}$$

where fixed $\operatorname{Re} p > s \geq 0$ is to the right of all singular points of $\mathcal{L}(f)(p)/P_n(p)$. Formula (17.28) can be simplified as follows. Since $a_0 \neq 0$, then by (17.26) we have

$$\mathcal{L}(v)(p) = \frac{1}{a_0}\frac{a_0}{P_n(p)}\mathcal{L}(f)(p) = \frac{1}{a_0}\frac{Q_{n-1}(p)}{P_n(p)}\mathcal{L}(f)(p)$$
$$= \frac{1}{a_0}\mathcal{L}(\varphi_{n-1})\mathcal{L}(f)(p) = \frac{1}{a_0}\mathcal{L}(\varphi_{n-1} * f)(p),$$

where φ_{n-1} is defined in (17.27). The inverse Laplace transform yields

$$v(t) = \frac{1}{a_0}\int_0^t \varphi_{n-1}(\tau)f(t-\tau)\mathrm{d}\tau. \tag{17.29}$$

Combining (17.23) and (17.29), we see that the solution of (17.21) is given by

$$y(t) = u(t) + v(t) = \sum_{j=0}^{n-1} y_j\varphi_j(t) + \frac{1}{a_0}\int_0^t \varphi_{n-1}(\tau)f(t-\tau)\mathrm{d}\tau. \tag{17.30}$$

Example 17.36. Let us solve the initial value problem,

$$y^{(4)}(t) + 2y''(t) + y(t) = 0, \quad y(0) = y'(0) = y''(0) = 0, \quad y'''(0) = 1.$$

Formula (17.30) leads in this case to the solution $y(t) = \varphi_3(t)$. But $\varphi_3(t)$ equals

$$\varphi_3(t) = \operatorname*{Res}_{p=\mathrm{i}} \frac{\mathrm{e}^{pt}}{p^4 + 2p^2 + 1} + \operatorname*{Res}_{p=-\mathrm{i}} \frac{\mathrm{e}^{pt}}{p^4 + 2p^2 + 1}$$
$$= \left(\mathrm{e}^{pt}\frac{1}{(p+\mathrm{i})^2}\right)'\Bigg|_{p=\mathrm{i}} + \left(\mathrm{e}^{pt}\frac{1}{(p-\mathrm{i})^2}\right)'\Bigg|_{p=-\mathrm{i}}$$
$$= t\mathrm{e}^{pt}\frac{1}{(p+\mathrm{i})^2}\Bigg|_{p=\mathrm{i}} - \mathrm{e}^{pt}\frac{2}{(p+\mathrm{i})^3}\Bigg|_{p=\mathrm{i}} + t\mathrm{e}^{pt}\frac{1}{(p-\mathrm{i})^2}\Bigg|_{p=-\mathrm{i}} - \mathrm{e}^{pt}\frac{2}{(p-\mathrm{i})^3}\Bigg|_{p=-\mathrm{i}}$$
$$= \frac{t\mathrm{e}^{\mathrm{i}t}}{(2\mathrm{i})^2} - \frac{2\mathrm{e}^{\mathrm{i}t}}{(2\mathrm{i})^3} + \frac{t\mathrm{e}^{-\mathrm{i}t}}{(-2\mathrm{i})^2} - \frac{2\mathrm{e}^{-\mathrm{i}t}}{(-2\mathrm{i})^3}$$
$$= -\frac{t\mathrm{e}^{\mathrm{i}t}}{4} + \frac{\mathrm{e}^{\mathrm{i}t}}{4\mathrm{i}} - \frac{t\mathrm{e}^{-\mathrm{i}t}}{4} - \frac{\mathrm{e}^{-\mathrm{i}t}}{4\mathrm{i}} = -\frac{t}{2}\cos t + \frac{1}{2}\sin t.$$

Example 17.37. Let us solve the initial value problem,

$$y''(t) + y(t) = \sin t, \quad y(0) = y'(0) = 0.$$

Formula (17.30) leads to the solution

$$y(t) = \int_0^t \varphi_1(\tau)\sin(t-\tau)\mathrm{d}\tau,$$

where

$$\varphi_1(t) = \operatorname*{Res}_{p=\mathrm{i}} \frac{e^{pt}}{p^2+1} + \operatorname*{Res}_{p=-\mathrm{i}} \frac{e^{pt}}{p^2+1} = \frac{e^{\mathrm{i}t}}{2\mathrm{i}} - \frac{e^{-\mathrm{i}t}}{2\mathrm{i}} = \sin t.$$

Thus,

$$\begin{aligned} y(t) &= \int_0^t \sin\tau \sin(t-\tau)\mathrm{d}\tau = -\frac{1}{2}\int_0^t (\cos t - \cos(2\tau - t))\mathrm{d}\tau \\ &= -\frac{t}{2}\cos t + \frac{1}{2}\frac{\sin(2\tau - t)}{2}\Big|_0^t = \frac{1}{2}\sin t - \frac{t}{2}\cos t. \end{aligned}$$

Example 17.38. Let us solve the initial value problem,

$$y''(t) + \omega^2 y(t) = \cos(\nu t), \quad y(0) = 0, \quad y'(0) = 1, \quad \nu, \omega \in \mathbb{C}.$$

Let first $\nu \neq \pm\omega$. Then (17.30) gives the solution as

$$y(t) = \varphi_1(t) + \int_0^t \varphi_1(\tau)\cos(\nu(t-\tau))\mathrm{d}\tau,$$

where $\varphi_1(t)$ is defined as

$$\varphi_1(t) = \operatorname*{Res}_{p=\mathrm{i}\omega} \frac{e^{pt}}{p^2+\omega^2} + \operatorname*{Res}_{p=-\mathrm{i}\omega} \frac{e^{pt}}{p^2+\omega^2} = \frac{e^{\mathrm{i}\omega t}}{2\mathrm{i}\omega} + \frac{e^{-\mathrm{i}\omega t}}{-2\mathrm{i}\omega} = \frac{\sin(\omega t)}{\omega}.$$

For $\omega = 0$, we have $\varphi_1(t) = t$. So, for $\omega \neq 0$ we get

$$y(t) = \frac{\sin(\omega t)}{\omega} + \frac{1}{\omega}\int_0^t \sin(\omega\tau)\cos(\nu(t-\tau))\mathrm{d}\tau.$$

Since $\nu \neq \pm\omega$, then the latter integral equals

$$\begin{aligned} &\frac{1}{2\omega}\int_0^t \left[\left(\frac{\cos((\omega-\nu)\tau + \nu t)}{\nu - \omega}\right)' - \left(\frac{\cos((\omega+\nu)\tau - \nu t)}{\nu + \omega}\right)'\right]\mathrm{d}\tau \\ &= \frac{1}{2}\left[\frac{\cos(\omega t)}{\nu-\omega} - \frac{\cos(\nu t)}{\nu - \omega} - \frac{\cos(\omega t)}{\nu+\omega} + \frac{\cos(\nu t)}{\nu+\omega}\right] = \frac{\cos(\omega t) - \cos(\nu t)}{\nu^2 - \omega^2}. \end{aligned}$$

Therefore, the solution is ($\nu \neq \pm\omega \neq 0$),

$$y(t) = \frac{\sin(\omega t)}{\omega} + \frac{\cos(\omega t) - \cos(\nu t)}{\nu^2 - \omega^2}.$$

If $\omega = 0$ and $\nu \neq \pm\omega$, then

$$y(t) = t + \frac{1 - \cos(\nu t)}{\nu^2}.$$

In the case $\nu = \pm\omega$, we may use the limiting process to obtain

$$y(t) = \frac{\sin(\omega t)}{\omega} + \frac{t}{2}\frac{\sin(\omega t)}{\omega}.$$

Problem 17.39. Solve the problems:

1. $y'(t) + by(t) = e^t, y(0) = y_0$,
2. $y'''(t) + y(t) = 1, y(0) = y'(0) = y''(0) = 0$,
3. $y''(t) + y(t) = \sin(\omega t), y(0) = 0, y'(0) = 1$,
4. $y^{(4)}(t) + 4y(t) = \sin t, y(0) = y'(0) = y''(0) = y'''(0) = 0$,
5. $y''(t) + 4y'(t) + 8y = 1, y(0) = y'(0) = 0$,
6. $y''(t) - y(t) = -2t(e^{-t} + 1), y(0) = 0, y'(0) = y_0$.

Example 17.40. Let us solve the integral equation

$$g(t) = f(t) + \lambda \int_0^t K(t - \tau)g(\tau)d\tau,$$

where $g, f, K \in \mathcal{F}^+$ with the corresponding growth indices. Applying the Laplace transform, we obtain

$$\mathcal{L}(g)(p) = \mathcal{L}(f)(p) + \lambda\mathcal{L}(K)(p)\mathcal{L}(g)(p).$$

So, we have (formally)

$$g(t) = \mathcal{L}^{-1}\left(\frac{\mathcal{L}(f)}{1 - \lambda\mathcal{L}(K)}\right)(t).$$

This formula can be simplified as follows (see Problem 17.15). We have

$$\begin{aligned} g(t) &= \mathcal{L}^{-1}\left(\mathcal{L}(f) + \lambda\frac{\mathcal{L}(K)}{1 - \lambda\mathcal{L}(K)}\mathcal{L}(f)\right)(t) \\ &= f(t) + \lambda\mathcal{L}^{-1}\left(\frac{\mathcal{L}(K)}{1 - \lambda\mathcal{L}(K)}\mathcal{L}(f)\right)(t) \\ &= f(t) + \lambda\int_0^t f(t - \tau)\mathcal{L}^{-1}\left(\frac{\mathcal{L}(K)}{1 - \lambda\mathcal{L}(K)}\right)(\tau)d\tau. \end{aligned}$$

This formula gives the solution with any kernel $K(t)$ of the integral equation. For example, if $K(t) = e^t$, then for $\operatorname{Re} p > 1$ we have

$$\mathcal{L}(K) = \frac{1}{p - 1}$$

and so we may conclude that

$$\mathcal{L}^{-1}\left(\frac{\mathcal{L}(K)}{1-\lambda\mathcal{L}(K)}\right)(t) = \mathcal{L}^{-1}\left(\frac{1}{p-\lambda-1}\right)(t) = \mathrm{e}^{(\lambda+1)t}.$$

Therefore, for this particular case the solution of the integral equation

$$g(t) = f(t) + \lambda \int_0^t \mathrm{e}^{t-\tau} g(\tau)\mathrm{d}\tau$$

is equal to

$$g(t) = f(t) + \lambda \int_0^t f(t-\tau)\mathrm{e}^{(\lambda+1)\tau}\mathrm{d}\tau = f(t) + \lambda \int_0^t f(\tau)\mathrm{e}^{(\lambda+1)(t-\tau)}\mathrm{d}\tau.$$

Problem 17.41. Solve the equations:

1. $f(t) = \int_0^t \mathrm{e}^{-(t-\tau)} g(\tau)\mathrm{d}\tau$
2. $g(t) = 1 - \int_0^t (t-\tau) g(\tau)\mathrm{d}\tau$
3. $f(t) = \int_0^t \sin^2(t-\tau) g(\tau)\mathrm{d}\tau$

Problem 17.42.

1. Generalize Problem 17.23 for the case $\nu > -1$. Namely, show that

$$\mathcal{L}(t^\nu)(p) = \frac{\Gamma(\nu+1)}{p^{\nu+1}}, \quad \nu > -1,$$

where $\mathcal{L}(t^\nu)(p)$ is understood as the limit

$$\mathcal{L}(t^\nu)(p) := \lim_{\delta\to+0} \int_\delta^\infty t^\nu \mathrm{e}^{-pt}\mathrm{d}t$$

which exists.

2. Using Part (1), solve the integral equation

$$g(t) = f(t) + \lambda \int_0^t \frac{g(\tau)}{(t-\tau)^\alpha}\mathrm{d}\tau, \quad 0 < \alpha < 1.$$

Problem 17.43 (*Abel's equation*). Let $0 < \alpha < 1$ and

$$f(t) = \int_0^t \frac{g(\tau)}{(t-\tau)^\alpha}\mathrm{d}\tau.$$

Show that

$$g(t) = \frac{\sin(\alpha\pi)}{\pi}\left(\frac{f(0)}{t^{1-\alpha}} + \int_0^t \frac{f'(\tau)\mathrm{d}\tau}{(t-\tau)^{1-\alpha}}\right)$$

is a solution of this equation. Hint: Use the first part of Problem 17.42 and the formula

$$\Gamma(\alpha)\Gamma(1-\alpha) = \frac{\pi}{\sin(\alpha\pi)}, \qquad 0 < \alpha < 1.$$

18 Special functions

18.1 Method of Frobenius

This method concerns the solving of the second-order differential equations of the form (and it has some connections with the complex analysis and special functions)

$$x^2y''(x) + xp(x)y'(x) + q(x)y(x) = 0 \tag{18.1}$$

under the assumption that all quantities in (18.1) are real-valued and the functions $p(x)$ and $q(x)$ are real analytic in the vicinity of $x = 0$. In that case, equation (18.1) can be solved by setting an ansatz

$$y(x) = \sum_{k=0}^{\infty} c_k x^{k+s}, \quad s \in \mathbb{C}, c_0 \neq 0, \tag{18.2}$$

where s and c_k are to be determined, so that equation (18.1) is satisfied. Further, we may assume without loss of generality that $c_0 = 1$.

Next, let

$$p(x) = \sum_{j=0}^{\infty} p_j x^j, \quad q(x) = \sum_{j=0}^{\infty} q_j x^j, \tag{18.3}$$

then substituting (18.2) and (18.3) into equation (18.1) we obtain

$$\sum_{k=0}^{\infty} c_k(k+s)(k+s-1)x^{k+s} + \sum_{j=0}^{\infty} p_j x^j \sum_{k=0}^{\infty} c_k(k+s)x^{k+s} + \sum_{j=0}^{\infty} q_j x^j \sum_{k=0}^{\infty} c_k x^{k+s} = 0.$$

This is equivalent to

$$\sum_{k=0}^{\infty} c_k(k+s)(k+s-1)x^{k+s} + \sum_{k,j=0}^{\infty} c_k((k+s)p_j + q_j)x^{k+s+j} = 0.$$

Introducing a new index of summation $k' := k + j$ and using then again index k instead of k', we obtain

$$\sum_{k=0}^{\infty} c_k(k+s)(k+s-1)x^{k+s} + \sum_{k=0}^{\infty}\left(\sum_{j=0}^{k} c_{k-j}((k-j+s)p_j + q_j)\right)x^{k+s} = 0.$$

Equating now the coefficients in front of x^{k+s}, we obtain equations for each $k = 0, 1, 2, \ldots$ as

$$c_k((k+s)(k+s-1) + (k+s)p_0 + q_0) + \sum_{j=1}^{k} c_{k-j}((k-j+s)p_j + q_j) = 0, \tag{18.4}$$

https://doi.org/10.1515/9783111632278-020

where the second term in this sum is excluded if $k = 0$. Since we assumed that $c_0 = 1$, we obtain (necessarily) the first equation from (18.4) in the form

$$s(s-1) + sp_0 + q_0 = 0. \tag{18.5}$$

This quadratic equation is said to be the *indicial equation* and it determines possible values for s. There are three possibilities:

1. The roots s_1, s_2 are real and ($s_1 \geq s_2$)
   (a) $s_1 - s_2$ does not belong to $\mathbb{N}_0$,
   (b) $s_1 - s_2 = n_0 \in \mathbb{N}_0$.
2. The roots s_1, s_2 are complex (complex conjugate since p_0 and q_0 are real).

Consider now first the case 1(a). Equations (18.4) (together with (18.5)) for the determination of coefficients c_k are recursive. Indeed, starting with (18.5), we rewrite (18.4) for $k = 1, 2, \ldots$ as ($l := k - j$),

$$c_k((k+s)(k+s-1)) + (c+k)p_0 + q_0) = -\sum_{l=0}^{k-1} c_l((l+s)p_{k-l} + q_{k-l}), \quad k = 1, 2, \ldots. \tag{18.6}$$

Since s satisfies (18.5) and for $k = 1, 2, \ldots$ the factor $(k+s)(k+s-1) + (k+s)p_0 + q_0$ is not equal to 0, the coefficients c_k are uniquely determined by (18.6) (separately with respect to s_1 and s_2). Hence, in case 1(a) the fundamental system of solutions (two linear independent solutions) for equation (18.1) is given by

$$y_1(x) = x^{s_1} \sum_{k=0}^{\infty} c_k(s_1)x^k, \quad y_2(x) = x^{s_2} \sum_{k=0}^{\infty} c_k(s_2)x^k. \tag{18.7}$$

Consider now the case 1(b). Let $s_1 - s_2 = n_0 \in \mathbb{N}_0$. In that case, the first linearly independent solution $y_1(x)$, which corresponds to s_1 can be defined as above (see (18.7)) since $s_1 + k, k = 1, 2, \ldots$ is not the root of the indicial equation (18.5). But the second solution $y_2(x)$ can be of the form

$$y_2(x) = Ay_1(x) \log x + x^{s_2} \sum_{k=0}^{\infty} b_k x^k, \tag{18.8}$$

where coefficients A and b_k are to be determined. Substituting ansatz (18.8) into equation (18.1) finally leads to the equality

$$\begin{aligned} &A \log x(x^2 y_1''(x) + xp(x)y_1'(x)q(x)y_1(x)) + 2Axy_1'(x) - Ay_1(x) + Ap(x)y_1(x) \\ &+ \sum_{k=0}^{\infty} b_k((k+s_2)(k+s_2-1) + p(x)(k+s_2) + q(x))x^{k+s_2} = 0. \end{aligned}$$

Taking now into account that $y_1(x)$ satisfies equation (18.1) and using assumptions (18.2) (see also (18.7)), we obtain

$$A\left(\sum_{k=0}^{\infty} c_k(s_1)\left(2(k+s_1)-1+\sum_{j=0}^{\infty} p_j x^j\right)x^{k+s_1}\right)$$
$$+\sum_{k=0}^{\infty} b_k\left((k+s_2)(k+s_2-1)+(k+s_2)\sum_{j=0}^{\infty} p_j x^j+\sum_{j=0}^{\infty} q_j x^j\right)x^{k+s_2}=0.$$

Since $s_1 = s_2 + n_0$, then changing the indexes of summation several times we obtain

$$A\sum_{k=n_0}^{\infty} c_{k-n_0}(s_1)(2(k+s_2)-1)x^{k+s_2}+A\sum_{k=n_0}^{\infty}\left(\sum_{l=0}^{k-n_0} c_l(s_1)p_{k-n_0-l}\right)x^{k+s_2}$$
$$+\sum_{k=0}^{\infty} b_k(k+s_2)(k+s_2-1)x^{k+s_2}+\sum_{k=0}^{\infty}\left(\sum_{l=0}^{k} b_l((l+s_2)p_{k-l}+q_{k-l})\right)x^{k+s_2}=0.$$

Equating now the coefficients in front of x^{k+s_2}, we obtain

$$A\left(c_{k-n_0}(s_1)(2(k+s_2)-1)+\sum_{l=0}^{k-n_0} c_l(s_1)p_{k-n_0-l}\right)$$
$$+b_k(k+s_2)(k+s_2-1)+\sum_{l=0}^{k} b_l((l+s_2)p_{k-l}+q_{k-l})=0. \tag{18.9}$$

Next, we consider two cases: $n_0 = 0$, i. e., $s_1 = s_2$, and $n_0 \in \mathbb{N}$. In the first case, due to (18.6), the coefficient in front of A in (18.9) for $k = 0$ is equal to 0, i. e., A is arbitrary (e. g., we may put $A = 1$), and for the coefficients b_k, $k = 1, 2, \ldots$ (further we also assume without loss of generality that $b_0 = 1$), we have the recursive equations

$$b_k((k+s_1)(k+s_1-1)+(k+s_1)p_0+q_0)$$
$$=-\sum_{l=0}^{k-1} b_l((l+s_1)p_{k-l}+q_{k-l})-c_k(s_1)(2(k+s_1)-1)-\sum_{l=0}^{k} c_l(s_1)p_{k-l}. \tag{18.10}$$

Since $k = 1, 2, \ldots$ the coefficient in front of b_k in (18.10) is not equal to 0 (see (18.5)), so the recursive equations (18.10) are uniquely solvable and, therefore, the first case when $n_0 = 0$ is completed. If now $n_0 \in \mathbb{N}$, then for $k = 0.1, 2, \ldots, n_0 - 1$ we have (see (18.9)) (only) the following recursive equations:

$$b_k((k+s_1)(k+s_1-1)+(k+s_1)p_0+q_0)=-\sum_{l=0}^{k-1} b_l((l+s_1)p_{k-l}+q_{k-l}). \tag{18.11}$$

For $k = n_0$, we have actually an equation for A ($c_0 = 1$),

$$A(2(n_0+s_2)-1+p_0))+\sum_{l=0}^{n_0-1} b_l((l+s_2)p_{n_0-l}+q_{n_0-l})=0, \tag{18.12}$$

where the coefficients $b_l, l = 0, 1, 2, \ldots, n_0 - 1$ are determined uniquely from the equations (18.11). Since $2n_0 + 2s_2 + p_0 - 1 = n_0 \neq 0$ ($s_1 = s_2 + n_0$), then A is uniquely determined. It can be mentioned here that A might be equal to 0.

For $k = n_0 + 1, n_0 + 2, \ldots$, we have the following recursive equations (A is determined already):

$$A\left(c_{k-n_0}(s_1)(2(k+s_2)-1) + \sum_{l=0}^{k-n_0} c_l(s_1)p_{k-n_0-l} \right)$$
$$+ b_k\big((k+s_2)(k+s_2-1) + (k+s_2)p_0 + q_0\big) + \sum_{l=0}^{k-1} b_l\big((l+s_2)p_{k-l} + q_{k-l}\big) = 0. \quad (18.13)$$

Equations (18.9)–(18.13) show that case 1 is completely considered.

In case 2, the roots of the indicial equation are complex conjugates, i. e., $s_1 = \alpha + \mathrm{i}\beta$, $s_2 = \alpha - \mathrm{i}\beta$, $\beta > 0$. Operating as in the case 1(a), we are looking for the solution of equation (18.1) in the form (18.2) with assumptions (18.3) such that

$$y(x) = x^{\alpha+\mathrm{i}\beta} \sum_{k=0}^{\infty} c_k x^k, \quad c_0 = 1, \quad (18.14)$$

where, as before, the coefficients $c_k(s_1)$ are uniquely determined from the following recursive equations:

$$c_k\big((k+s)(k+s-1)\big) + (c+k)p_0 + q_0\big) = -\sum_{l=0}^{k-1} c_l\big((l+s)p_{k-l} + q_{k-l}\big), \quad k = 1, 2, \ldots. \quad (18.15)$$

The coefficients c_k here might be complex since s is complex in this case. It can be mentioned also that $c_k(s_1) = \overline{c_k(s_2)}$ due to real p_k and q_k. Denoting now $c_k, k = 1, 2, \ldots$ from (18.15) as $c_k = \alpha_k + \mathrm{i}\beta_k$ we have that

$$y_1(x) := \operatorname{Re} y(x) = \cos(\beta \log x) \sum_{k=0}^{\infty} \alpha_k x^{k+\alpha} - \sin(\beta \log x) \sum_{k=0}^{\infty} \alpha_k x^{k+\alpha} \quad (18.16)$$

and

$$y_2(x) := \operatorname{Im} y(x) = \cos(\beta \log x) \sum_{k=0}^{\infty} \beta_k x^{k+\alpha} + \sin(\beta \log x) \sum_{k=0}^{\infty} \beta_k x^{k+\alpha} \quad (18.17)$$

are linearly independent and, therefore, solutions (18.16) and (18.17) form the fundamental system of solutions for equation (18.1). It completes case 2.

Next, we consider some examples.

Example 18.1. Let us solve by the method of Frobenius the differential equation

$$x^2 y''(x) - xy'(x) + (1-x)y(x) = 0.$$

Since $p(x) = -1$ and $q(x) = 1 - x$, we have that (see (18.3)) $p_0 = -1, p_1 = p_2 = \cdots = 0$, $q_0 = 1, q_1 = -1, q_2 = q_3 = \cdots = 0$. Hence, the indicial equation is equal to

$$s(s-1) - s + 1 = 0 \equiv (s-1)^2 = 0,$$

i. e., $s_1 = s_2 = 1$ and we are in the case 1(b) with $n_0 = 0$. That is why the first independent solution is given by

$$y_1(x) = x \sum_{k=0}^{\infty} c_k x^k, \quad c_0 = 1,$$

and (see (18.6)) for $k = 1, 2, \ldots$ we have

$$c_k(k^2 + k - k - 1 + 1) = -\sum_{l=0}^{k-1} c_l((l+1)p_{k-l} + q_{k-l}).$$

It is equivalent in this case to

$$k^2 c_k = c_{k-1} \equiv c_k = \frac{c_{k-1}}{k^2},$$

so that $c_1 = 1, c_2 = \frac{1}{4}, c_3 = \frac{1}{36}, \ldots$ and

$$y_1(x) = x + x^2 + \frac{x^3}{4} + \frac{x^4}{36} + \cdots$$

whereas the second linearly independent solution is equal to

$$y_2(x) = y_1(x) \log x + x \sum_{k=0}^{\infty} b_k x^k,$$

where $b_0 = 1$ and for $k = 1, 2, \ldots$ we have that (see (18.9), (18.10) with $A = 1$)

$$c_k(2k+1) - c_k + k^2 b_k - b_{k-1} = 0 \equiv b_k = \frac{b_{k-1}}{k^2} - \frac{2c_k}{k}, \quad k = 1, 2, \ldots.$$

These formulae yield that

$$b_0 = 1, \quad b_1 = -1, \quad b_2 = -\frac{1}{2}, \quad b_3 = -\frac{2}{27}, \quad \ldots$$

so that

$$y_2(x) = \log x\left(x + x^2 + \frac{x^3}{4} + \frac{x^4}{36} + \cdots\right) + x - x^2 - \frac{x^3}{2} - \frac{2x^4}{27} + \cdots.$$

Example 18.2. Let us solve the differential equation

$$2x^2y''(x) - xy'(x) + (1-x)y(x) = 0.$$

Since $p(x) = -\frac{1}{2}$ and $q(x) = \frac{1-x}{2}$, we have that (see (18.3)) $p_0 = -\frac{1}{2}$, $p_1 = p_2 = \cdots = 0$, $q_0 = \frac{1}{2}$, $q_1 = -\frac{1}{2}$, $q_2 = q_3 = \cdots = 0$. Hence, the indicial equation is equal to

$$s(s-1) - \frac{s}{2} + \frac{1}{2} = 0 \equiv s^2 - \frac{3}{2}s + \frac{1}{2} = 0,$$

i. e., $s_1 = 1$, $s_2 = \frac{1}{2}$ and we are in the case 1(a). That is why two linearly independent solutions are given by ($x \geq 0$),

$$y_1(x) = x\sum_{k=0}^{\infty} c_k(s_1)x^k, \quad c_0 = 1$$

and

$$y_2(x) = \sqrt{x}\sum_{k=0}^{\infty} c_k(s_2)x^k, \quad c_0 = 1.$$

For s_1, we have the following recursive equations (see (18.9), (18.10) with $A = 1$):

$$c_k\left(k^2 + \frac{k}{2}\right) = \frac{c_{k-1}}{2} \equiv c_k = \frac{c_{k-1}}{2k^2 + k}, \quad k = 1, 2, \ldots.$$

For s_2, we have, respectively,

$$c_k\left(k^2 - \frac{k}{2}\right) = \frac{c_{k-1}}{2} \equiv c_k = \frac{c_{k-1}}{2k^2 - k}, \quad k = 1, 2, \ldots.$$

Problem 18.3. Solve the following differential equations by the method of Frobenius:
1. $x^2y''(x) + xy'(x) + (x^2 - \frac{1}{4})y(x) = 0,$
2. $x^2y''(x) + \frac{x}{2}y'(x) + \frac{x}{4}y(x) = 0,$
3. $x^2y''(x) + xy'(x) + (1-x)y(x) = 0.$

18.2 Bessel functions

The following differential equation is said to be *Bessel's equation*:

$$x^2y''(x) + xy'(x) + (x^2 - \nu^2)y(x) = 0, \quad \nu \in \mathbb{R} \tag{18.18}$$

and the solutions of this equation are called *Bessel's functions of order* $\pm\nu$.

Using the method of Frobenius (see Section 18.1), we may conclude that the indicial equation (see (18.5)) has the roots $s_1 = \nu$, $s_2 = -\nu$, $\nu > 0$, and we have the following possibilities:

(a) $s_1 - s_2 = 2\nu$ does not belong to $\mathbb{N}_0$.
(b) $s_1 - s_2 = 2\nu \in \mathbb{N}_0$, i. e., $\nu = \frac{n}{2}$, $n \in \mathbb{N}_0$.

Considering first the case (a), we obtain the recursive equations for the coefficients c_k (see (18.6)),

$$c_k((\nu+k)^2 - \nu^2) = -\sum_{l=0}^{k-1} c_l((l+\nu)p_{k-l} + q_{k-l}), \quad c_0 \neq 0,$$

where $p_1 = p_2 = \cdots = 0$, $q_1 = 0$, $q_2 = 1$, $q_3 = q_4 = \cdots = 0$ (see (18.18)). This is equivalent to the equations

$$c_1 = c_3 = \cdots = 0, c_k = -\frac{c_{k-2}}{k^2 + 2k\nu}, \quad k = 2, 4, \ldots. \tag{18.19}$$

Solving by induction the second equation in (18.19), we obtain

$$c_{2m} = \frac{(-1)^m c_0 \Gamma(\nu+1)}{2^{2m} m! \Gamma(\nu+m+1)}, \quad m = 1, 2, \ldots,$$

where Γ is Euler's gamma function. Choosing now $c_0 := \frac{1}{2^\nu \Gamma(\nu+1)}$, we obtain the following two linearly independent solutions (Bessel functions) of Bessel's equation (18.18):

$$J_\nu(x) = \sum_{k=0}^{\infty} \frac{(-1)^k}{k!\Gamma(k+\nu+1)} \left(\frac{x}{2}\right)^{2k+\nu}, \quad J_{-\nu}(x) = \sum_{k=0}^{\infty} \frac{(-1)^k}{k!\Gamma(k-\nu+1)} \left(\frac{x}{2}\right)^{2k-\nu}, \tag{18.20}$$

where $x > 0$ and $\nu \neq n, n = 1, 2, \ldots$. For the particular case $\nu = 0$, formulae (18.20) are valid and give us only one independent solution of Bessel's equation (18.18). Moreover, if $\nu > 0$ then the first formula in (18.20) is still valid, but it gives us only the first independent solution of (18.18) if $\nu = 1, 2, \ldots$ and another one will be introduced for this case later.

Proposition 18.4.
1. *Let $\nu \geq 0$. Then as $x \to 0$, we have*

$$J_\nu(x) = \left(\frac{x}{2}\right)^\nu \frac{1}{\Gamma(\nu+1)} + O(x^{\nu+2}), \quad J'_\nu(x) = \frac{\nu}{2}\left(\frac{x}{2}\right)^{\nu-1} \frac{1}{\Gamma(\nu+1)} + O(x^{\nu+1}).$$

2. *Let $\nu > 0, \nu \neq 1, 2, \ldots$. Then as $x \to 0$, we have*

$$J_{-\nu}(x) = \left(\frac{x}{2}\right)^{-\nu} \frac{1}{\Gamma(-\nu+1)} + O(x^{-\nu+2}),$$

$$J'_{-\nu}(x) = -\frac{\nu}{2}\left(\frac{x}{2}\right)^{-\nu-1} \frac{1}{\Gamma(-\nu+1)} + O(x^{-\nu+1}).$$

Proof. It follows immediately from representation (18.20) and possibility of the differentiation of the power series term by term. □

Proposition 18.5. *Let $\nu > 0, \nu \neq 1, 2, \ldots$. The Wronskian of J_ν and $J_{-\nu}$ is equal to*

$$W(J_\nu, J_{-\nu}) = -\frac{2\sin(\nu\pi)}{\pi x}, \quad x > 0.$$

Proof. By the definition of the Wronskian, we have

$$W(J_\nu, J_{-\nu}) = J_\nu(x) J'_{-\nu}(x) - J'_\nu(x) J_\nu(x).$$

Next, since

$$x^2 J''_\nu(x) + x J'_\nu(x) + (x^2 - \nu^2) J_\nu(x) = 0$$

and

$$x^2 J''_{-\nu}(x) + x J'_{-\nu}(x) + (x^2 - \nu^2) J_{-\nu}(x) = 0$$

then multiplying the first equation by $J_{-\nu}$ and the second by J_ν, and considering their difference we obtain

$$x^2 (J''_\nu(x) J_{-\nu}(x) - J_\nu(x) J''_{-\nu}(x)) - x W(J_\nu, J_{-\nu}) = 0.$$

Hence, we obtain the differential equation for W as

$$x^2 W' + xW = 0 \quad \text{or} \quad \frac{W'}{W} = -\frac{1}{x} \quad \text{or} \quad W(x) = \frac{C}{x},$$

where constant C is to be determined. Indeed, using Proposition 18.4 we have as $x \to 0$ that

$$\begin{aligned} W(x) &\approx -\left(\frac{x}{2}\right)^{\nu} \frac{1}{\Gamma(\nu+1)} \frac{\nu}{2} \left(\frac{x}{2}\right)^{-\nu-1} \frac{1}{\Gamma(-\nu+1)} \\ &\quad - \left(\frac{x}{2}\right)^{-\nu} \frac{1}{\Gamma(-\nu+1)} \frac{\nu}{2} \left(\frac{x}{2}\right)^{\nu-1} \frac{1}{\Gamma(\nu+1)} \\ &= -\frac{2\nu}{x\Gamma(\nu+1)\Gamma(-\nu+1)} = -\frac{2}{x\Gamma(\nu)\Gamma(-\nu+1)} = -\frac{2\sin(\nu\pi)}{\pi x}. \end{aligned}$$

In the latter calculations, we have used the properties of Euler's gamma function (see Theorem 17.19 and Theorem 17.27). Thus, we may conclude that the required constant is $C = -\frac{2}{\pi}\sin(\nu\pi)$ and this proves the proposition. □

Corollary 18.6. *If $\nu > 0, \nu \neq 1, 2, \ldots$, then Bessel's functions $J_\nu, J_{-\nu}$ from* (18.20) *are linearly independent.*

Consider now $\nu = n \in \mathbb{N}_0$ and Bessel's functions J_n and J_{-n}, i. e.,

$$J_n(x) = \sum_{k=0}^{\infty} \frac{(-1)^k}{k!\Gamma(k+n+1)} \left(\frac{x}{2}\right)^{2k+n}, \quad J_{-n}(x) = \sum_{k=0}^{\infty} \frac{(-1)^k}{k!\Gamma(k-n+1)} \left(\frac{x}{2}\right)^{2k-n}.$$

Since for $k = 0, 1, 2, \ldots, n-1$, we have $\Gamma(k-n+1) = \infty$ (see Corollary 17.20) then actually

$$
\begin{aligned}
J_{-n}(x) &= \sum_{k=n}^{\infty} \frac{(-1)^k}{k!\Gamma(k-n+1)} \left(\frac{x}{2}\right)^{2k-n} \\
&= \sum_{k'=0}^{\infty} \frac{(-1)^{k'+n}}{(k'+n)!\Gamma(k'+1)} \left(\frac{x}{2}\right)^{2k'+n} = (-1)^n J_n(x).
\end{aligned} \tag{18.21}
$$

Hence, J_n and J_{-n} are linearly dependent and, therefore, we need to find the second solution of (18.18) (which is linearly independent to $J_n, n = 0, 1, 2, \ldots$). To do it, we introduce for $\nu \neq 0, 1, 2, \ldots$ a new function $Y_\nu(x)$ as

$$
\begin{aligned}
Y_\nu(x) &:= \frac{\cos(\nu\pi)J_\nu(x) - J_{-\nu}(x)}{\sin(\nu\pi)}, \quad Y_{-\nu}(x) := \frac{J_\nu(x) - \cos(\nu\pi)J_{-\nu}(x)}{\sin(\nu\pi)}, \\
Y_{\pm n}(x) &:= \lim_{\nu \to \pm n, \nu \neq \pm n} Y_\nu(x), \quad n \in \mathbb{N}_0, \quad x > 0.
\end{aligned} \tag{18.22}
$$

This function $Y_\nu(x)$ is said to be *Neumann's function.*

It can be checked that for any $\nu \in \mathbb{R}$ the Wronskian of J_ν and Y_ν is equal to

$$
W(J_\nu, Y_\nu) = \frac{2}{\pi x}, \quad x > 0. \tag{18.23}
$$

Problem 18.7. Prove the equality (18.23).

The equality (18.23) shows that for any $\nu \in \mathbb{R}$ the functions $J_\nu(x)$ and $Y_\nu(x)$ are linearly independent.

Proposition 18.8. *For any $n = 0, 1, 2, \ldots$, Neumann's function $Y_n(x)$ can be calculated as*

$$
Y_n(x) = \frac{1}{\pi}(J_\nu(x))'\bigg|_{\nu=n} + \frac{(-1)^n}{\pi}(J_\nu(x))'\bigg|_{\nu=-n}, \tag{18.24}
$$

where the derivatives are taken with respect to ν.

Proof. Equation (18.22) and equality (18.21) lead to the equality

$$
\begin{aligned}
Y_n(x) &= \lim_{\nu \to n} \frac{\cos(\nu\pi)J_\nu(x) - (-1)^n J_n(x) + J_{-n}(x) - J_{-\nu}(x)}{\sin(\nu\pi)} \\
&= \lim_{\tau \to 0} \left(\frac{\cos(\tau\pi)J_{\tau+n}(x) - J_n(x)}{\sin(\tau\pi)} + (-1)^n \frac{J_{-n}(x) - J_{-\tau-n}(x)}{\sin(\nu\pi)} \right) \\
&= \frac{1}{\pi}(J_\nu(x))'\bigg|_{\nu=n} + \frac{(-1)^n}{\pi}(J_\nu(x))'\bigg|_{\nu=-n}
\end{aligned}
$$

since $\lim_{\tau\to 0} \frac{\sin(\tau\pi)}{\tau} = \pi$ and $\lim_{\tau \to 0} \frac{\cos(\tau\pi)-1}{\sin(\tau\pi)} = 0$. The proposition is proved. □

Corollary 18.9. *For $n = 0, 1, 2, \dots$, the following equality holds:*

$$Y_{-n}(x) = (-1)^n Y_n(x).$$

Proof. It follows immediately from the latter proposition, the definition of $Y_n(x)$ from (18.22), and identity (18.21). □

Example 18.10. Returning to the Laplace transform, we show that ($\operatorname{Re} p > 0$)

$$\mathcal{L}^{-1}\left(\frac{1}{\sqrt{1+p^2}}\right) = J_0(t), \quad t > 0.$$

Indeed, the single-valued function $F(p) = \frac{1}{\sqrt{1+p^2}}$ for $|p| > 1$ and with $\operatorname{Re} p > 0$ has the Laurent's expansion

$$F(p) = \frac{1}{p}\frac{1}{\sqrt{\frac{1}{p^2}+1}} = \sum_{k=0}^{\infty}(-1)^k \frac{(2k)!}{2^{2k}(k!)^2}\frac{1}{p^{2k+1}}.$$

Applying now Example 17.17, we obtain that

$$\mathcal{L}^{-1}(F(p))(t) = \sum_{k=0}^{\infty}(-1)^k \frac{(2k)!}{2^{2k}(k!)^2}\frac{t^{2k}}{(2k)!} = \sum_{k=0}^{\infty}\frac{(-1)^k}{(k!)^2}\left(\frac{t}{2}\right)^{2k} = J_0(t).$$

The latter equality follows from the definition of Bessel function J_0 (see, e. g., (18.20)). As an immediate consequence of this fact, we can get the very useful identity

$$\int_0^t J_0(\tau)J_0(t-\tau)\mathrm{d}\tau = \sin t, \quad t > 0.$$

To show this, we use first (17.10) and then (17.8), and obtain

$$\mathcal{L}\left(\int_0^t J_0(\tau)J_0(t-\tau)\mathrm{d}\tau\right) = \mathcal{L}^2(J_0) = \frac{1}{1+p^2} = \mathcal{L}(\sin t).$$

Similar to the latter example, we can get that (compare with Problem 17.23)

$$\mathcal{L}\left(\frac{1}{\sqrt{p+a}}\right) = \frac{\mathrm{e}^{-ia}}{\sqrt{\pi t}}, \quad t > 0, a \in \mathbb{C},$$

and

$$\mathcal{L}\left(\frac{1}{p}\mathrm{e}^{-\frac{1}{p}}\right) = J_0(2\sqrt{t}).$$

Proposition 18.11. *Let $\nu \in \mathbb{R}$ and $x > 0$. Then the following recursive formulae hold:*
1. $(x^\nu J_\nu(x))' = x^\nu J_{\nu-1}(x)$,
2. $(x^{-\nu} J_\nu(x))' = -x^{-\nu} J_{\nu+1}(x)$, *in particular,* $J_0'(x) = -J_1(x)$,
3. $xJ_\nu'(x) - \nu J_\nu(x) = -xJ_{\nu+1}(x)$,
4. $xJ_\nu'(x) + \nu J_\nu(x) = xJ_{\nu-1}(x)$,
5. $xJ_{\nu+1}(x) + xJ_{\nu-1}(x) = 2\nu J_\nu(x)$,
6. $J_{\nu-1}(x) - J_{\nu+1}(x) = 2J_\nu'(x)$,

where derivatives are taken with respect to x.

Proof. We will prove (1) and (2) and all other equalities can be proved similar to (1) and (2). Indeed, using (18.20) and differentiating the corresponding series term by term, we obtain

$$
\begin{aligned}
(x^\nu J_\nu(x))' &= \sum_{k=0}^{\infty} \frac{(-1)^k (k+\nu)}{k!\Gamma(k+\nu+1)} \left(\frac{x}{2}\right)^{2k+\nu-1} x^\nu \\
&= x^\nu \sum_{k=0}^{\infty} \frac{(-1)^k}{k!\Gamma(k+1+\nu-1)} \left(\frac{x}{2}\right)^{2k+\nu-1} = x^\nu J_{\nu-1}(x).
\end{aligned}
$$

This proves (1). To prove (2), we first write (see (18.20))

$$
\begin{aligned}
\left(\frac{J_\nu(x)}{x^\nu}\right)' &= \sum_{k=1}^{\infty} \frac{(-1)^k 2k}{k!\Gamma(k+\nu+1)} \frac{x^{2k-1}}{2^{2k+\nu}} \\
&= -\sum_{k'=0}^{\infty} \frac{(-1)^{k'}(k'+1)}{(k'+1)!\Gamma(k'+\nu+2)} \frac{x^{2k'+1}}{2^{2k'+1+\nu}} \\
&= -x^{-\nu} \sum_{k'=0}^{\infty} \frac{(-1)^{k'}}{k'!\Gamma(k'+1+\nu+1)} \frac{x^{2k'+1+\nu}}{2^{2k'+1+\nu}} = -x^{-\nu} J_{\nu+1}(x).
\end{aligned}
$$

This proves (2). It can be mentioned here that (6) is obtained from (3) and (4). The proposition is proved. □

Remark. Since for $n \in \mathbb{N}_0$ (see (5) from Proposition 18.11),

$$
J_{n+1}(x) = \frac{2n}{x} J_n(x) - J_{n-1}(x),
$$

and since $J_1(x) = -J_0'(x)$, then any Bessel's function $J_n(x)$ of the integer order can be expressed via $J_0(x)$ and its derivative only. This explains the importance of the zero-order Bessel function $J_0(x)$.

Example 18.12 (Bessel's functions of the half-integer order). Let us first calculate Bessel's and Neumann's functions $J_{\frac{1}{2}}(x)$ and $Y_{\frac{1}{2}}(x)$. By (18.20), we have (using the properties of gamma function Γ, see Theorem 17.19)

$$J_{\frac{1}{2}}(x) = \sum_{k=0}^{\infty} \frac{(-1)^k}{k!\Gamma(k+\frac{3}{2})}\left(\frac{x}{2}\right)^{2k+\frac{1}{2}} = \frac{1}{\sqrt{x}} \sum_{k=0}^{\infty} \frac{(-1)^k x^{2k+1}}{k!(2k+1)(2k-1)\cdots 1\Gamma(\frac{1}{2})} \frac{2^k}{2^{2k+\frac{1}{2}}}$$
$$= \sqrt{\frac{2}{\pi x}} \sum_{k=0}^{\infty} \frac{(-1)^k x^{2k+1}}{k!(2k+1)(2k-1)\cdots 1} \frac{1}{2^{k+1}}$$
$$= \sqrt{\frac{2}{\pi x}} \sum_{k=0}^{\infty} \frac{(-1)^k x^{2k+1}}{(2k)!!(2k+1)!!} = \sqrt{\frac{2}{\pi x}} \sin x.$$

Similarly, we obtain

$$J_{-\frac{1}{2}}(x) = \sqrt{\frac{2}{\pi x}} \cos x.$$

Using now (18.22), we obtain

$$Y_{\frac{1}{2}}(x) = \frac{\cos \frac{\pi}{2} J_{\frac{1}{2}}(x) - J_{-\frac{1}{2}}(x)}{\sin \frac{\pi}{2}} = -J_{-\frac{1}{2}}(x) = -\sqrt{\frac{2}{\pi x}} \cos x.$$

We may prove even more. Namely, for any $n \in \mathbb{N}$, the following is true:

$$J_{n-\frac{1}{2}}(x) = \sqrt{\frac{2}{\pi x}} x^n \left(-\frac{1}{x}\frac{\mathrm{d}}{\mathrm{d}x}\right)^n \cos x. \tag{18.25}$$

To prove this equality, we use induction with respect to n. Indeed, for $n = 1$, this is true (see formula for $J_{\frac{1}{2}}(x)$ from above). Let us assume that for any $n \in \mathbb{N}$ formula (18.25) holds. Then our task is to show that

$$J_{n+\frac{1}{2}}(x) = \sqrt{\frac{2}{\pi x}} x^{n+1} \left(-\frac{1}{x}\frac{\mathrm{d}}{\mathrm{d}x}\right)^{n+1} \cos x.$$

Using the induction hypothesis, we will have

$$\sqrt{\frac{2}{\pi x}} x^{n+1} \left(-\frac{1}{x}\frac{\mathrm{d}}{\mathrm{d}x}\right)^{n+1} \cos x = \sqrt{\frac{2}{\pi x}} x^{n+1} \left(-\frac{1}{x}\frac{\mathrm{d}}{\mathrm{d}x}\right)\left(-\frac{1}{x}\frac{\mathrm{d}}{\mathrm{d}x}\right)^n \cos x$$
$$= \sqrt{\frac{2}{\pi x}} x^{n+1} \left(-\frac{1}{x}\frac{\mathrm{d}}{\mathrm{d}x}\right)\left(\frac{J_{n-\frac{1}{2}}(x)}{\sqrt{\frac{2}{\pi x}} x^n}\right)$$
$$= -x^{n-\frac{1}{2}} \left(\frac{J_{n-\frac{1}{2}}(x)}{x^{n-\frac{1}{2}}}\right)' = J_{n+\frac{1}{2}}(x)$$

due to the recursive formulae (see (2) from Proposition 18.11). This proves (18.25) by the induction.

Problem 18.13. Prove that for any $n = 0, 1, 2, \ldots$ it holds that:

1.

$$Y_{n+\frac{1}{2}}(x) = (-1)^{n+1} J_{-n-\frac{1}{2}}(x), \quad Y_{-n-\frac{1}{2}}(x) = (-1)^{n+1} J_{n+\frac{1}{2}}(x).$$

2.

$$Y_{n-\frac{1}{2}}(x) = \sqrt{\frac{2}{\pi x}} x^n \left(-\frac{1}{x}\frac{\mathrm{d}}{\mathrm{d}x}\right)^n \sin x.$$

Definition 18.14. Let $\nu \in \mathbb{R}$ and $x > 0$. The functions

$$H_\nu^{(1)}(x) := J_\nu(x) + \mathrm{i}Y_\nu(x), \quad H_\nu^{(2)}(x) := J_\nu(x) - \mathrm{i}Y_\nu(x)$$

are called *Hankel functions* of order ν and of the first and of the second kind, respectively.

It is clear that $H_\nu^{(1)}(x) = \overline{H_\nu^{(2)}(x)}$. Moreover,

$$H_\nu^{(1)}(x) = \frac{J_{-\nu}(x) - \mathrm{e}^{-\mathrm{i}\nu\pi} J_\nu(x)}{\mathrm{i}\sin(\nu\pi)}, \quad H_\nu^{(2)}(x) = -\frac{J_{-\nu}(x) - \mathrm{e}^{\mathrm{i}\nu\pi} J_\nu(x)}{\mathrm{i}\sin(\nu\pi)}$$

and

$$H_\nu^{(1)}(x) = \mathrm{e}^{-\mathrm{i}\nu\pi} H_{-\nu}^{(1)}(x), \quad H_\nu^{(2)}(x) = \mathrm{e}^{\mathrm{i}\nu\pi} H_{-\nu}^{(2)}(x)$$

and (see Example 18.12 and Problem 18.13)

$$H_{n-\frac{1}{2}}^{(1,2)}(x) = \sqrt{\frac{2}{\pi x}} x^n \left(-\frac{1}{x}\frac{\mathrm{d}}{\mathrm{d}x}\right)^n \mathrm{e}^{\pm \mathrm{i}x}, \quad n = 0, 1, 2, \ldots,$$

respectively.

The Hankel functions admit for $x > 0$ the following integral representation:

$$H_\nu^{(1)}(x) = \frac{1}{\mathrm{i}\pi} \int_{-\infty}^{+\infty+\mathrm{i}\pi} \mathrm{e}^{-\nu t + x\sinh t} \mathrm{d}t, \quad H_\nu^{(2)}(x) = -\frac{1}{\mathrm{i}\pi} \int_{-\infty}^{+\infty-\mathrm{i}\pi} \mathrm{e}^{-\nu t + x\sinh t} \mathrm{d}t, \tag{18.26}$$

where the integration occurs along the curve that can be chosen as follows: from $-\infty$ to 0 along the negative real axis, then from 0 to $\pm \mathrm{i}\pi$ along the imaginary axis, and then from $\pm \mathrm{i}\pi$ to $+\infty \pm \mathrm{i}\pi$ along a line parallel to the real axis, respectively.

We return now to the asymptotic expansions of Bessel's, Neumann's, and Hankel functions for small argument. Proposition 18.4 yields that when $x \to 0$,

$$J_\nu(x) \approx \left(\frac{x}{2}\right)^\nu \frac{1}{\Gamma(\nu + 1)}, \quad \nu \geq 0$$

and

$$J_{-\nu}(x) \approx \left(\frac{x}{2}\right)^{-\nu} \frac{1}{\Gamma(-\nu+1)}, \qquad \nu > 0, \nu \neq 1, 2, \dots.$$

Hence, for any $\nu > 0, \nu \neq 1, 2, \dots$ when $x \to 0$ (see (18.22)),

$$\begin{aligned} Y_\nu(x) &\approx \cot(\nu\pi)\left(\frac{x}{2}\right)^{\nu} \frac{1}{\Gamma(\nu+1)} - \left(\frac{x}{2}\right)^{-\nu} \frac{1}{\sin(\nu\pi)\Gamma(-\nu+1)} \\ &\approx -\left(\frac{x}{2}\right)^{-\nu} \frac{1}{\sin(\nu\pi)\Gamma(-\nu+1)} = -\frac{\Gamma(\nu)}{\pi}\left(\frac{x}{2}\right)^{-\nu}. \end{aligned} \tag{18.27}$$

Problem 18.15. Obtain the analogue of the approximation (18.27) for Neumann's functions $Y_{-\nu}(x)$ (see (18.22)) with $\nu > 0, \nu \neq 1, 2, \dots$.

Let us consider now the expansions when $x \to 0$ for Neumann's functions $Y_n(x), n \in \mathbb{N}_0$. Due to (18.24), we have

$$\begin{aligned} Y_0(x) &= \frac{2}{\pi} \sum_{k=0}^{\infty} \frac{(-1)^k x^{2k}}{2^{2k} k!} \left(\frac{(\frac{x}{2})^\nu}{\Gamma(\nu+k+1)} \right)' \Bigg|_{\nu=0} = \frac{2}{\pi} \sum_{k=0}^{\infty} \frac{(-1)^k x^{2k}}{2^{2k} k!} \frac{\log \frac{x}{2}}{\Gamma(k+1)} \\ &- \frac{2}{\pi} \sum_{k=0}^{\infty} \frac{(-1)^k x^{2k}}{2^{2k} k!} \frac{\Gamma'(k+1)}{\Gamma^2(k+1)} \approx \frac{2}{\pi} \log \frac{x}{2}, \qquad x \to 0. \end{aligned} \tag{18.28}$$

This implies, e. g. (combining with Proposition 18.4) that

$$H_0^{(1)}(x) \approx 1 + \mathrm{i}\frac{2}{\pi} \log \frac{x}{2}, \quad H_0^{(2)}(x) \approx 1 - \mathrm{i}\frac{2}{\pi} \log \frac{x}{2}, \quad x \to 0.$$

Similar to (18.28) and due to (18.24), we obtain that for $n = 1, 2, \dots$,

$$\begin{aligned} Y_n(x) &= \frac{1}{\pi} \sum_{k=0}^{\infty} \frac{(-1)^k x^{2k+n}}{2^{2k+n} k!} \frac{1}{\Gamma(n+k+1)} \log \frac{x}{2} \\ &- \frac{1}{\pi} \sum_{k=0}^{\infty} \frac{(-1)^k x^{2k+n}}{2^{2k+n} k!} \frac{\Gamma'(n+k+1)}{\Gamma^2(n+k+1)} \\ &+ \frac{(-1)^n}{\pi} \sum_{k=0}^{\infty} \frac{(-1)^k x^{2k-n}}{2^{2k-n} k!} \frac{1}{\Gamma(-n+k+1)} \log \frac{x}{2} \\ &- \frac{(-1)^n}{\pi} \sum_{k=0}^{\infty} \frac{(-1)^k x^{2k-n}}{2^{2k-n} k!} \frac{\Gamma'(-n+k+1)}{\Gamma^2(-n+k+1)}. \end{aligned}$$

Since $\Gamma(-n+k+1) = \infty$ for $k = 0, 1, 2, \dots, n-1$, then we obtain from the latter representation that

$$Y_n(x) \approx \frac{(-1)^{n+1}}{\pi} \left(\frac{x}{2}\right)^{-n} \frac{\Gamma'(-n+1)}{\Gamma^2(-n+1)} = -\frac{(n-1)!}{\pi} \left(\frac{x}{2}\right)^{-n}, \qquad x \to 0, \tag{18.29}$$

where $n = 1, 2, \dots$. Since $Y_{-n}(x) = (-1)^n Y_n(x)$ (see Corollary 18.9), then we obtain from (18.29) the corresponding asymptotic for $Y_{-n}(x)$ as well.

Problem 18.16. Prove that (see (18.29))

$$(-1)^n \frac{\Gamma'(-n+1)}{\Gamma^2(-n+1)} = \Gamma(n), \quad n = 1, 2, \ldots.$$

Hint: Use Corollary 17.20 and Problem 17.22, and the continuity of the gamma function $\Gamma(\nu)$ with respect to ν.

Problem 18.17. Based on Proposition 18.4, (18.22), (18.27), and (18.29), construct the asymptotic approximation for Hankel functions $H_\nu^{(1,2)}(x)$ as $x \to 0$.

Problem 18.18. Show that two linearly independent solutions $j_n(x)$ and $k_n(x)$ of the differential equation

$$x^2 y(x) + 2xy'(x) + (x^2 - n(n+1))y(x) = 0, \quad x > 0, n = 0, 1, 2, \ldots$$

are given by

$$j_n(x) = \sqrt{\frac{\pi}{2x}} J_{n+\frac{1}{2}}(x), \quad k_n(x) = \sqrt{\frac{\pi}{2x}} Y_{n+\frac{1}{2}}(x). \tag{18.30}$$

These functions are called the *spherical Bessel's and Neumann's functions*, respectively. Hint: Use the changes of variable $u(x) = x^a y(x)$ and use then Bessel's equation (18.18) for $u(x)$.

Problem 18.19. Based on Proposition 18.11(1) and (2), Example 18.12 and (18.30), and using induction prove the following identities:

$$j_n(x) = (-x)^n \left(\frac{\mathrm{d}}{x\mathrm{d}x}\right)^n \frac{\sin x}{x}, \quad k_n(x) = -(-x)^n \left(\frac{\mathrm{d}}{x\mathrm{d}x}\right)^n \frac{\cos x}{x}, \tag{18.31}$$

where $n \in \mathbb{N}_0$.

Generating functions

Definition 18.20. A complex-valued function $G(z, x)$ of complex variable z (with real parameter x) is said to be a *generating function* of the functional sequence $\{g_n(x)\}_{n=-\infty}^{\infty}$ if

$$G(z, x) = \sum_{n=-\infty}^{\infty} g_n(x) z^n \tag{18.32}$$

and this Laurent's expansion converges in the corresponding region with respect to z.

Theorem 18.21. *The function* $G(z, x) := \mathrm{e}^{\frac{x}{2}(z - z^{-1})}$ *is a generating function for the sequence* $\{J_n(x)\}_{n=-\infty}^{\infty}$.

Proof. The function $e^{\frac{x}{2}z}$ with respect to z has the following Taylor's expansion:

$$e^{\frac{x}{2}z} = \sum_{k=0}^{\infty} \frac{x^k}{2^k k!} z^k, \quad |z| < \infty$$

and the function $e^{-\frac{x}{2}z^{-1}}$ with respect to z has the following Laurent's expansion:

$$e^{-\frac{x}{2}z^{-1}} = \sum_{j=0}^{\infty} (-1)^j \frac{x^j}{2^j j!} z^{-j}, \quad |z| > 0.$$

Due to unsymmetrical property of $G(z,x)$ with respect to z, we have that $g_n(x) = (-1)^n g_{-n}(x)$, $n \in \mathbb{Z}$. Then the following expansion holds:

$$\begin{aligned}
e^{\frac{x}{2}(z-z^{-1})} &= \sum_{k,j=0}^{\infty} \frac{(-1)^j}{k!j!} \left(\frac{x}{2}\right)^{k+j} z^{k-j} \\
&= \sum_{n=0}^{\infty} \left(\sum_{j=0}^{\infty} \frac{(-1)^j}{j!(n+j)!} \left(\frac{x}{2}\right)^{2n+j} \right) z^n + \sum_{n=1}^{\infty} (-1)^n \left(\sum_{j=0}^{\infty} \frac{(-1)^j}{j!(n+j)!} \left(\frac{x}{2}\right)^{2n+j} \right) z^{-n} \\
&= \sum_{n=-\infty}^{\infty} J_n(x) z^n.
\end{aligned}$$

We have used here the fact that $J_n = (-1)^n J_{-n}$. This proves the theorem. □

If we consider $z = e^{i\theta}$ and $z = ie^{i\theta}$, then we obtain the following useful identities.

Corollary 18.22. *We have*

$$e^{ix \sin\theta} = \sum_{n=-\infty}^{\infty} J_n(x) e^{in\theta}, \quad e^{ix\cos\theta} = \sum_{n=-\infty}^{\infty} i^n J_n(x) e^{in\theta}, \tag{18.33}$$

in particular

$$\cos(x \sin\theta) = \sum_{n=-\infty}^{\infty} J_n(x)\cos(n\theta), \quad \sin(x\sin\theta) = \sum_{n=-\infty}^{\infty} J_n(x)\sin(n\theta)$$

and

$$\cos(x\cos\theta) = \sum_{n=-\infty}^{\infty} J_n(x)\cos(n\theta + n\pi/2), \quad \sin(x\cos\theta) = \sum_{n=-\infty}^{\infty} J_n(x)\sin(n\theta + n\pi/2).$$

If we consider $\theta = \frac{\pi}{2}$, then we have the following.

Corollary 18.23.

$$\cos x = J_0(x) + 2\sum_{k=1}^{\infty} (-1)^k J_{2k}(x), \quad \sin x = 2\sum_{k=0}^{\infty} (-1)^k J_{2k+1}(x). \tag{18.34}$$

Problem 18.24. Show that

$$J_0(x) + 2\sum_{k=1}^{\infty} J_{2k}(x) \equiv 1.$$

Problem 18.25. Show that

$$J_0(x) = \sum_{k=0}^{\infty}\left(\frac{1-x^2}{2}\right)^k J_k(1).$$

Problem 18.26. Show that

$$\sin x = J_1(x) + 2\sum_{k=1}^{\infty}(-1)^{k+1}J'_{2k}(x), \quad \cos x = 2\sum_{k=0}^{\infty}(-1)^k J'_{2k+1}(x). \tag{18.35}$$

Problem 18.27. Prove that:
1. $\cos(m\theta - x\sin\theta) = \sum_{n=-\infty}^{\infty} J_n(x)\cos((m-n)\theta)$, $m \in \mathbb{R}$.
2. $J_n(x) = \frac{1}{\pi}\int_0^{\pi}\cos(n\theta - x\sin\theta)\mathrm{d}\theta$, $n \in \mathbb{N}_0$.
   Hint: Integrate (1) with respect to x under the condition that $m \in \mathbb{Z}$.
3. $J_0(x) = \frac{2}{\pi}\int_0^1 \frac{\cos(x\xi)}{\sqrt{1-\xi^2}}\mathrm{d}\xi, J_1(x) = \frac{2x}{\pi}\int_0^1 \frac{\sin(x\xi)}{\sqrt{1-\xi^2}}\mathrm{d}\xi.$
4. $\frac{x}{1-x} = 2\sum_{n=1}^{\infty} J_n(nx)$, in particular $\sum_{n=1}^{\infty} J_n(\frac{n}{2}) = \frac{1}{2}$.
5. $|J_n^{(k)}(x)| \le 1$, $k \in \mathbb{N}_0$, $n \in \mathbb{N}_0$.
6. $J_{n+m}(x+y) = \sum_{n=-\infty}^{\infty} J_n(x)J_{m-n}(y)$, $m \in \mathbb{N}_0$.
7. $Y_0(x) = \frac{1}{\pi}\int_0^{\pi}\sin(x\sin\theta)\mathrm{d}\theta - \frac{2}{\pi}\int_0^{\infty}\mathrm{e}^{-x\sinh t}\mathrm{d}t.$
   Hint: Use (2) and the definition of $Y_0(x)$.
8. Generalize (7) showing that

$$Y_n(x) = \frac{1}{\pi}\int_0^{\pi}\sin(x\sin\theta - n\theta)\mathrm{d}\theta - \frac{2}{\pi}\int_0^{\infty}\frac{\mathrm{e}^{nt} + (-1)^n\mathrm{e}^{-nt}}{2}\mathrm{e}^{-x\sinh t}\mathrm{d}t.$$

9. Prove the integral representation for the Hankel functions (18.26).
10. $\mathcal{L}^{-1}\left(\frac{(\sqrt{1+p^2}-p)^n}{\sqrt{1+p^2}}\right) = J_n(t).$
    Hint: Use (2) and Mellin's formula.
11. $J_0(x) = \frac{2}{\pi}\int_0^{\infty}\sin(x\cosh t)\mathrm{d}t$, $Y_0(x) = -\frac{2}{\pi}\int_0^{\infty}\cos(x\cosh t)\mathrm{d}t$.
12. $\int_0^x tJ_\nu(\alpha t)J_\nu(\beta t)\mathrm{d}t = \frac{x}{\alpha^2-\beta^2}(J_\nu(\alpha x)(J_\nu(\beta x))' - (J_\nu(\alpha x))'J_\nu(\beta x))$, $\alpha \ne \beta$, $\nu > -1$.

Problem 18.28. Prove the orthogonality conditions for Bessel's functions:
1.

$$\int_0^{\infty} J_\alpha(x)J_\beta(x)\frac{\mathrm{d}x}{x} = \frac{2}{\pi}\frac{\sin(\frac{\pi}{2}(\alpha-\beta))}{\alpha^2-\beta^2}, \quad \alpha,\beta > 0,$$

in particular,

$$\int_0^\infty J_\alpha^2(x)\frac{\mathrm{d}x}{x} = \frac{1}{2\alpha}, \quad \alpha > 0.$$

2.

$$\int_0^1 tJ_\nu(\alpha t)J_\nu(\beta t)\mathrm{d}t = 0, \quad \alpha \neq \beta, \quad J_\nu(\alpha) = J_\nu(\beta) = 0, \quad \nu > -1,$$

$$\int_0^1 t(J_\nu(\alpha t))^2\mathrm{d}t = \frac{1}{2}(J_{\nu+1}(\alpha))^2, \quad \nu > -1.$$

Remark. Based on the latter problem (see (2)), we can consider the so-called *Fourier–Bessel expansion*. Namely, let $\alpha_j, j = 1, 2, \dots$ be positive roots of the Bessel function $J_\nu(x), \nu > -1$ and let $f(x)$ be integrable on the interval $[0, 1]$. Then $f(x)$ can be represented as the *Fourier–Bessel expansion*

$$f(x) \approx \sum_{j=1}^\infty A_j J_\nu(x\alpha_j),$$

where the coefficients are defined

$$A_j = \frac{2}{(J_{\nu+1}(\alpha_j))^2}\int_0^1 xf(x)J_\nu(x\alpha_j)\mathrm{d}x.$$

The convergence of this expansion is not discussed here.

Remark. There is not so trivial observation of the Bessel's function, which is concerned to the Dirac δ-function. Namely, in the sense of distributions (i. e., in the sense of some integral identities), the following equality holds:

$$\int_0^\infty J_\nu(tx)J_\nu(ts)t\mathrm{d}t = \frac{\delta(x-s)}{s} = \frac{\delta(s-x)}{x}, \quad \nu > -1,$$

where the symbol δ denotes a one-dimensional Dirac δ-function. As a consequence of this equality, it can be proved that for any integrable function $f(x)$ on the interval $(0, \infty)$ we have

$$f(x) \approx \int_0^\infty J_\nu(tx)\mathrm{d}t \int_0^\infty f(s)J_\nu(ts)s\mathrm{d}s,$$

which can be also understood in the sense of distributions.

Example 18.29. Returning to the Bessel's function of order zero, we want to show that

$$\int_0^\infty J_0(x)\mathrm{d}x = 1, \quad \int_0^\infty J_1(x)\mathrm{d}x = 1.$$

The second equality follows immediately from the relation $J_1(x) = -J_0'(x)$. To prove the first equality, we first show that for any $\epsilon > 0$,

$$\int_0^\infty \mathrm{e}^{-\epsilon x} J_0(x)\mathrm{d}x = \frac{1}{\sqrt{1+\epsilon^2}}.$$

Indeed, using part (3) from Problem 18.27, we have that

$$\begin{aligned}\int_0^\infty \mathrm{e}^{-\epsilon x} J_0(x)\mathrm{d}x &= \frac{2}{\pi}\int_0^1 \frac{\mathrm{d}\xi}{\sqrt{1-\xi^2}}\int_0^\infty \mathrm{e}^{-\epsilon x}\cos(x\xi)\mathrm{d}x \\ &= \operatorname{Re}\frac{2}{\pi}\int_0^1 \frac{\mathrm{d}\xi}{\sqrt{1-\xi^2}}\int_0^\infty \mathrm{e}^{-\epsilon x}\mathrm{e}^{\mathrm{i}x\xi}\mathrm{d}x \\ &= \operatorname{Re}\frac{2}{\pi}\int_0^1 \frac{\mathrm{d}\xi}{\sqrt{1-\xi^2}}\int_0^\infty \mathrm{e}^{-x(\epsilon-\mathrm{i}\xi)}\mathrm{d}x \\ &= \operatorname{Re}\frac{2}{\pi}\int_0^1 \frac{1}{\sqrt{1-\xi^2}(\epsilon-\mathrm{i}\xi)}\mathrm{d}\xi = \frac{2}{\pi}\int_0^1 \frac{\epsilon}{\sqrt{1-\xi^2}(\epsilon^2+\xi^2)}\mathrm{d}\xi.\end{aligned}$$

Using the change of variable $\xi = \sin t$, the latter integral becomes

$$\frac{2\epsilon}{\pi}\int_0^{\pi/2} \frac{\mathrm{d}t}{\epsilon^2+\sin^2 t}\mathrm{d}t = \frac{\epsilon}{2\pi}\int_{-\pi}^{\pi} \frac{\mathrm{d}t}{\epsilon^2+\sin^2 t}.$$

The latter integral can be calculated as the trigonometric integral (see Section 13.1) if we use a new variable $z = \mathrm{e}^{\mathrm{i}t}$, $t \in [-\pi, \pi]$. In that case, it will be equal to

$$-\frac{4\epsilon}{2\pi\mathrm{i}}\int_{|z|=1} \frac{z\mathrm{d}z}{z^4 - 2z^2(1+2\epsilon^2)+1}.$$

The singular points of the integrand function that belong to the unit disk are

$$z_1 = \sqrt{1+\epsilon^2} - \epsilon, \quad z_2 = -\sqrt{1+\epsilon^2} + \epsilon.$$

Calculating now the residues of the integrand with respect to these singular points z_1 and z_2 and using the Cauchy's residue theorem (see Theorem 11.10), we obtain that the latter value will be equal to

$$-4\epsilon\left(\frac{z_1}{4z_1^3-4z_1(1+2\epsilon^2)}+\frac{z_2}{4z_2^3-4z_2(1+2\epsilon^2)}\right)=-2\epsilon\left(\frac{1}{-2\epsilon\sqrt{1+\epsilon^2}}\right).$$

This proves the needed statement. Finally, we see that

$$\int_0^\infty J_0(x)\mathrm{d}x := \lim_{\epsilon\to 0}\int_0^\infty \mathrm{e}^{-\epsilon x}J_0(x)\mathrm{d}x = \lim_{\epsilon\to 0}\frac{1}{\sqrt{1+\epsilon^2}} = 1.$$

Problem 18.30. Prove that

$$\int_0^\infty \frac{\sin x}{x}J_0(ax)\mathrm{d}x = \frac{\pi}{2}, \quad 0 < a < 1,$$

$$\int_0^\infty \frac{\sin x}{x}J_1(ax)\mathrm{d}x = \frac{1}{a}, \quad a > 1.$$

Bessel's functions of the complex argument

Bessel's and Neumann's functions (see (18.20), (18.22), and (18.24)) can be defined for complex arguments as well, and an important special case is that of a purely imaginary, i. e., $z = \mathrm{i}x$. In that case, the solutions to the Bessel's equation are called *modified Bessel's and Neumann's functions* and they are defined as

$$I_\nu(x) := \mathrm{i}^{-\nu}J_\nu(\mathrm{i}x) = \sum_{k=0}^\infty \frac{1}{k!\Gamma(k+\nu+1)}\left(\frac{x}{2}\right)^{2k+\nu}, \quad x > 0 \tag{18.36}$$

and

$$K_\nu(x) := \frac{\pi}{2}\frac{I_{-\nu}(x)-I_\nu(x)}{\sin(\nu\pi)}, \quad x > 0, \tag{18.37}$$

when ν is not an integer. When ν is an integer, then the limit (analogue to (18.24)) is used. These are chosen to be real-valued for real and positive argument x. The modified Neumann's function $K_\nu(x)$ is often called the *MacDonald function of order* ν.

Problem 18.31. Prove that the modified Bessel's and Neumann's functions $I_\nu(x)$ and $K_\nu(x)$ are two linearly independent solutions of the equation

$$x^2y''(x) + xy'(x) - (x^2+\nu^2)y(x) = 0, \quad x > 0.$$

Problem 18.32. Prove that for any $n \in \mathbb{N}_0$ the following are true (modified Bessel's and Neumann's function of the half-integer order)

$$I_{n-\frac{1}{2}}(x) = \sqrt{\frac{2}{\pi x}} x^n \left(\frac{1}{x}\frac{\mathrm{d}}{\mathrm{d}x}\right)^n \cosh x, \quad K_{n-\frac{1}{2}}(x) = \sqrt{\frac{2}{\pi x}} x^n \left(-\frac{1}{x}\frac{\mathrm{d}}{\mathrm{d}x}\right)^n \mathrm{e}^{-x}.$$

Remark. We present the Zommerfeld integral representation (very useful for applications) for the MacDonald function

$$K_\nu(x) = \frac{x^\nu}{2^{\nu+1}} \int_0^\infty \mathrm{e}^{-t-x^2/4t} t^{-\nu-1} \mathrm{d}t, \quad x > 0, \nu \in \mathbb{R}. \tag{18.38}$$

Also useful in applications is the Poisson integral representation for the MacDonald functions

$$K_\nu(x) = \frac{\sqrt{\pi}(2x)^\nu \mathrm{e}^{-x}}{\Gamma(\nu + \frac{1}{2})} \int_0^\infty \mathrm{e}^{-2xt}(t^2 + t)^{\nu-1/2} \mathrm{d}t, \quad x > 0, \nu > -\frac{1}{2}.$$

The Zommerfeld representation of the MacDonald function (18.38) allows us to evaluate many useful integrals.

Example 18.33 (Sonine–Gegenbauer-type integral). Let us consider the integral

$$I := \int_0^\infty K_\mu(ax) J_\nu(bx) x^{\nu-\mu+1} \mathrm{d}x, \quad a, b > 0, \nu - \mu > -1.$$

We want to show that

$$I = 2^{\nu-\mu} \frac{b^\nu \Gamma(\nu - \mu + 1)}{a^\mu (a^2 + b^2)^{\nu-\mu+1}}.$$

Using (18.38), we get

$$I = \frac{a^\mu}{2^{\mu+1}} \int_0^\infty J_\nu(bx) x^{\nu+1} \mathrm{d}x \int_0^\infty \mathrm{e}^{-t-(ax)^2/4t} t^{-\mu-1} \mathrm{d}t.$$

Changing the order of integration and using the power series for the Bessel's function (see (18.20)), we obtain

$$\begin{aligned} I &= \frac{a^\mu}{2^{\mu+1}} \int_0^\infty \frac{\mathrm{e}^{-t}}{t^{\mu+1}} \mathrm{d}t \int_0^\infty J_\nu(bx) x^{\nu+1} \mathrm{e}^{-(ax)^2/4t} \mathrm{d}x \\ &= \frac{a^\mu}{2^{\mu+1}} \int_0^\infty \frac{\mathrm{e}^{-t}}{t^{\mu+1}} \mathrm{d}t \sum_{k=0}^\infty \frac{(-1)^k}{k!\Gamma(k + \nu + 1)} \left(\frac{b}{2}\right)^{2k+\nu} \int_0^\infty x^{2k+2\nu+1} \mathrm{e}^{-(ax)^2/4t} \mathrm{d}x \end{aligned}$$

$$= \left(\text{changing variable } u := \frac{(ax)^2}{4t}\right)$$

$$= 2^{\nu-\mu}\frac{b^\nu}{a^{2\nu-\mu+2}}\int_0^\infty \frac{e^{-t}}{t^{\mu-\nu}}\sum_{k=0}^\infty \frac{(-1)^k}{k!\Gamma(k+\nu+1)}\left(\frac{b\sqrt{t}}{a}\right)^{2k}\Gamma(k+\nu+1)dt$$

$$= 2^{\nu-\mu}\frac{b^\nu}{a^{2\nu-\mu+2}}\int_0^\infty \frac{e^{-t(1+b^2/a^2)}}{t^{\mu-\nu}}dt = 2^{\nu-\mu}\frac{b^\nu\Gamma(\nu-\mu+1)}{a^\mu(a^2+b^2)^{\nu-\mu+1}}.$$

This proves the claim. In particular, for any $\nu \in \mathbb{R}$ we have

$$\int_0^\infty K_\nu(ax)J_\nu(bx)x\,dx = \frac{b^\nu}{a^\nu(a^2+b^2)}.$$

Taking into account the asymptotic of the MacDonald function (see (18.37), (18.27) and Proposition 18.4) as $z \to 0$, i. e.,

$$K_\mu(z) \approx \frac{\pi}{2}\left(\frac{z}{2}\right)^{-\nu}\frac{1}{\sin(\mu\pi)\Gamma(\mu+1)} = \frac{\Gamma(\mu)}{2}\left(\frac{z}{2}\right)^{-\mu},$$

we obtain that (letting $a \to 0$)

$$\int_0^\infty J_\nu(bx)x^{\nu-2\mu+1}dx = \frac{2^{\nu-2\mu+1}\Gamma(\nu-\mu+1)}{b^{\nu-2\mu+2}\Gamma(\mu)}, \quad \mu-1<\nu<2\mu-\frac{1}{2}.$$

In particular,

$$\int_0^\infty J_\nu(bx)x^{1-\nu}dx = \frac{2^{1-\nu}}{b^{2-\nu}\Gamma(\nu)}, \quad \nu > \frac{1}{2}.$$

Problem 18.34. Let $a, b > 0$. Using the same technique as in Example 18.33, prove that:

1.

$$\int_0^\infty \frac{K_\mu(a\sqrt{1+x^2})J_0(bx)x}{(\sqrt{1+x^2})^\mu}dx = \frac{(\sqrt{a^2+b^2})^{\mu-1}K_{\mu-1}(\sqrt{a^2+b^2})}{a^\mu}.$$

In particular, letting $a \to 0$,

$$K_{\mu-1}(b) = \left(\frac{2}{b}\right)^{\mu-1}\Gamma(\mu)\int_0^\infty \frac{J_0(bx)x}{(1+x^2)^\mu}dx, \quad \mu > \frac{1}{4}.$$

2.

$$\int_0^\infty e^{-a^2x^2}J_\nu(bx)x^{\nu+1}dx = \frac{b^\nu}{2^{\nu+1}a^{2\nu+2}}e^{-b^2/4a^2}.$$

In particular,

$$\int_0^\infty e^{-a^2x^2} J_0(bx)x\mathrm{d}x = \frac{1}{2a^2} e^{-b^2/4a^2}.$$

3.

$$\int_0^\infty e^{-ax} J_\nu(bx)x^\nu \mathrm{d}x = \frac{(2b)^\nu \Gamma(\nu + 1/2)}{\sqrt{\pi}(a^2 + b^2)^{\nu+1/2}}.$$

In particular,

$$\int_0^\infty e^{-ax} J_0(bx)\mathrm{d}x = \frac{1}{\sqrt{a^2 + b^2}}.$$

Returning to the Bessel's and Neumann's functions, we can define them not only for a purely imaginary argument (see (18.36) and (18.37) from above) but also for any complex argument $z \in \mathbb{C}$ and this extension will lead to the entire functions on the complex plane $\mathbb{C}$. For these purposes, one can use, e. g., the continuation of the power series (18.20) to the complex variable $z \in \mathbb{C}$. But we will do it differently (actually equivalently) as follows. We prove first the following integral representation for the Bessel's functions:

$$J_\nu(x) = \frac{(\frac{x}{2})^\nu}{\sqrt{\pi}\Gamma(\nu + \frac{1}{2})} \int_{-1}^1 (1 - t^2)^{\nu - \frac{1}{2}} \cos(xt)\mathrm{d}t, \quad \nu > -\frac{1}{2}, x > 0. \tag{18.39}$$

Indeed, using the Taylor's expansion for the real-valued function cosine, i. e.,

$$\cos(xt) = \sum_{k=0}^\infty \frac{(-1)^k (xt)^{2k}}{(2k)!}, \quad xt \in \mathbb{R},$$

we obtain that

$$\begin{aligned}
&\frac{(\frac{x}{2})^\nu}{\sqrt{\pi}\Gamma(\nu + \frac{1}{2})} \int_{-1}^1 (1 - t^2)^{\nu - \frac{1}{2}} \cos(xt)\mathrm{d}t \\
&\quad = \frac{(\frac{x}{2})^\nu}{\sqrt{\pi}\Gamma(\nu + \frac{1}{2})} \sum_{k=0}^\infty \frac{(-1)^k x^{2k}}{(2k)!} 2 \int_0^1 (1 - t^2)^{\nu - \frac{1}{2}} t^{2k} \mathrm{d}t.
\end{aligned}$$

The latter integral can be evaluated using the new variable and Part (3) from Theorem 17.19 as

$$2\int_0^1 (1-t^2)^{\nu-\frac{1}{2}} t^{2k}\mathrm{d}t = \int_0^1 (1-\xi)^{\nu-\frac{1}{2}}\xi^{k-\frac{1}{2}}\mathrm{d}\xi = B\left(\nu+\frac{1}{2}, k+\frac{1}{2}\right)$$
$$= \frac{\Gamma(\nu+\frac{1}{2})\Gamma(k+\frac{1}{2})}{\Gamma(\nu+k+1)} = \frac{\Gamma(\nu+\frac{1}{2})\sqrt{\pi}(2k-1)!!}{2^k\Gamma(\nu+k+1)}.$$

Substituting this integral to the latter sum, we obtain that it is equal to

$$\frac{(\frac{x}{2})^\nu}{\sqrt{\pi}\Gamma(\nu+\frac{1}{2})}\sum_{k=0}^{\infty}\frac{(-1)^k x^{2k}}{(2k)!}\frac{\Gamma(\nu+\frac{1}{2})\sqrt{\pi}(2k-1)!!}{2^k\Gamma(\nu+k+1)} = \sum_{k=0}^{\infty}\frac{(-1)^k(\frac{x}{2})^{\nu+2k}}{k!\Gamma(\nu+k+1)} = J_\nu(x).$$

This proves the equality (18.39).

Let now the complex variable $z \in \mathbb{C}$ belong to the complex plane with the cut along the negative real axis, i. e., $z \in \mathbb{C}$, $|\arg z| < \pi$. This restriction is necessary for the univalence of multivalued function z^ν with noninteger ν. Now we define the Bessel's function J_ν for such complex z (and for complex ν) as (see (18.39))

$$J_\nu(z) := \frac{(\frac{z}{2})^\nu}{\sqrt{\pi}\Gamma(\nu+\frac{1}{2})}\int_{-1}^{1}(1-t^2)^{\nu-\frac{1}{2}}\cos(zt)\mathrm{d}t, \quad \operatorname{Re}\nu > -\frac{1}{2}, z \in \mathbb{C}. \tag{18.40}$$

Since $\operatorname{Re}\nu > -\frac{1}{2}$ and $|\cos(zt)| \le \mathrm{e}^{|z|}$, $|t| \le 1$, the integral in (18.40) converges uniformly (together with its derivative with respect to z) for $\operatorname{Re}\nu \ge -\frac{1}{2}+\delta$, $\delta > 0$ and $|z| \le R$. This implies that the Bessel's function defined by (18.39) is analytic for each of the arguments $z, |\arg z| < \pi$, and $\nu, \operatorname{Re}\nu > -\frac{1}{2}$. This function can be interpreted as the entire function of z of order 1.

Using now the Taylor's expansion for the complex-valued function $\cos(\cdot)$ of the complex variable, the integral representation (18.39) can be used (somehow in reverse order) to show that the series (18.20) for the Bessel's functions can be extended for the complex variable $z, |\arg z| < \pi$, so that

$$J_\nu(z) = \sum_{k=0}^{\infty}\frac{(-1)^k}{k!\Gamma(k+\nu+1)}\left(\frac{z}{2}\right)^{2k+\nu}, \quad J_{-\nu}(z) = \sum_{k=0}^{\infty}\frac{(-1)^k}{k!\Gamma(k-\nu+1)}\left(\frac{z}{2}\right)^{2k-\nu}, \tag{18.41}$$

where $\nu \ne -1, -2, \ldots$. It allows us to define the Neumann's functions (see (18.22) and (18.24)) also for the complex variable $z, |\arg z| < \pi$, and for all $\nu \ne -1, -2, \ldots$. For the values $\nu = -1, -2, \ldots$, we will use again (see (18.21) and Corollary 18.9) the following relations ($z, |\arg z| < \pi$):

$$J_{-n}(z) = (-1)^n J_n(z), \quad Y_{-n}(z) = (-1)^n Y_n(z), \quad n \in \mathbb{N}.$$

Definition 18.14 of the Hankel functions of the real variable can be extended now to the complex variable $z, |\arg z| < \pi$ such that

$$H_\nu^{(1,2)}(z) = J_\nu(z) \pm \mathrm{i}Y_\nu(z),$$

$$H_\nu^{(1)}(z) = \frac{J_{-\nu}(z) - \mathrm{e}^{-\mathrm{i}\nu\pi}J_\nu(z)}{\mathrm{i}\sin(\nu\pi)}, \quad H_\nu^{(2)}(z) = -\frac{J_{-\nu}(z) - \mathrm{e}^{\mathrm{i}\nu\pi}J_\nu(z)}{\mathrm{i}\sin(\nu\pi)}, \quad \nu \in \mathbb{C}.$$

It can be also checked that the Bessel's functions satisfy the differential equation (see (18.18))

$$z^2y''(z) + zy'(z) + (z^2 - \nu^2)y(z) = 0, \quad z \in \mathbb{C}, \nu \in \mathbb{C}.$$

Problem 18.35. Show that for any $\nu \in \mathbb{C}$ and $z \in \mathbb{C}$ the function:

1.
$$y(z) := \int_0^\pi \cos(\nu\theta - z\theta)\mathrm{d}\theta$$

   satisfies the differential equation

$$z^2y''(z) + zy'(z) + (z^2 - \nu^2)y(z) = \sin(\nu\pi)(z - \nu)$$

   which reduces to the Bessel's equation (18.18) if $\nu \in \mathbb{Z}$.

2.
$$y(z) := \int_0^\pi \mathrm{e}^{\nu z\cos\theta}(A + B\log(z\sin^2\theta))\mathrm{d}\theta$$

   satisfies the differential equation

$$zy''(z) + y'(z) - \nu^2 zy(z) = 0,$$

   where A and B are arbitrary constants, and $\log(\cdot)$ is any branch of the logarithmic function. Hint: Use the symmetricity of the real-valued functions $\cos(\cdot)$ and $\sin^2(\cdot)$ with respect to $\frac{\pi}{2}$.

The orthogonality condition (see (2) of Problem 18.28) and the integral representation (18.40) for the Bessel's functions allow us to investigate the question of their zeros. More precisely, we will show that the zeros other than 0 of the Bessel's function are real and of the multiplicity 1. If $z = x + \mathrm{i}y$, then the integral representation (18.39) yields ($\nu > -\frac{1}{2}$)

$$J_\nu(z) = \frac{(\frac{z}{2})^\nu}{\sqrt{\pi}\Gamma(\nu + \frac{1}{2})}\int_{-1}^{1}(1 - t^2)^{\nu-\frac{1}{2}}\cos(zt)\mathrm{d}t$$
$$= \frac{(\frac{z}{2})^\nu}{\sqrt{\pi}\Gamma(\nu + \frac{1}{2})}\int_{-1}^{1}(1 - t^2)^{\nu-\frac{1}{2}}(\cos(xt)\cosh(yt) - \mathrm{i}\sin(xt)\sinh(yt))\mathrm{d}t$$

and

$$J_\nu(\overline{z}) = \frac{(\frac{\overline{z}}{2})^\nu}{\sqrt{\pi}\Gamma(\nu+\frac{1}{2})}\int_{-1}^{1}(1-t^2)^{\nu-\frac{1}{2}}(\cos(xt)\cosh(yt)+\mathrm{i}\sin(xt)\sinh(yt))\mathrm{d}t.$$

These equalities (since ν is real) lead to the important property $\overline{J_\nu(z)} = J_\nu(\overline{z})$. The latter equality implies then that if z, $\operatorname{Im} z \neq 0$, is a zero of $J_\nu(\cdot)$, then its complex conjugate $\overline{z} \neq z$ is also a zero of $J_\nu(\cdot)$. Therefore, the orthogonality condition (see (2) of Problem 18.28) reads as

$$\int_0^1 tJ_\nu(zt)J_\nu(\overline{z}t)\mathrm{d}t = \int_0^1 t|J_\nu(zt)|^2\mathrm{d}t = 0.$$

But due to the continuity of the Bessel's functions, it is equivalent to the fact that $J_\nu(\cdot) \equiv 0$. This contradiction shows that the zeros cannot be complex, i. e., if the zeros exist then they are necessarily real. The existence of the zeros of $J_\nu(x)$ with real x and $\nu > -\frac{1}{2}$ follows from the main property of continuous functions, crossing zero. Namely, the integral representation (18.39) and the oscillatory nature of the function $\cos(xt)$ shows that at the infinitely many points x the right-hand side changes the sign. This implies the existence of infinitely many zeros of the Bessel's function $J_\nu(x)$. We may prove even more.

Proposition 18.36. *Between any two consecutive real zeros of the function $x^{-\nu}J_\nu(x)$, there is one and only one zero of the function $x^{-\nu}J_{\nu+1}(x)$.*

Proof. Using the recursive relation (2) of Proposition 18.11 above, we have that

$$x^{-\nu}J_{\nu+1}(x) = -(x^{-\nu}J_\nu(x))'_x.$$

Thus, the application of the classical Rolle's theorem gives that between any two consecutive real zeros of the function $x^{-\nu}J_\nu(x)$ there is at least one zero of the function $x^{-\nu}J_{\nu+1}(x)$. Similarly, the recursive relation (1) of Proposition 18.11 gives that between any two consecutive real zeros of the function $x^{\nu+1}J_{\nu+1}(x)$ there is at least one zero of the function $x^{\nu+1}J_\nu(x)$. Next, the functions $x^{-\nu}J_\nu(x)$ and $(x^{-\nu}J_\nu(x))'_x$ have no common zeros since the first function of these two satisfy the equation

$$xy''(x) + (2\nu+1)y'(x) + xy(x) = 0.$$

If $y(x)$ and $y'(x)$ are equal to 0 at some point $x \neq 0$, then by the differentiation of this equation we obtain by induction that all derivatives $y^{(n)}(x), n = 0, 1, 2, \ldots$ will be equal to 0 at this point and it implies that $y(x) \equiv 0$.

To complete the proof of the proposition, it remains to show that there is no other zeros between 0 and the minimal zero $x_0 \neq 0$ of the function $x^{-\nu}J_{\nu+1}(x)$. It might be mentioned here that 0 is a zero for the function $x^{-\nu}J_{\nu+1}(x)$. Indeed, if we assume that there is another zero x_1 such that $0 < x_1 < x_0$, then it will contradict the fact that between 0 and x_0 there are no zeros of the function $x^{\nu+1}J_\nu(x)$. Hence, the proposition is completely proved. □

Remark. It might be mentioned that the latter proposition can be also obtained by the Laguerre's theorem (see Theorem 15.18) if one takes into account that $J_\nu(z)$ is an entire function of order 1.

Proposition 18.37. *If $\nu > -\frac{1}{2}$, then the Bessel's functions $J_\nu(z)$ can be represented as*

$$J_\nu(z) = \frac{(\frac{z}{2})^\nu}{\Gamma(\nu+1)} \prod_{n=1}^{\infty}\left(1 - \frac{z^2}{(z_n^{(\nu)})^2}\right),$$

where $0 < z_1^{(\nu)} < z_2^{(\nu)} < \cdots$ are positive real zeros of $J_\nu(z)$, which are accumulating only at the infinity.

Proof. Together with the positive real zeros $z_n^{(\nu)}, n \in \mathbb{N}$, there will be the negative real zeros such that $z_{-n}^{(\nu)} = -z_n^{(\nu)}, n \in \mathbb{N}$. This fact follows from the equality:

$$J_\nu(-z) = e^{i\pi\nu} J_\nu(z).$$

Next, since for $\nu > -\frac{1}{2}$ the Bessel's function is an entire function of order 1 (see (18.40)) for $z, |\arg z| < \pi$, then Hadamard's theorem (see Theorem 15.13) yields

$$J_\nu(z) = Cz^\nu e^{az} \prod_{n=-\infty, n\neq 0}^{\infty}\left(1 - \frac{z}{z_n^{(\nu)}}\right) e^{z/z_n^{(\nu)}}.$$

Using the property of the zeros $z_n^{(\nu)}$, we obtain

$$J_\nu(z) = Cz^\nu e^{az} \prod_{n=1}^{\infty}\left(1 - \frac{z^2}{(z_n^{(\nu)})^2}\right).$$

Now, using the asymptotic of $J_\nu(z)$ as $z \to 0$ (see Proposition 18.4) and its generalization for complex z (18.40), we first obtain that the constant $C = \frac{1}{2^\nu \Gamma(\nu+1)}$ and, second, the constant $a = 0$ in view of the property $J_\nu(-z) = e^{i\pi\nu} J_\nu(z)$. Hence, the proposition is proved. □

18.3 Elliptic functions

Let ω_1, ω_2 be two arbitrary numbers (complex in general) such that the fraction $\frac{\omega_1}{\omega_2}$ is not real. Let $f(z)$ be a complex-valued function of the complex variable z such that

$$f(z + 2\omega_1) = f(z), \quad f(z + 2\omega_2) = f(z). \tag{18.42}$$

This type of function is called the *doubly periodic* with periods $2\omega_1$ and $2\omega_2$.

Definition 18.38. Let function $f(z)$ be doubly periodic satisfying (18.42). If this function $f(z)$ is analytic and all possible singular points are poles and they are finite, then $f(z)$ is said to be an *elliptic function.*

Remark. It is clear that if $f(z)$ is doubly periodic, then $f'(z)$ is also doubly periodic with the same periods as $f(z)$. Consequently, if function $f(z)$ is elliptic, then $f'(z)$ is also elliptic.

Consider the parallelogram P shown in Figure 18.1. It is called the *basic parallelogram of periods* if for any $\omega \in P \setminus \{2\omega_1, 2\omega_2, 2\omega_1 + 2\omega_2\}$ and for all z, we have

$$f(z + \omega) \neq f(z).$$

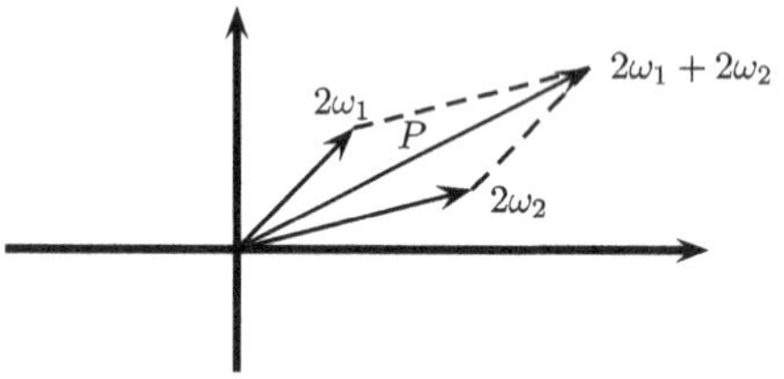

Figure 18.1: Basic parallelogram of periods.

It is clear that the whole complex plane $\mathbb{C}$ can be covered by the family of the parallelograms, which are congruent to the basic parallelogram P such that each of the points $2m\omega_1 + 2n\omega_2, n, m \in \mathbb{Z}$ are the vertices of these parallelograms. It is also clear that for all $z \in \mathbb{C}$ the points

$$z, \quad z + 2\omega_1, \quad z + 2\omega_2, \quad z + 2\omega_1 + 2\omega_1, \quad \ldots, \quad z + 2m\omega_1 + 2n\omega_2, \quad \ldots$$

have the same positions in the corresponding parallelograms and they are called *comparable*. The comparability of such two points z and z' is denoted as

$$z' = z(\mathrm{mod} 2\omega_1, 2\omega_2).$$

Hence, the elliptic function takes the same values at the comparable points, i. e., its values at any parallelogram are repeated by the values of this function in the basic parallelogram of the periods P. The interior of these parallelograms is called a *cell*. We may assume without loss of generality in the future that all poles are located inside of each cell. All these facts and the analyticity of elliptic functions yield the following properties.

Proposition 18.39. *Let $f(z)$ be elliptic function. Then:*
1. *The number of poles is finite in each cell.*
2. *The number of zeros is finite in each cell.*
3. *The sum of the residues with respect to poles in each cell is equal to* 0.

4. *Assuming that there are no poles in the cell yields* $f(z) \equiv$ const.
5. *The number of roots of the equation*

$$f(z) - C = 0, \quad C \in \mathbb{C},$$

   in the cell is independent on this constant C and this number is said to be the order of elliptic function *and it is equal to the number of poles in the cell taking into account their multiplicity.*
6. *The order of the elliptic functions cannot be less than 2.*
7. *The sum of the zeros of the elliptic function inside of any cell is comparable with the sum of its poles there.*

Proof. Parts (1) and (2) are quite evident. In the first case, if we assume that there are infinitely many poles in the cell, then a limiting point will be an essential singularity, which contradicts the ellipticity of this function. In the second case, the assumption of an infinite number of zeros leads to the fact that this elliptic function is identically equal to 0.

Let now P be any cell, then $2\pi\mathrm{i}$ times the sum of the residues there will be equal to

$$\begin{aligned}\int_{\partial P} f(z)\mathrm{d}z &= \left(\int_{\zeta}^{\zeta+2\omega_1} + \int_{\zeta+2\omega_1}^{\zeta+2\omega_1+2\omega_2} + \int_{\zeta+2\omega_1+2\omega_2}^{\zeta+2\omega_2} + \int_{\zeta+2\omega_1}^{\zeta}\right) f(z)\mathrm{d}z \\ &= \int_{\zeta}^{\zeta+2\omega_1} f(z)\mathrm{d}z + \int_{\zeta}^{\zeta+2\omega_2} f(z+2\omega_1)\mathrm{d}z \\ &\quad - \int_{\zeta}^{\zeta+2\omega_1} f(z+2\omega_2)\mathrm{d}z - \int_{\zeta}^{\zeta+2\omega_2} f(z)\mathrm{d}z = 0\end{aligned}$$

due to the doubly periodicity of $f(z)$. This proves (3).

If an elliptic function has no poles in the cell (due to periodicity it has no poles anywhere), then it is analytic and bounded in the whole complex plane. It is implied by the Liouville's theorem that $f(z) \equiv$ constant. This proves (4).

To prove (5), we use the principle of argument (see Theorem 12.4). Due to this, the difference between the zeros and the poles of the function $f(z) - C$ in the cell P is equal to

$$\frac{1}{2\pi\mathrm{i}} \int_{\partial P} \frac{f'(z)}{f(z) - C}\mathrm{d}z.$$

But since $f'(z)$ is also doubly periodic (with the same periods as $f(z)$), then using the procedure as in the proof of (3), we obtain that the latter integral is equal to 0. Hence, the number of zeros of $f(z) - C$ is equal to the number of the poles of $f(z) - C$, i. e., of $f(z)$. And this number is independent on C. This proves (5).

The order of elliptic function cannot be less than 2. If it turned out that the order is equal to 1, then there will be only one (at most) pole of order 1. But in that case the residue of $f(z)$ at this pole is not equal to 0, which contradicts with (3).

To prove (7), we first prove the formula, which has an independent important interest. Namely, the following is true:

$$\frac{1}{2\pi \mathrm{i}}\int_{\partial P}\frac{zf'(z)}{f(z)}dz = \sum_{j=1}^{k} r_j z_j - \sum_{j=1}^{l} s_j \tilde{z}_j, \tag{18.43}$$

where r_j is the multiplicity of the zero z_j and s_j is the order of the pole $\tilde{z}_j$ for this function $f(z)$. Indeed, by the Cauchy's residue theorem, we have

$$\frac{1}{2\pi \mathrm{i}}\int_{\partial P}\frac{zf'(z)}{f(z)}dz = \sum_{z=z_j, z=\tilde{z}_j} \operatorname{Res}\left(\frac{zf'(z)}{f(z)}\right).$$

Next, since near the point z_j we have that

$$\begin{aligned}\frac{zf'(z)}{f(z)} &= \frac{zr_j(A_j(z-z_j)^{r_j-1} + O((z-z_j)^{r_j})}{A_j(z-z_j)^{r_j} + o((z-z_j)^{r_j})}\\ &= \frac{zr_j}{z-z_j} + zr_j o(1) = \frac{z_j r_j}{z-z_j} + r_j + zr_j o(1), \quad z \to z_j,\end{aligned}$$

then this representation yields that

$$\operatorname*{Res}_{z=z_j}\left(\frac{zf'(z)}{f(z)}\right) = r_j z_j.$$

Similarly, we obtain that

$$\frac{zf'(z)}{f(z)} = -\frac{\tilde{z}_j s_j}{z-\tilde{z}_j} - s_j + o(1), \quad z \to \tilde{z}_j,$$

and consequently,

$$\operatorname*{Res}_{z=\tilde{z}_j}\left(\frac{zf'(z)}{f(z)}\right) = -s_j \tilde{z}_j.$$

These two facts prove the required formula (18.43). Further, using the doubly periodicity of $f(z)$, we have that

$$
\begin{aligned}
&\frac{1}{2\pi\mathrm{i}}\int_{\partial P} f(z)\mathrm{d}z \\
&= \frac{1}{2\pi\mathrm{i}}\left(\int_{\zeta}^{\zeta+2\omega_1} + \int_{\zeta+2\omega_1}^{\zeta+2\omega_1+2\omega_2} + \int_{\zeta+2\omega_1+2\omega_2}^{\zeta+2\omega_2} + \int_{\zeta+2\omega_1}^{\zeta}\right)\frac{zf'(z)}{f(z)}\mathrm{d}z \\
&= \frac{1}{2\pi\mathrm{i}}\left(-2\omega_2\int_{\zeta}^{\zeta+2\omega_1}\frac{f'(z)}{f(z)}\mathrm{d}z + 2\omega_1\int_{\zeta}^{\zeta+2\omega_2}\frac{f'(z)}{f(z)}\mathrm{d}z\right) \\
&= \frac{1}{2\pi\mathrm{i}}\left(-2\omega_2\log f(z)\big|_{\zeta}^{\zeta+2\omega_1} + 2\omega_1\log f(z)\big|_{\zeta}^{\zeta+2\omega_2}\right) \\
&= \frac{1}{2\pi\mathrm{i}}\left(-2\mathrm{i}\omega_2(\arg f(\zeta+2\omega_1) - \arg f(\zeta)) + 2\mathrm{i}\omega_1(\arg f(\zeta+2\omega_2) - \arg f(\zeta))\right) \\
&= -2\omega_2 n + 2\omega_1 m, \quad n, m \in \mathbb{Z}
\end{aligned}
$$

due to the properties of the multivalued function $\log(\cdot)$. Combining all these facts, we finally obtain that

$$
\sum_{j=1}^{k} r_j z_j - \sum_{j=1}^{l} s_j \tilde{z}_j = 2\omega_1 m + 2\omega_2 n,
$$

which is what was needed to be shown. The proposition is completely proved now. □

The simplest elliptic functions are the functions of order 2. They can be divided into two classes: there is only one pole of order 2 with the residue equaling to 0, and there are two simple poles with the residues of the opposite sign (with equal modulii). The functions of the first class are called *Weierstrass elliptic functions*, and of the second class are called *Jacobi elliptic functions*. We will show that any elliptic function can be expressed through the Weierstrass or Jacobi elliptic functions. We focus our attention mainly on the Weierstrass elliptic functions.

Weierstrass elliptic function

Definition 18.40. We define the Weierstrass elliptic function $\wp(z)$ by the equality

$$
\wp(z) := \frac{1}{z^2} + \sum_{\substack{n,m=-\infty \\ n^2+m^2>0}}^{\infty}\left(\frac{1}{(z-2m\omega_1-2n\omega_2)^2} - \frac{1}{(2m\omega_1+2n\omega_2)^2}\right), \tag{18.44}
$$

where ω_1 and ω_2 satisfy the conditions (18.42).

It is clear that the series (18.44) converges absolutely and uniformly in z outside of the points $2m\omega_1 + 2n\omega_2$ and defines therefore an analytic function such that the points $2m\omega_1 + 2n\omega_2$ are the poles of order 2.

Problem 18.41. Show that

$$\wp(z) = \left(\frac{\pi}{2\omega_1}\right)^2\left(-\frac{1}{3} + \sum_{n=-\infty}^{\infty} \frac{1}{\sin^2(\pi\frac{z-2n\omega_2}{2\omega_1})} - \sum_{n=-\infty,n\neq 0}^{\infty} \frac{1}{\sin^2(\pi\frac{n\omega_2}{\omega_1})}\right)$$

$$= \left(\frac{\pi}{2\omega_1}\right)^2\left(-\frac{1}{3} + \sum_{n=-\infty}^{\infty} \frac{1}{\sin^2(\pi\frac{z-2n\omega_2}{2\omega_1})} - 2\sum_{n=1}^{\infty} \frac{1}{\sin^2(\pi\frac{n\omega_2}{\omega_1})}\right).$$

Hint: Use the representation for the function $\frac{1}{\sin^2(\cdot)}$ in Example 11.19 and Problem 11.20(3).

Since (18.44) is the series of analytic functions and the series which is differentiated term by term converges uniformly outside of the poles, then we obtain

$$\wp'(z) := -2 \sum_{n,m=-\infty}^{\infty} \frac{1}{(z - 2m\omega_1 - 2n\omega_2)^3}.$$

The latter equality and (18.44) immediately imply that (using the fact that $\wp$ is even)

$$\wp(-z) = \wp(z), \quad \wp'(-z) = -\wp'(z).$$

Further, we have

$$\wp'(z + 2\omega_1) := -2 \sum_{n,m=-\infty}^{\infty} \frac{1}{(z - 2m\omega_1 - 2n\omega_2 - 2\omega_1)^3}$$

$$= -2 \sum_{n,m'=-\infty}^{\infty} \frac{1}{(z - 2m'\omega_1 - 2n\omega_2)^3} = \wp'(z).$$

Similarly, we have that

$$\wp'(z + 2\omega_2) = \wp'(z),$$

i. e., $\wp'(z)$ is doubly periodic with the periods $2\omega_1$ and $2\omega_2$. Integrating now the latter equalities with respect to z, we get

$$\wp(z + 2\omega_1) = \wp(z) + A_1, \quad \wp(z + 2\omega_2) = \wp(z) + A_2$$

with some constants A_1, A_2. If we put here $z = -\omega_1$ and $z = -\omega_2$, respectively, we obtain that $A_1 = 0, A_2 = 0$. This means then that $\wp(z)$ is doubly periodic with periods $2\omega_1$ and $2\omega_2$ having only poles in the complex plane $\mathbb{C}$ of order 2. Therefore, the equality (18.44) defines an elliptic function.

Remark. The doubly periodicity of the Weierstrass function $\wp(z)$ can be obtained straightforwardly from definition (18.44).

We may now obtain the ordinary differential equation satisfied by the Weierstrass elliptic function. Since $\wp(z) - \frac{1}{z^2}$ is analytic in the neighborhood of the point $z = 0$ and since $\wp(z)$ is even, then the Taylor's expansion near $z = 0$ gives the representation

$$\wp(z) - \frac{1}{z^2} = \tilde{g}_2 z^2 + \tilde{g}_3 z^4 + O(z^6),$$

and so

$$\wp'(z) + \frac{2}{z^3} = 2\tilde{g}_2 z + 4\tilde{g}_3 z^3 + O(z^5).$$

It follows that

$$\wp^3(z) = \frac{1}{z^6} + 3\tilde{g}_2 \frac{1}{z^2} + 3\tilde{g}_3 + O(z^2)$$

and

$$(\wp'(z))^2 = \frac{4}{z^6} - 8\tilde{g}_2 \frac{1}{z^2} - 16\tilde{g}_3 + O(z^2).$$

Consequently, we have that

$$(\wp'(z))^2 - 4\wp^3(z) = -20\tilde{g}_2 \frac{1}{z^2} - 28\tilde{g}_3 + O(z^2).$$

The latter representation can be rewritten equivalently as

$$(\wp'(z))^2 - 4\wp^3(z) = -20\tilde{g}_2 \wp(z) - 28\tilde{g}_3 + O(z^2).$$

This means that the function

$$(\wp'(z))^2 - 4\wp^3(z) + 20\tilde{g}_2 \wp(z) + 28\tilde{g}_3$$

is elliptic and analytic near $z = 0$, i. e., it is an elliptic function without singular points (taking into account its periodicity) and, therefore, it is identically equal to the constant (see Proposition 18.39, Part (4)). Letting $z \to 0$, we obtain that this constant is equal to 0, and thus $\wp(z)$ satisfies the nonlinear differential equation

$$(\wp'(z))^2 = 4\wp^3(z) - 20\tilde{g}_2 \wp(z) - 28\tilde{g}_3.$$

Denoting $20\tilde{g}_2$ by g_2, and $28\tilde{g}_3$ by g_3, we get finally that

$$(\wp'(z))^2 = 4\wp^3(z) - g_2 \wp(z) - g_3, \tag{18.45}$$

where g_2 and g_3 are said to be the *invariants of the elliptic function* $\wp(z)$ and defined as

$$g_2 = \sum_{\substack{n,m=-\infty \\ n^2+m^2>0}}^{\infty} \frac{60}{(2m\omega_1 + 2n\omega_2)^4}, \quad g_3 = \sum_{\substack{n,m=-\infty \\ n^2+m^2>0}}^{\infty} \frac{140}{(2m\omega_1 + 2n\omega_2)^6}. \tag{18.46}$$

Problem 18.42.

1. Prove the formulas (18.46).
2. Show that near $z = 0$ the following representation holds:

$$\wp(z) = \frac{1}{z^2} + \sum_{j=1}^{\infty} c_{2j} z^{2j},$$

where $c_2 = \frac{g_2}{20}$, $c_4 = \frac{g_3}{28}$, $c_6 = \frac{g_2^2}{80\cdot 15}$, $c_8 = \frac{3g_2g_3}{80\cdot 77}, \dots$.

Conversely to (18.45), if there is a differential equation

$$(u'(z))^2 = 4u^3(z) - g_2u(z) - g_3$$

and if it is possible to define the numbers ω_1 and ω_2 such that g_2 and g_3 satisfy (18.46), then the general solution of this differential equation will be the function

$$u(z) = \wp(\pm z + \alpha),$$

where α is a constant of integration. Indeed, introducing a new variable ξ satisfying $u(z) = \wp(\xi)$, one can see that

$$(u'(z))^2 \equiv (\wp'(\xi))^2 (\xi'(z))^2 = 4\wp^3(z) - g_2\wp(z) - g_3,$$

which evidently implies that $(\xi'(z))^2 = 1$, i. e., $\xi'(z) = \pm 1$ or $\xi(z) = \pm z + \alpha$. Since $\wp(z)$ is even, then we obtain the needed statement.

The Weierstrass elliptic function has an integral representation. Let us consider the equality

$$z = \int_{\zeta}^{\infty} \frac{1}{\sqrt{4t^3 - g_2t - g_3}} \, dt,$$

where the integration goes over arbitrary curve connecting ζ and ∞ bypassing the roots of the polynomial $4t^3 - g_2t - g_3$. Differentiating this integral with respect to z, we obtain

$$(\zeta'(z))^2 = 4\zeta^3(z) - g_2\zeta(z) - g_3,$$

and consequently (as was shown earlier),

$$\zeta(z) = \wp(z + \alpha).$$

But if $\zeta \to \infty$, then $z \to 0$, and hence a is a pole of the Weierstrass function $\wp(\cdot)$, i. e.,

$$a = 2m\omega_1 + 2n\omega_2, \quad m, n \in \mathbb{Z}.$$

It implies that

$$\zeta(z) = \wp(z + 2m\omega_1 + 2n\omega_2) = \wp(z)$$

and, therefore,

$$z = \int_{\wp(z)}^{\infty} \frac{1}{\sqrt{4t^3 - g_2 t - g_3}} dt. \tag{18.47}$$

This integral representation might be considered as an equivalent definition of the Weierstrass elliptic function $\wp(z)$ with the invariants g_2 and g_3.

Proposition 18.43 (Addition formula). *Let $\wp(\cdot)$ be the Weierstrass elliptic function with periods $2\omega_1$ and $2\omega_2$, and let $z, y \in \mathbb{C}$ be arbitrary points such that $z \neq y(\bmod 2\omega_1, 2\omega_2)$. Then*

$$\wp(z + y) = \frac{1}{4}\left(\frac{\wp'(z) - \wp'(y)}{\wp(z) - \wp(y)}\right)^2 - \wp(z) - \wp(y). \tag{18.48}$$

Proof. Since $z \neq y(\bmod 2\omega_1, 2\omega_2)$, then $\wp(z) \neq \wp(y)$. Consider (under this condition) the system of equations

$$\wp'(z) = A\wp(z) + B, \quad \wp'(y) = A\wp(y) + B,$$

where the values A and B (depending on z and y) are to be determined. It is easy to check that the solutions A and B are equal to

$$A = \frac{\wp'(z) - \wp'(y)}{\wp(z) - \wp(y)}, \quad B = \frac{\wp(z)\wp'(y) - \wp(y)\wp'(z)}{\wp(z) - \wp(y)}.$$

Next, considering with these A and B, the function

$$\wp'(\zeta) - A\wp(\zeta) - B,$$

we may conclude that it is an elliptic function with the pole at $\zeta = 0$ of order 3. Hence, by the property (7) of Proposition 18.39, this function has only three zeros whose sum is comparable with some $2m\omega_1 + 2n\omega_2$. But $\zeta = z$ and $\zeta = y$ are the zeros of this function, and then the third zero is comparable with $\zeta = -z - y$, i. e.,

$$\wp'(-z - y) = A\wp(-z - y) + B$$

or, equivalently,

$$-\wp'(z+y) = A\wp(z+y) + B.$$

Taking now into account that $\wp(\cdot)$ is the Weierstrass elliptic function, we obtain actually that

$$(\wp'(\zeta))^2 = (A\wp(\zeta) + B)^2 = 4\wp^3(\zeta) - g_2\wp(\zeta) - B.$$

This implies that

$$4\wp^3(\zeta) - A^2\wp^2(\zeta) - (g_2 + 2AB)\wp(\zeta) - g_3 - B = 0.$$

This equation as the polynomial equation of degree 3 has exactly three roots $\wp(z)$, $\wp(y)$, and $\wp(z+y)$, and they are different from each other. Using the property of the coefficients of the cubic polynomials, we have that

$$\wp(z) + \wp(y) + \wp(z+y) = \frac{1}{4}A^2.$$

Substituting instead of A, its expression from above yields (18.48). □

Corollary 18.44 (Doubling formula). *Suppose that $2z$ is not a period of the Weierstrass function $\wp(\cdot)$. Then the following is true:*

$$\wp(2z) = \frac{1}{4}\left(\frac{\wp''(z)}{\wp'(z)}\right)^2 - 2\wp(z). \tag{18.49}$$

Proof. Letting $y \to z$ in (18.48) yields (since $2z$ is not a period)

$$\wp(2z) = \frac{1}{4}\lim_{y\to z}\left(\frac{\wp'(z) - \wp'(y)}{\wp(z) - \wp(y)}\right)^2 - 2\wp(z).$$

Applying L'Hôspital's rule for the limit we will have

$$\lim_{y\to z}\frac{\wp'(z) - \wp'(y)}{\wp(z) - \wp(y)} = \frac{\wp''(z)}{\wp'(z)}.$$

This proves the corollary. □

Proposition 18.45. *Let $\wp(z)$ be the Weierstrass elliptic function with periods $2\omega_1$ and $2\omega_2$. Then the values*

$$e_1 := \wp(\omega_1), \quad e_2 := \wp(\omega_2), \quad e_3 := \wp(\omega_1 + \omega_2) \tag{18.50}$$

are not equal to each other and they are the roots of the polynomial equation

$$4w^3 - g_2 w - g_3 = 0.$$

Proof. Consider first the value $\wp'(\omega_1)$. Then

$$\wp'(\omega_1) = -\wp'(-\omega_1) = -\wp'(2\omega_1 - \omega_1) = -\wp'(\omega_1),$$

and consequently, $\wp'(\omega_1) = 0$. Similarly, we obtain that

$$\wp'(\omega_2) = 0, \quad \wp'(\omega_1 + \omega_2) = 0.$$

Since $\wp'(z)$ is also periodic (as $\wp(z)$) with the same periods $2\omega_1$ and $2\omega_2$, then $\wp'(z)$ has three and only three zeros, i. e., these values ω_1, ω_2, and $\omega_3 = -\omega_1 - \omega_2$ (or the values which are comparable with them). Considering the function $\wp(z) - e_1$ yields that $z = \omega_1$ is a double zero ($\wp'(\omega_1) = 0$) of this function and all other zeros must be comparable with ω_1. The same is true for the functions $\wp(z) - e_2$ and $\wp(z) - e_3$ with their zeros ω_2 and ω_3, respectively. We can obtain now from here that the values e_1, e_2, and e_3 are not equal to each other. Indeed, if we assume, e. g., that $e_1 = e_2$, and then function $\wp(z) - e_1$ would have a zero at $z = \omega_2$, i. e., at the point that is not comparable with ω_1. This contradiction shows that the values e_1, e_2, e_3 are not equal to each other. In addition, since $\wp'(z) = 0$ at the points $\omega_1, \omega_2, \omega_3$ and

$$(\wp'(z))^2 = 4\wp^3(z) - g_2\wp(z) - g_3$$

then the right-hand side is equal to 0 if and only if $\wp(z) = e_1, e_2$, or e_3. It means that e_1, e_2, e_3 are the roots of the equation

$$4w^3 - g_2 w - g_3 = 0.$$

This proves the proposition. □

Remark. The latter proposition implies that

$$(\wp'(z))^2 = 4(\wp(z) - e_1)(\wp(z) - e_2)(\wp(z) - e_3)$$

and

$$\begin{cases} e_1 + e_2 + e_3 = 0, \\ e_1e_2 + e_2e_3 + e_3e_1 = -\frac{1}{4}g_2, \\ e_1e_2e_3 = \frac{1}{4}g_3. \end{cases} \tag{18.51}$$

Recalling the Cardano's formulae (see Chapter 2), we have that the discriminant Δ of the Weierstrass elliptic function is equal to

$$\Delta = g_2^3 - 27g_3^2 \tag{18.52}$$

and if g_2 and g_3 are real and $\Delta > 0$, then (see Problem 2.12) all three zeros e_1, e_2, e_3 are real and can be put in the following order: $e_1 > e_2 > e_3$. In that case, we can obtain that (see (18.47))

$$\begin{cases} \omega_1 = \int_{e_1}^{\infty} \frac{1}{\sqrt{4t^3-g_2t-g-3}} dt, \\ \omega_3 = -i \int_{-\infty}^{e_3} \frac{1}{\sqrt{-4t^3+g_2t+g_3}} dt, \end{cases} \tag{18.53}$$

so that ω_1 is real and ω_3 is pure imaginary.

Problem 18.46. Find the formula for ω_2 similar to (18.53).

The addition formula (18.48) and the first equality of (18.51) lead to

$$\begin{aligned} \wp(z+\omega_1) + \wp(z) + \wp(\omega_1) &= \frac{1}{4}\left(\frac{\wp'(z) - \wp'(\omega_1)}{\wp(z) - \wp(\omega_1)}\right)^2 = \frac{(\wp'(z))^2}{(\wp(z) - e_1)^2} \\ &= \frac{(\wp(z) - e_1)(\wp(z) - e_2)(\wp(z) - e_3)}{(\wp(z) - e_1)^2} \\ &= \frac{(\wp(z) - e_2)(\wp(z) - e_3)}{\wp(z) - e_1}. \end{aligned}$$

It can be rewritten as

$$\begin{aligned} \wp(z+\omega_1) &= \frac{(\wp(z) - e_2)(\wp(z) - e_3)}{\wp(z) - e_1} - \wp(z) - e_1 \\ &= \frac{2\wp(z)e_1 + e_2e_3}{\wp(z) - e_1} - e_1 = e_1 + \frac{(e_1 - e_2)(e_1 - e_3)}{\wp(z) - e_1}. \end{aligned} \tag{18.54}$$

The latter equality is an addition formula for the half-period. Similarly, we have

$$\wp(z+\omega_2) = e_2 + \frac{(e_2 - e_1)(e_2 - e_3)}{\wp(z) - e_2}.$$

Problem 18.47. Using (18.48) and (18.54), show that:

1. $\wp^2(z) = \frac{\wp''(z)}{6} + \frac{g_2}{12}$.
2. $\wp(z+y) - \wp(z-y) = -\frac{\wp'(z)\wp'(y)}{(\wp(z)-\wp(y))^2}$.
3. $\wp(\frac{\omega_1}{2}) = e_1 \pm \sqrt{(e_1 - e_2)(e_1 - e_3)}$.
4. $\wp(\omega_2 + \frac{\omega_1}{2}) = e_1 \mp \sqrt{(e_1 - e_2)(e_1 - e_3)}$.
5. $\wp'(\frac{\omega_1}{2}) = -2(\sqrt{(e_1 - e_2)^2(e_1 - e_3)} + \sqrt{(e_1 - e_2)(e_1 - e_3)^2})$.
6. $\frac{\wp'(z+\omega_1)}{\wp'(z)} = -(\frac{\wp(\frac{\omega_1}{2})-\wp(\omega_1)}{\wp(z)-\wp(\omega_1)})^2$.
7. The discriminant Δ of the equation $4t^3 - g_2t - g_3$ is equal to

$$\wp'(z)\wp'(z+\omega_1)\wp'(z+\omega_2)\wp'(z+\omega_3) = 16(e_1 - e_2)^2(e_2 - e_3)^2(e_3 - e_1)^2 = g_2^3 - 27g_3^2.$$

The Weierstrass elliptic functions may help in the integration (not in the elementary functions) of some specific integrals. Moreover, they help very much in solving of some nonlinear ordinary differential equations in terms of the Weierstrass functions. More precisely, let

$$R(f) := \alpha f^4 + 4\beta f^3 + 6\gamma f^2 + 4\delta f + \epsilon$$

be a polynomial of degree 4. We are interested in the evaluation of the integral

$$z := \int_{f_0}^{f(z)} \frac{1}{\sqrt{R(g)}} \mathrm{d}g, \tag{18.55}$$

where f_0 is an arbitrary constant for the moment. Further, we assume that $R(f)$ has no multiple zeros, otherwise the integral (18.55) can be evaluated in elementary functions using the Euler's substitutions. As known from the courses of linear algebra, the invariants of this polynomial $R(f)$ are equal to

$$g_2 = \alpha\epsilon - 4\beta\delta + 3\gamma^2, \quad g_3 = \alpha\gamma\epsilon + 2\beta\gamma\delta - \gamma^3 - \alpha\delta^2 - \beta^2\epsilon. \tag{18.56}$$

We will show that $f(z)$ from (18.55) can be presented as a rational function of the Weierstrass elliptic function $\wp(z) \equiv \wp(z; g_2, g_3)$.

Proposition 18.48. *Let f_0 from* (18.55) *be a simple root of $R(f)$. Then*

$$f(z) = f_0 + \frac{R'(f_0)}{4\wp(z; g_2, g_3) - \frac{R''(f_0)}{6}}, \tag{18.57}$$

where $'$ here denotes the derivative with respect to f. Moreover, this function $f(z)$ solves nonlinear ordinary differential equation with initial data f_0,

$$(f'(z))^2 = R(f), \quad f(0) = f_0.$$

Proof. Let f_0 be a simple root of $R(\cdot)$. Then by the Taylor's expansion near f_0, we have

$$\begin{aligned} R(g) &= R'(f_0)(g - f_0) + \frac{R''(f_0)}{2!}(g - f_0)^2 + \frac{R'''(f_0)}{3!}(g - f_0)^3 + \frac{R^{(iv)}(f_0)}{4!}(g - f_0)^4 \\ &\equiv \alpha(g - f_0)^4 + 4\tilde{\beta}(g - f_0)^3 + 6\tilde{\gamma}(g - f_0)^2 + 4\tilde{\delta}(g - f_0), \end{aligned}$$

where

$$\tilde{\beta} = \alpha f_0 + \beta, \quad \tilde{\gamma} = \alpha f_0^2 + 2\beta f_0 + \gamma, \quad \tilde{\delta} = \alpha f_0^3 + 3\beta f_0^2 + 3\gamma f_0 + \delta. \tag{18.58}$$

Changing now the variable of integration in (18.55) as $\tau := \frac{1}{g - f_0}$ and denoting by $\xi := \frac{1}{f - f_0}$, we obtain

$$z = \int_{\xi}^{\infty} \frac{1}{\sqrt{4\tilde{\delta}\tau^3 + 6\tilde{\gamma}\tau^2 + 4\tilde{\beta}\tau + \alpha}} \, d\tau.$$

Let us introduce here a new variable (again) as $\eta := \tilde{\delta}\tau + \frac{\tilde{\gamma}}{2}$. And denoting by $\phi := \tilde{\delta}\xi + \frac{\tilde{\gamma}}{2}$, we obtain that

$$z = \int_{\phi}^{\infty} \frac{1}{\sqrt{4\eta^3 - (3\tilde{\gamma}^2 - 4\tilde{\beta}\tilde{\delta})\eta - (2\tilde{\beta}\tilde{\gamma}\tilde{\delta} - \tilde{\gamma}^2 - \alpha\tilde{\delta}^2)}} \, d\eta.$$

It must be mentioned that $\tilde{\delta} \neq 0$, otherwise f_0 will be multiple zero for $R(\cdot)$. Also, it is easy to check that (due to the fact that f_0 is a simple root) the coefficients of the integrand in the latter integral are actually equal to g_2 and g_3 (see (18.56) and (18.58)), respectively, i. e.,

$$3\tilde{\gamma}^2 - 4\tilde{\beta}\tilde{\delta} = g_2, \quad 2\tilde{\beta}\tilde{\gamma}\tilde{\delta} - \tilde{\gamma}^2 - \alpha\tilde{\delta}^2 = g_3,$$

and consequently, $\phi = \wp(z; g_2, g_3)$. This implies that

$$\xi = \frac{1}{f - f_0} = \frac{\wp(z; g_2, g_3)}{\tilde{\delta}}$$

or, equivalently,

$$f(z) = f_0 + \frac{\tilde{\delta}}{\wp(z; g_2, g_3) - \frac{\tilde{\gamma}}{2}},$$

and finally

$$f(z) = f_0 + \frac{R'(f_0)}{4\wp(z; g_2, g_3) - \frac{R''(f_0)}{6}},$$

which proves (18.57) and the first part of the proposition. To prove the last part, we differentiate (18.55) with respect to f (considering z as a function of f) and obtain

$$\frac{dz}{df} = \frac{1}{\sqrt{R(f)}} \quad \text{or} \quad \frac{df}{dz} = \sqrt{R(f(z))}.$$

This proves the last part of the proposition. □

Problem 18.49 (*Weierstrass formula*). Show that if

$$z := \int_{f_0}^{f(z)} \frac{1}{\sqrt{R(g)}} \, dg,$$

where f_0 is an arbitrary constant and the polynomial $R(g)$ has no multiple zeros, then

$$f(z) = f_0 + \frac{\frac{d\wp(z)}{dz}\sqrt{R(f_0)} + \frac{R'(f_0)}{2}\left(\wp(z) - \frac{R''(f_0)}{24}\right) + \frac{R(f_0)R'''(f_0)}{24}}{(4\wp(z) - \frac{R''(f_0)}{6})^2 - \alpha\frac{R(f_0)}{2}}, \tag{18.59}$$

where $\wp(z) = \wp(z; g_2, g_3)$ with g_2 and g_3 from (18.56). Show that this function $f(z)$ solves the nonlinear ordinary differential equation

$$\left(\frac{df(z)}{dz}\right)^2 = R(f(z))$$

with initial data $f(0) = f_0$.

Problem 18.50. Show that under the conditions of Proposition 18.48 the following is true:

1.

$$\sqrt{R(f(z))} = -\frac{R'(f_0)\frac{d\wp(z)}{dz}}{2\wp(z) - \frac{R''(f_0)}{12}},$$

2.

$$\wp(z) = \frac{R'(f_0)}{4(f(z) - f_0)} + \frac{R''(f_0)}{24},$$

where $\wp(z) = \wp(z; g_2, g_3)$.

Problem 18.51. Let f and z be connected as

$$z = \int_{-\infty}^{f(z)} \frac{1}{\sqrt{t^4 + 6ct^2 + e^2}} dt.$$

Show that

$$f(z) = \frac{\frac{d\wp(z)}{dz}}{2(\wp(z) + c)},$$

where $\wp(z)$ is the Weierstrass elliptic function constructed by the roots $e_1 = -c$, $e_2 = \frac{c+e}{2}$, $e_3 = \frac{c-e}{2}$.

Quasiperiodic zeta and sigma functions

The *zeta function* $\zeta(z)$ is defined via the Weierstrass elliptic function as

$$\zeta'(z) = -\wp(z)$$

with the uniqueness condition

$$\lim_{z\to 0}\left(\zeta(z)-\frac{1}{z}\right)=0.$$

This definition implies that

$$\begin{aligned}\zeta(z)&=\frac{1}{z}-\int_0^z\left(\wp(t)-\frac{1}{t^2}\right)\mathrm{d}t\\&=\frac{1}{z}-\sum_{\substack{n,m=-\infty\\ m^2+n^2>0}}^{\infty}\int_0^z\left(\frac{1}{(t-2m\omega_1-2n\omega_2)^2}-\frac{1}{(2m\omega_1+2n\omega_2)^2}\right)\mathrm{d}t\\&=\frac{1}{z}+\sum_{\substack{n,m=-\infty\\ m^2+n^2>0}}^{\infty}\left(\frac{1}{z-2m\omega_1-2n\omega_2}+\frac{1}{2m\omega_1+2n\omega_2}+\frac{z}{(2m\omega_1+2n\omega_2)^2}\right).\end{aligned}\tag{18.60}$$

This formula immediately gives that $\zeta(-z)=-\zeta(z)$, i. e., the zeta function is odd and has a pole of order 1 at any point comparable with 0. It also gives that the residue at the pole is equal to 1, and thus the zeta function is not elliptic (and also is not doubly periodic). This function however, is *quasiperiodic* in the sense that

$$\begin{cases}\zeta(z+2\omega_1)=\zeta(z)+2\zeta(\omega_1),\\ \zeta(z+2\omega_2)=\zeta(z)+2\zeta(\omega_2).\end{cases}\tag{18.61}$$

Indeed, since $\wp(z+2\omega_1)=\wp(z)$ then integrating this equality yields

$$\zeta(z+2\omega_1)=\zeta(z)+C.$$

Considering now $z=-\omega_1$ and using oddness of $\zeta(\cdot)$, we obtain

$$\zeta(\omega_1)=-\zeta(\omega_1)+C.$$

This proves the first equality from (18.61). Similarly, we obtain the second equality from (18.61). There is some interesting connection between $\zeta(\omega_1)$ and $\zeta(\omega_2)$. Namely,

$$\zeta(\omega_1)\omega_2-\zeta(\omega_2)\omega_1=\frac{\pi\mathrm{i}}{2}.\tag{18.62}$$

To check (18.62), consider the following integral:

$$\int_{\partial P}\zeta(z)\mathrm{d}z,$$

where P is any cell (as for the Weierstrass elliptic function). Since inside of any cell there is only one pole of order 1 with the residue 1, we have

$$\int_{\partial P} \zeta(z)\mathrm{d}z = 2\pi \mathrm{i}.$$

On the other hand (using (18.61)), we will have that

$$2\pi \mathrm{i} = \int_{\partial P} \zeta(z)\mathrm{d}z = \int_{w}^{w+2\omega_1} (\zeta(z) - \zeta(z + 2\omega_2))\mathrm{d}z - \int_{w}^{w+2\omega_2} (\zeta(z) - \zeta(z + 2\omega_1))\mathrm{d}z$$
$$= -4\zeta(\omega_2)\omega_1 + 4\zeta(\omega_1)\omega_2,$$

which is what needed to be shown.

Problem 18.52. Prove that for any x, y, z with $x + y + z = 0$, the following is true:

$$(\zeta(x) + \zeta(y) + \zeta(z))^2 + \zeta'(x) + \zeta'(y) + \zeta'(z) = 0.$$

There is another function, which is closely connected with the Weierstrass elliptic function. The *sigma function* $\sigma(z)$ is defined as

$$(\log \sigma(z))' = \zeta(z)$$

with the uniqueness condition

$$\lim_{z\to 0} \frac{\sigma(z)}{z} = 1.$$

Integrating here and using (18.60), we obtain

$$\sigma(z) = C \exp(\log z) \exp\Bigg(\sum_{\substack{n,m=-\infty \\ m^2+n^2>0}}^{\infty} \int_0^z \Bigg(\frac{1}{t - 2m\omega_1 - 2n\omega_2} + \frac{1}{2m\omega_1 + 2n\omega_2}$$
$$+ \frac{t}{(2m\omega_1 + 2n\omega_2)^2} \Bigg)\mathrm{d}t \Bigg)$$
$$= zC \exp\Bigg(\sum_{\substack{n,m=-\infty \\ m^2+n^2>0}}^{\infty} \Bigg(\log(z - 2m\omega_1 - 2n\omega_2) + \frac{z}{2m\omega_1 + 2n\omega_2}$$
$$+ \frac{z^2}{2(2m\omega_1 + 2n\omega_2)^2} \Bigg)\Bigg),$$

and consequently,

$$\frac{\sigma(z)}{z} = C \prod_{\substack{n,m=-\infty \\ m^2+n^2>0}}^{\infty} \left((z - 2m\omega_1 - 2n\omega_2)\mathrm{e}^{\frac{z}{2m\omega_1+2n\omega_2} + \frac{z^2}{2(2m\omega_1+2n\omega_2)^2}}\right).$$

Since $\lim_{z\to 0} \frac{\sigma(z)}{z} = 1$, then the constant of integration C is equal to

$$C = \prod_{\substack{n,m=-\infty \\ m^2+n^2>0}}^{\infty} \left(-\frac{1}{2m\omega_1 + 2n\omega_2}\right).$$

Hence, we obtain finally that

$$\sigma(z) = z \prod_{\substack{n,m=-\infty \\ m^2+n^2>0}}^{\infty} \left(\left(1 - \frac{z}{2m\omega_1 + 2n\omega_2}\right) e^{\frac{z}{2m\omega_1+2n\omega_2} + \frac{z^2}{2(2m\omega_1+2n\omega_2)^2}}\right). \tag{18.63}$$

This representation shows that $\sigma(z)$ is an entire function with simple zeros at the points $z = 2m\omega_1 + 2n\omega_2, m, n \in \mathbb{Z}$. It is also clear that $\sigma(z)$ is odd. The sigma function $\sigma(z)$ is quasiperiodic in the sense that

$$\sigma(z + 2\omega_1) = -e^{2\zeta(\omega_1)(z+\omega_1)}\sigma(z), \quad \sigma(z + 2\omega_2) = -e^{2\zeta(\omega_2)(z+\omega_2)}\sigma(z).$$

Indeed, since

$$\zeta(z + 2\omega_1) = \zeta(z) + 2\zeta(\omega_1), \quad \zeta(z + 2\omega_2) = \zeta(z) + 2\zeta(\omega_2)$$

then the definition of $\sigma(z)$ yields

$$\sigma(z + 2\omega_1) = Ce^{2\zeta(\omega_1)z}\sigma(z), \quad \sigma(z + 2\omega_2) = C'e^{2\zeta(\omega_2)z}\sigma(z),$$

where C, C' are the constants of integration. Putting $z = -\omega_1$ and $z = -\omega_2$, respectively, and using that $\sigma(z)$ is odd we get

$$\sigma(\omega_1) = -Ce^{-2\zeta(\omega_1)\omega_1}\sigma\omega_1, \quad \sigma(\omega_2) = -C'e^{-2\zeta(\omega_2)\omega_2}\sigma(\omega_2)$$

and further,

$$C = -e^{2\zeta(\omega_1)\omega_1}, \quad C' = -e^{2\zeta(\omega_2)\omega_2}.$$

Substituting these constants, we obtain the quasiperiodicity of the sigma function $\sigma(z)$.

Problem 18.53.

1. Show that

$$\frac{\sigma(2z)}{\sigma^4(z)} = -\wp'(z), \quad \frac{\sigma(3z)}{\sigma^9(z)} = 3\wp(z)(\wp'(z))^2 - \frac{(\wp''(z))^2}{4},$$

   where $\wp(z)$ is the Weierstrass elliptic function.

2. Let $e_1 > e_2 > e_3$, where $e_1 = \wp(\omega_1)$, $e_2 = \wp(\omega_2)$, $e_3 = \wp(\omega_1 + \omega_2)$, and $\wp(z)$ is the Weierstrass elliptic function with periods $2\omega_1, 2\omega_2$. What one can say about the values of the function

$$\psi(z) := \zeta(z) - \frac{z\zeta(\omega_1 + \omega_2)}{\omega_1 + \omega_2}$$

when z goes along the parallelogram with vertices $-\omega_1, \omega_1, -\omega_2, -2\omega_1 - \omega_2$?

Problem 18.54. Show that

$$\int \frac{1}{((x^2 - a)(x^2 - b))^{1/4}} \mathrm{d}x = -\frac{1}{2} \log \frac{\sigma(z - z_0)}{\sigma(z + z_0)} + \frac{\mathrm{i}}{2} \log \frac{\sigma(z - \mathrm{i}z_0)}{\sigma(z + \mathrm{i}z_0)},$$

where

$$x^2 = a + \frac{1}{6(\wp(z) - \wp(z_0))}, \quad \wp^2(z_0) = \frac{1}{6(a - b)}$$

and $\wp(\cdot)$ is the Weierstrass elliptic function with the invariants

$$g_2 = \frac{2b}{3a(a - b)}, \quad g_3 = 0.$$

We are in position now to show that any elliptic function with periods $2\omega_1$ and $2\omega_2$ can be expressed in terms of the Weierstrass elliptic function $\wp(z)$ and its derivative $\wp'(z)$ with the same periods. Since the following identity holds,

$$f(z) = \frac{1}{2}(f(z) + f(-z)) + \frac{(f(z) - f(-z))\wp'(z)}{2\wp'(z)},$$

and since $\wp'(z)$ is odd, then the problem would be solved if we express any even elliptic function $\psi(z)$ through $\wp(z)$ and $\wp'(z)$. Indeed, if $a_1, a_2, \dots, a_k$ are the zeros of $\psi(z)$ in some cell, then $-a_1, -a_2, \dots, -a_k$ (comparable with $a_1, a_2, \dots, a_k$) are also the zeros in this cell. Similarly, if $b_1, b_2, \dots, b_k$ are the poles, then $-b_1, -b_2, \dots, -b_k$ are also the poles there. Considering the function

$$F(z) := \frac{1}{\psi(z)} \frac{\prod_{j=1}^{k}(\wp(z) - \wp(a_j))}{\prod_{j=1}^{k}(\wp(z) - \wp(b_j))}$$

we conclude that $F(z)$ is an elliptic function with periods $2\omega_1$ and $2\omega_2$. Moreover, this function has no poles since the zeros of $\psi(z)$ are exactly the same as the zeros of the product in the numerator, and the zeros of the product in the denominator are exactly the poles of $\psi(z)$. It means that $F(z)$ is a bounded (in this cell) analytic function, and thus, by the Liouville's theorem, $F(z) \equiv$ constant. It implies that

$$\psi(z) = C\frac{\prod_{j=1}^{k}(\wp(z) - \wp(a_j))}{\prod_{j=1}^{k}(\wp(z) - \wp(b_j))}.$$

Applying this procedure to each of the functions,

$$f(z) + f(-z), \quad \frac{f(z) - f(-z)}{\wp'(z)},$$

we obtain the required result. Moreover, we obtain that this relation is a rational function of $\wp(z)$ and $\wp'(z)$.

Problem 18.55. Show that any elliptic function $f(z)$ with periods $2\omega_1$ and $2\omega_2$ can be represented (in some cell) as

$$f(z) = C + \sum_{j=1}^{k}\sum_{l=1}^{r_j}\frac{(-1)^{l-1}}{(l-1)!}c_{jl}\zeta^{(l-1)}(z - a_j),$$

where C is some constant, $\zeta(\cdot)$ is the zeta function with quasiperiods $2\omega_1, 2\omega_2$ (see (18.61)), and $a_1, a_2, \dots, a_k$ are the poles of $f(z)$ in this cell with the main part of the Laurent's expansion near these poles as

$$\frac{c_{j1}}{z - a_j} + \frac{c_{j2}}{(z - a_j)^2} + \cdots + \frac{c_{jr_j}}{(z - a_j)^{r_j}}, \quad j = 1, 2, \dots, k.$$

Problem 18.56. Prove that any elliptic function $f(z)$ with periods $2\omega_1$ and $2\omega_2$ can be represented (in some cell) as

$$f(z) = C\prod_{j=1}^{k}\frac{\sigma(z - a_j)}{\sigma(z - b_j)},$$

where C is some constant, $a_1, a_2, \dots, a_k$ are the zeros of $f(z)$, and $b_1, b_2, \dots, b_k$ are the poles of $f(z)$ such that

$$a_1 + a_2 + \cdots + a_k = b_1 + b_2 + \cdots + b_k$$

and $\sigma(\cdot)$ is the sigma quasiperiodic function constructed via the Weierstrass elliptic function with periods $2\omega_1$ and $2\omega_2$.

Problem 18.57.

1. Show that

$$\wp(z) - \wp(y) = -\frac{\sigma(z + y)\sigma(z - y)}{\sigma^2(z)\sigma^2(y)},$$

where $\wp(\cdot)$ is the Weierstrass elliptic function and $\sigma(\cdot)$ is the corresponding sigma function.

2. Differentiating the equality from (1), show that

$$\frac{1}{2}\frac{\wp'(z)-\wp'(y)}{\wp(z)-\wp(y)} = \zeta(z+y)-\zeta(z)-\zeta(y),$$

where $\zeta(\cdot)$ is the corresponding zeta function.

Jacobi elliptic functions

As it was mentioned earlier, the Jacobi elliptic functions can be defined as the elliptic functions of order 2 having (in cell) two simple poles with the residues in them equal in magnitude but opposite in sign. A more useful definition of the basic elliptic functions $\mathrm{sn}(\cdot)$, $\mathrm{cn}(\cdot)$, $\mathrm{dn}(\cdot)$ is connected with the inversion of *elliptic integrals of the first kind*, i. e., the integrals of the form

$$z := \int_0^y \frac{1}{\sqrt{(1-t^2)(1-k^2t^2)}}\,dt, \tag{18.64}$$

where the complex k^2 does not belong to $(-\infty, 0] \cup [1, \infty)$ and $y = y(z)$ is a function of z everywhere except the singular points, which are simple poles. Specifically, the Jacobi sn-function, $y = \mathrm{sn}(z; k)$ is defined via the integral (18.64). This definition implies immediately that $\mathrm{sn}(\cdot)$ is odd. The Jacobi $\mathrm{cn}(\cdot)$ and $\mathrm{dn}(\cdot)$-functions might be defined equivalently, using the following identities:

$$\begin{cases} \mathrm{sn}^2(z;k) + \mathrm{cn}^2(z;k) = 1, \\ k^2\,\mathrm{sn}^2(z;k) + \mathrm{dn}^2(z;k) = 1, \\ \mathrm{cn}(0;k) = \mathrm{dn}(0;k) = 1. \end{cases} \tag{18.65}$$

Problem 18.58.

1. Show that the identities (18.65) define $\mathrm{cn}(z;k)$ and $\mathrm{dn}(z;k)$ uniquely.
2. Show that $\mathrm{cn}(z;k)$ and $\mathrm{dn}(z;k)$ are even functions.
3. Show that if

$$z = \int_{y_2}^1 \frac{1}{\sqrt{(1-t^2)(1-k^2(1-t^2))}}\,dt, \quad z = \int_{y_3}^1 \frac{1}{\sqrt{(1-t^2)(t^2-1+k^2)}}\,dt, \tag{18.66}$$

then $y_2 = \mathrm{cn}(z;k)$ and $y_3 = \mathrm{dn}(z;k)$, respectively.

4. Show that

$$\begin{cases} (\mathrm{sn}(z;k))'_z = \mathrm{cn}(z;k)\cdot\mathrm{dn}(z;k), \\ (\mathrm{cn}(z;k))'_z = -\,\mathrm{sn}(z;k)\cdot\mathrm{dn}(z;k), \\ (\mathrm{dn}(z;k))'_z = -k^2\,\mathrm{sn}(z;k)\cdot\mathrm{cn}(z;k). \end{cases}$$

5. Show that $y_1 = \operatorname{sn}(z;k)$, $y_2 = \operatorname{cn}(z;k)$, and $y_3 = \operatorname{dn}(z;k)$ might be also considered as the solutions of the following differential equations:

$$\begin{cases} (y_1'(z))^2 = (1 - y_1^2(z))(1 - k^2 y_1^2(z)), & y_1(0) = 0, \\ (y_2'(z))^2 = (1 - y_2^2(z))(1 - k^2(1 - y_2^2(z))), & y_2(0) = 1, \\ (y_3'(z))^2 = (1 - y_3^2(z))(y_3^2(z) - 1 + k^2), & y_3(0) = 1. \end{cases} \tag{18.67}$$

6. Show that for $k = 0$, we have

$$\operatorname{sn}(z;0) = \sin z, \quad \operatorname{cn}(z;0) = \cos z, \quad \operatorname{dn}(z;0) \equiv 1.$$

7. Prove the addition formula for elliptic functions $\operatorname{sn}(\cdot)$, $\operatorname{cn}(\cdot)$, $\operatorname{dn}(\cdot)$:

$$\begin{cases} \operatorname{sn}(z+y;k) = \frac{\operatorname{sn}(z;k)\operatorname{cn}(y;k)\operatorname{dn}(y;k)+\operatorname{sn}(y;k)\operatorname{cn}(z;k)\operatorname{dn}(z;k)}{1-k^2\operatorname{sn}^2(z;k)\operatorname{sn}^2(y;k)}, \\ \operatorname{cn}(z+y;k) = \frac{\operatorname{cn}(z;k)\operatorname{cn}(y;k)-\operatorname{sn}(y;k)\operatorname{sn}(y;k)\operatorname{dn}(z;k)\operatorname{dn}(y;k)}{1-k^2\operatorname{sn}^2(z;k)\operatorname{sn}^2(y;k)}, \\ \operatorname{dn}(z+y;k) = \frac{\operatorname{dn}(z;k)\operatorname{dn}(y;k)-k^2\operatorname{sn}(y;k)\operatorname{sn}(y;k)\operatorname{cn}(z;k)\operatorname{cn}(y;k)}{1-k^2\operatorname{sn}^2(z;k)\operatorname{sn}^2(y;k)}. \end{cases}$$

There are some special constants with respect to Jacobi functions, which have independent interest. The first one is defined as

$$K := \int_0^1 \frac{1}{\sqrt{(1-t^2)(1-k^2t^2)}}\,dt. \tag{18.68}$$

Consequently,

$$\operatorname{sn}(K;k) = 1, \quad \operatorname{cn}(K;k) = 0, \quad \operatorname{dn}(K;k) = \sqrt{1-k^2},$$

where a square root is equal to $+\sqrt{1-k^2}$ for $0 < k < 1$. These equalities and the addition formula imply (see Problem 18.58, Part (7)) that

$$\begin{cases} \operatorname{sn}(z+K;k) = \frac{\operatorname{cn}(z;k)}{\operatorname{dn}(z;k)}, \\ \operatorname{cn}(z+K;k) = -\sqrt{1-k^2}\frac{\operatorname{sn}(z;k)}{\operatorname{dn}(z;k)}, \\ \operatorname{dn}(z+K;k) = \frac{\sqrt{1-k^2}}{\operatorname{dn}(z;k)}. \end{cases}$$

Repeating this process once more and using the latter equalities, one can obtain

$$\operatorname{sn}(z+2K;k) = -\operatorname{sn}(z;k), \quad \operatorname{cn}(z+2K;k) = -\operatorname{cn}(z;k), \quad \operatorname{dn}(z+2K;k) = \operatorname{dn}(z;k).$$

Consequently,

$$\operatorname{sn}(z+4K;k) = \operatorname{sn}(z;k), \quad \operatorname{cn}(z+4K;k) = \operatorname{cn}(z;k),$$

i. e., the value $4K$ is the period for elliptic functions $\operatorname{sn}(z;k)$ and $\operatorname{cn}(z;k)$, and the value $2K$ is the period for $\operatorname{dn}(z;k)$.

Problem 18.59. Show that

$$\operatorname{sn}\left(\frac{K}{2};k\right)=\frac{1}{k}\sqrt{1-\sqrt{1-k^2}},\quad \operatorname{cn}\left(\frac{K}{2};k\right)=\frac{1}{k}\sqrt{k^2-1+\sqrt{1-k^2}},$$
$$\operatorname{dn}\left(\frac{K}{2};k\right)=(1-k^2)^{1/4}.$$

Another important constant is K', which is defined as

$$K':=\int_0^1\frac{1}{\sqrt{(1-t^2)(k^2t^2+1-t^2)}}\mathrm{d}t, \tag{18.69}$$

where (as before) the complex k^2 does not belong to $(-\infty,0]\cup[1,\infty)$. We want to show that actually this constant K' is equal to (compare with (18.68))

$$K'=\int_1^{1/k}\frac{1}{\sqrt{(t^2-1)(1-k^2t^2)}}\mathrm{d}t.$$

Indeed, let us first assume that $0<k<1$ and consider a new variable

$$s:=\frac{1}{\sqrt{1-(1-k^2)t^2}},$$

which leads to the equalities

$$\sqrt{s^2-1}=\frac{t\sqrt{1-k^2}}{\sqrt{1-(1-k^2)t^2}},\qquad \sqrt{1-k^2s^2}=\frac{\sqrt{(1-k^2)(1-t^2)}}{\sqrt{1-(1-k^2)t^2}},$$
$$\frac{\mathrm{d}s}{\mathrm{d}t}=\frac{t(1-k^2)}{(1-(1-k^2)t^2)^{3/2}}.$$

Substituting these equalities into (18.69) yield the required result for the case $0<k<1$. For this real value k, we consider now the integral

$$\int_0^{1/k}\frac{1}{\sqrt{(1-t^2)(1-k^2t^2)}}\mathrm{d}t,$$

where the curve of integration from 0 to $\frac{1}{k}$ goes under the point 1 such that it consists of the lines on the real axis: 0 to $1-\delta$, $1+\delta$ to $\frac{1}{k}$, with a small positive δ; and a semicircle with radius δ centered at 1. Then we have

$$\int_0^{1/k} \frac{1}{\sqrt{(1-t^2)(1-k^2t^2)}}dt = \int_0^{1-\delta} \frac{1}{\sqrt{(1-t^2)(1-k^2t^2)}}dt + \int_{|\zeta-1|=\delta,\operatorname{Im}\zeta>0} \frac{1}{\sqrt{(1-\zeta^2)(1-k^2\zeta^2)}}d\zeta + \int_{1+\delta}^{1/k} \frac{1}{\sqrt{(1-t^2)(1-k^2t^2)}}dt.$$

Letting $\delta \to +0$, we can see that the integral over semicircle tends to 0, and hence, we obtain that

$$\int_0^{1/k} \frac{1}{\sqrt{(1-t^2)(1-k^2t^2)}}dt = \int_0^{1} \frac{1}{\sqrt{(1-t^2)(1-k^2t^2)}}dt + \int_1^{1/k} \frac{1}{\sqrt{(1-t^2)(1-k^2t^2)}}dt$$
$$= K + \mathrm{i}\int_1^{1/k} \frac{1}{\sqrt{(t^2-1)(1-k^2t^2)}}dt = K + \mathrm{i}K'. \tag{18.70}$$

Considering now the analytical continuation of these values as the functions of the complex k into the complex plane with cut of along the real axis from $-\infty$ to 0 and from 1 to $+\infty$, we conclude that the formula (18.70) holds for all such complex k. Moreover, (18.70) together with (18.65) lead to

$$\operatorname{sn}(K+\mathrm{i}K';k) = \frac{1}{k}, \quad \operatorname{cn}(K+\mathrm{i}K';k) = -\sqrt{1-\frac{1}{k^2}}, \quad \operatorname{dn}(K+\mathrm{i}K';k) = 0.$$

Problem 18.60. Show that

1.
$$\begin{cases} \operatorname{sn}(z+4K+4\mathrm{i}K';k) = \operatorname{sn}(z;k), \\ \operatorname{cn}(z+2K+2\mathrm{i}K';k) = \operatorname{cn}(z;k), \\ \operatorname{dn}(z+4K+4\mathrm{i}K';k) = \operatorname{dn}(z;k), \end{cases}$$

   i. e., the functions $\operatorname{sn}(\cdot)$, $\operatorname{dn}(\cdot)$ have a period $4K+4\mathrm{i}K'$, and the function $\operatorname{cn}(\cdot)$ has a period $2K+2\mathrm{i}K'$.

2.
$$\begin{cases} \operatorname{sn}(z+2\mathrm{i}K';k) = \operatorname{sn}(z;k), \\ \operatorname{cn}(z+4\mathrm{i}K';k) = \operatorname{cn}(z;k), \\ \operatorname{dn}(z+4\mathrm{i}K';k) = \operatorname{dn}(z;k), \end{cases} \tag{18.71}$$

   i. e., the values $2\mathrm{i}K'$ and $4\mathrm{i}K'$ are the periods of $\operatorname{sn}(\cdot)$, $\operatorname{cn}(\cdot)$, and $\operatorname{dn}(\cdot)$, respectively.

   Since (see Problem 18.58, Part (4))

$$(\operatorname{sn}(z;k))'_z = \operatorname{cn}(z;k)\operatorname{dn}(z;k),$$

and

$$\begin{aligned}(\operatorname{sn}(z;k))_z''' &= 4k^2\operatorname{sn}^2(z;k)\operatorname{cn}(z;k)\operatorname{dn}(z;k)\\ &\quad - \operatorname{cn}(z;k)\operatorname{dn}(z;k)(\operatorname{dn}^2(z;k)+k^2\operatorname{cn}^2(z;k))\end{aligned}$$

then the Taylor's expansions near $z = 0$ (taking into account the oddness of $\operatorname{sn}(\cdot)$ and some similar formulas for the derivative of even functions $\operatorname{cn}(\cdot)$ and $\operatorname{dn}(\cdot)$) of these functions are

$$\begin{cases}\operatorname{sn}(z;k) = z - \frac{1+k^2}{6}z^3 + O(z^5),\\ \operatorname{cn}(z;k) = 1 - \frac{1}{2}z^2 + O(z^4),\\ \operatorname{dn}(z;k) = 1 - \frac{k^2}{2}z^2 + O(z^4).\end{cases} \tag{18.72}$$

Problem 18.61. Using similar procedure as in (18.72), prove that

$$\begin{cases}\operatorname{sn}(z+\mathrm{i}K';k) = \frac{1}{kz} + \frac{1+k^2}{6k}z + O(z^3),\\ \operatorname{cn}(z+\mathrm{i}K';k) = -\frac{\mathrm{i}}{kz} + \frac{2k^2-1}{6k}\mathrm{i}z + O(z^3),\\ \operatorname{dn}(z+\mathrm{i}K';k) = -\frac{\mathrm{i}}{z} + \frac{2-k^2}{6}\mathrm{i}z + O(z^3).\end{cases} \tag{18.73}$$

Remark. These formulas (18.73) show that the elliptic Jacobi functions $\operatorname{sn}(\cdot)$, $\operatorname{cn}(\cdot)$, and $\operatorname{dn}(\cdot)$ have a singularity at the point $z = \mathrm{i}K'$ and this singularity is a simple pole with the residues $\frac{1}{k}$, $-\frac{\mathrm{i}}{k}$, and $-\mathrm{i}$, respectively.

Problem 18.62. Prove that:

1.

$$\operatorname{sn}\left(\frac{\mathrm{i}K'}{2};k\right) = \frac{\mathrm{i}}{\sqrt{k}},\quad \operatorname{cn}\left(\frac{\mathrm{i}K'}{2};k\right) = \sqrt{1+\frac{1}{k}},\quad \operatorname{dn}\left(\frac{\mathrm{i}K'}{2};k\right) = \sqrt{1+k},$$

2.

$$\begin{aligned}\operatorname{sn}\left(\frac{K+\mathrm{i}K'}{2};k\right) &= \frac{\sqrt{1+k}+\mathrm{i}\sqrt{1-k}}{\sqrt{2k}},\\ \operatorname{cn}\left(\frac{K+\mathrm{i}K'}{2};k\right) &= \frac{(1-\mathrm{i})(1-k^2)^{1/4}}{\sqrt{2k}},\\ \operatorname{dn}\left(\frac{K+\mathrm{i}K'}{2};k\right) &= \frac{(1-k^2)^{1/4}(\sqrt{1+\sqrt{1-k^2}} - \mathrm{i}\sqrt{1-\sqrt{1-k^2}})}{\sqrt{2}}.\end{aligned}$$

There is an important connection between the Jacobi and Weierstrass elliptic functions. The following proposition holds.

Proposition 18.63. *Let e_1, e_2, e_3 be three different numbers (complex, in general) such that $e_1+e_2+e_3 = 0$. Then*

$$\wp(z; g_2, g_3) = e_3 + \frac{e_1 - e_3}{\operatorname{sn}^2(z\sqrt{e_1 - e_3}; \sqrt{\frac{e_2-e_3}{e_1-e_3}})}, \tag{18.74}$$

where the invariants of the Weierstrass elliptic function $\wp(\cdot)$ are equal to

$$g_2 = -4(e_1e_2 + e_2e_3 + e_1e_3), \quad g_3 = 4e_1e_2e_3.$$

Proof. Consider the following function:

$$F(z) := e_3 + \frac{e_1 - e_3}{\operatorname{sn}^2(\lambda z; k)},$$

where $\operatorname{sn}(\cdot)$ is the Jacobi elliptic function with the parameters λ, k to be determined. Next,

$$F'(z) = -\frac{2\lambda(e_1 - e_3)\operatorname{cn}(\lambda z; k)\operatorname{dn}(\lambda z; k)}{\operatorname{sn}^3(\lambda z; k)}$$

and, therefore,

$$\begin{aligned}
(F'(z))^2 &= \frac{4\lambda^2(e_1 - e_3)^2\operatorname{cn}^2(\lambda z; k)\operatorname{dn}^2(\lambda z; k)}{\operatorname{sn}^6(\lambda z; k)} \\
&= \frac{4\lambda^2(e_1 - e_3)^2(1 - \frac{1}{\operatorname{sn}^2(\lambda z;k)})(k^2 - \frac{1}{\operatorname{sn}^2(\lambda z;k)})}{\operatorname{sn}^2(\lambda z; k)} \\
&= \frac{4\lambda^2(e_1 - e_3)^2(\frac{1}{\operatorname{sn}^2(\lambda z;k)} - 1)(\frac{1}{\operatorname{sn}^2(\lambda z;k)} - k^2)}{\operatorname{sn}^2(\lambda z; k)} \\
&= 4\lambda^2(e_1 - e_3)^2\left(\frac{F(z) - e_3}{e_1 - e_3} - 1\right)\left(\frac{F(z) - e_3}{e_1 - e_3} - k^2\right)\left(\frac{F(z) - e_3}{e_1 - e_3}\right).
\end{aligned}$$

Finally, we get

$$(F'(z))^2 = \frac{4\lambda^2}{(e_1 - e_3)}(F(z) - e_1)(F(z) - e_3)(F(z) - e_3 - k^2(e_1 - e_3)).$$

Choosing $\lambda^2 := e_1 - e_3$ and $k^2 := \frac{e_2-e_3}{e_1-e_3}$, we obtain that

$$(F'(z))^2 = 4F^3(z) - g_2F(z) - g_3,$$

where g_2 and g_3 are the invariants of the Weierstrass elliptic function chosen as above. This proves the proposition. □

Remark. The equality (18.74) can be rewritten as

$$\operatorname{sn}^2(\zeta; k) = \frac{e_1 - e_3}{\wp(\frac{\zeta}{\sqrt{e_1-e_3}}; g_2, g_3) - e_3}, \tag{18.75}$$

where g_2 and g_3 are as in (18.74), and $\frac{e_2-e_3}{e_1-e_3} = k^2$.

The Jacobi elliptic functions sn(⋅), cn(⋅), dn(⋅), and others can be expanded into a trigonometric Fourier series (see Proposition 18.105 and Theorem 18.106 below). Consider, e. g., the Jacobi function $\operatorname{sn}(\frac{2Kx}{\pi};k)$ for real $x \in [-\pi,\pi]$. This function is odd and periodic with period 2π (since $4K$ is a period of sn(⋅)). So, we have

$$\operatorname{sn}\left(\frac{2Kx}{\pi};k\right)=\sum_{j=1}^{\infty} b_j \sin(jx), \quad b_j=\frac{1}{\pi}\int_{-\pi}^{\pi}\operatorname{sn}\left(\frac{2Kx}{\pi};k\right)\sin(jx)\mathrm{d}x$$

or, equivalently,

$$\mathrm{i}\pi b_j=\int_{-\pi}^{\pi}\operatorname{sn}\left(\frac{2Kx}{\pi};k\right)\mathrm{e}^{\mathrm{i}jx}\mathrm{d}x.$$

To evaluate the latter integral, we consider it in the complex plane with respect to the variable z instead of x, along the parallelogram with vertices $-\pi, \pi, \mathrm{i}\pi\tau, -2\pi+\mathrm{i}\pi\tau$, where $\tau=\frac{K'}{K}$ under the assumption $\frac{K'}{K}>0$. By the Cauchy's residue theorem, this integral is equal to the sum of the residues of the function $\operatorname{sn}(\frac{2Kz}{\pi};k)\mathrm{e}^{\mathrm{i}jz}$ at the points $z=\frac{1}{2}\mathrm{i}\pi\tau=\frac{\mathrm{i}\pi K'}{2K}$ and $z=-\pi+\frac{1}{2}\mathrm{i}\pi\tau=-\pi+\frac{\mathrm{i}\pi K'}{2K}$, i. e., this integral is equal to (see (18.73))

$$2\pi\mathrm{i}\left(\frac{1}{k}\frac{\pi}{2K}\mathrm{e}^{\frac{\mathrm{i}\pi jK'}{2K}}-\frac{1}{k}\frac{\pi}{2K}\mathrm{e}^{-\mathrm{i}\pi j+\frac{\mathrm{i}\pi jK'}{2K}}\right)=\frac{\pi^2\mathrm{i}}{Kk}(1-(-1)^j)\mathrm{e}^{-\frac{K'j}{2K}}.$$

But on the other hand, this integral is equal to

$$\begin{aligned}
&\int_{-\pi}^{\pi}\operatorname{sn}\left(\frac{2Kx}{\pi};k\right)\mathrm{e}^{\mathrm{i}jx}\mathrm{d}x+\int_{\pi}^{\mathrm{i}\pi\tau}\operatorname{sn}\left(\frac{2Kz}{\pi};k\right)\mathrm{e}^{\mathrm{i}jz}\mathrm{d}z\\
&\quad+\int_{\mathrm{i}\pi\tau}^{-2\pi+\mathrm{i}\pi\tau}\operatorname{sn}\left(\frac{2Kz}{\pi};k\right)\mathrm{e}^{\mathrm{i}jz}\mathrm{d}z+\int_{-2\pi+\mathrm{i}\pi\tau}^{-\pi}\operatorname{sn}\left(\frac{2z}{\pi};k\right)\mathrm{e}^{\mathrm{i}jz}\mathrm{d}z\\
&=(1+(-1)^j\mathrm{e}^{-\frac{K'j}{K}})\int_{-\pi}^{\pi}\operatorname{sn}\left(\frac{2Kx}{\pi};k\right)\mathrm{e}^{\mathrm{i}jx}\mathrm{d}x
\end{aligned}$$

since the second and the fourth integrals due to the periodicity cancel each other. We have used also a new variable in the second integral $z'=z-\pi+\mathrm{i}\pi\tau$. Combining these results all together, we obtain

$$\mathrm{i}\pi b_j=\int_{-\pi}^{\pi}\operatorname{sn}\left(\frac{2Kx}{\pi};k\right)\mathrm{e}^{\mathrm{i}jx}\mathrm{d}x=\frac{\frac{\pi^2\mathrm{i}}{Kk}(1-(-1)^j)\mathrm{e}^{-\frac{K'j}{2K}}}{(1+(-1)^j\mathrm{e}^{-\frac{K'j}{K}})}.$$

This yields that $b_j=0$ if j is even and if j is odd, then

$$b_j = \frac{2\pi e^{-\frac{K'j}{2K}}}{Kk(1 - e^{-\frac{K'j}{K}})}.$$

Thus, the trigonometric Fourier series for the elliptic Jacobi function sn(·) for real x is equal to

$$\operatorname{sn}\left(\frac{2Kx}{\pi}; k\right) = \frac{2\pi}{Kk} \sum_{j=1}^{\infty} \frac{e^{-\frac{K'j}{2K}}}{1 - e^{-\frac{K'j}{K}}} \sin((2j-1)x).$$

It can be mentioned here that the latter equality can be extended to the complex x in the region $|\operatorname{Im} x| < \frac{\pi K'}{2K}$.

Problem 18.64. Similar to the trigonometric Fourier series for sn(·) show that:

1.
$$\operatorname{cn}\left(\frac{2Kx}{\pi}; k\right) = \frac{2\pi}{Kk} \sum_{j=0}^{\infty} \frac{e^{-\frac{K'(2j+1)}{2K}}}{1 + e^{-\frac{K'(2j-1)}{K}}} \cos((2j+1)x),$$

2.
$$\operatorname{dn}\left(\frac{2Kx}{\pi}; k\right) = \frac{\pi}{2K} + \frac{2\pi}{K} \sum_{j=1}^{\infty} \frac{e^{-\frac{K'j}{K}}}{1 + e^{-\frac{2K'j}{K}}} \cos(2xj).$$

18.4 Orthogonal polynomials

Definition 18.65. A function $f(x)$ (possibly complex-valued), defined on a closed interval $[a, b]$, is said to be *piecewise continuous* if there is partition of $[a, b]$ such that:
1. $a = x_0 < x_1 < x_2 < \cdots < x_n = b$.
2. $f(x)$ is continuous on each interval $(x_{j-1}, x_j), j = 1, 2, \ldots, n$.
3. There is $\lim_{x \to x_j \pm 0} f(x), j = 0, 1, 2, \ldots, n$.

The space of all such functions is denoted as $PC[a, b]$.

In the frame of Definition 18.65, we say that a function $f(x)$ is *piecewise smooth* if instead of (2) and (3), we have that:
1. $f'(x)$ is continuous on each interval $(x_{j-1}, x_j), j = 1, 2, \ldots, n$.
2. There is $\lim_{x \to x_j \pm 0} f'(x), j = 0, 1, 2, \ldots, n$.

The space of all such functions is denoted as $PS[a, b]$.

Example 18.66. Consider the following two functions:
1. If $f(x) = |x|, x \in [-1.1]$, then $f \in PS[-1, 1]$.

2. If $f(x) = \sqrt{x}$, $x \in [0,1]$ and $f(x) = -x$, $x \in [-1,0]$, then $f(x)$ is continuous on the interval $[-1,1]$ but $f(x)$ does not belong to $PS[-1,1]$.

Definition 18.67. A linear space E is said to be a *Euclidean space* if there is a function $(\cdot,\cdot)_E : E \times E \to \mathbb{C}$ (it is called an *inner product*) such that:
1. For any $f \in E$, it follows that $(f,f)_E \geq 0$ and $(f,f)_E = 0$ if and only if $f = 0$.
2. For any $f, g \in E$, we have that $(f,g)_E = \overline{(g,f)_E}$.
3. For any $\alpha_1, \alpha_2 \in \mathbb{C}$, $f_1, f_2, g \in E$, the following equality holds: $(\alpha_1 f_1 + \alpha_2 f_2, g)_E = \alpha_1 (f_1, g)_E + \alpha_2 (f_2, g)_E$.

Remark. The properties (2) and (3) of Definition 18.67 imply that for any $\alpha \in \mathbb{C}, f, g \in E$, we have that $(f, \alpha g)_E = \overline{\alpha}(f,g)_E$.

Example 18.68. Consider the linear space of functions, defined on an interval (a,b), which satisfy $\int_a^b |f(x)|^2 w(x)dx < \infty$, where $w(x) \geq 0$. We denote this linear space by $L_w^2(a,b)$. If the inner product is defined as

$$(f,g)_{L_w^2(a,b)} := \int_a^b f(x)\overline{g(x)}w(x)dx,$$

then $L_w^2(a,b)$ becomes a Euclidean space.

Definition 18.69. Let F be a linear space. A mapping $\|\cdot\|_F : F \to \mathbb{R}$ is said to be a *norm* and F is said to be a *normed space* if:
1. For any $f \in F$, it follows that $\|f\|_F \geq 0$ and $\|f\|_F = 0$ if and only if $f = 0$.
2. For any $\alpha \in \mathbb{C}, f \in F$, we have that $\|\alpha f\|_F = |\alpha|\|f\|_F$.
3. For any $f, g \in F$, the *triangle inequality* $\|f + g\|_F \leq \|f\|_F + \|g\|_F$ holds.

If E is a Euclidean space, then its inner product $(\cdot,\cdot)_E$ induces a norm by

$$\|f\|_E := \sqrt{(f,f)_E}, \quad f \in E$$

and it is called the *induced norm.*

Problem 18.70. Show that the induced norm satisfies all conditions of Definition 18.69.

Theorem 18.71 (Cauchy–Schwarz–Bunjakovskii inequality). *Let E be a Euclidean space with inner product $(\cdot,\cdot)_E$. Then the following inequality holds:*

$$|(f,g)_E| \leq \|f\|_E \|g\|_E, \quad f, g \in E, \tag{18.76}$$

where $\|\cdot\|_E$ is the induced norm. This inequality is called the Cauchy–Schwarz–Bunjakovskii inequality and the equality here holds if and only if $f = \mu g$ for some $\mu \in \mathbb{C}$.

Proof. Due to Definition 18.67 for any $f, g \in E$, $\lambda \in \mathbb{C}$, we have that

$$(f + \lambda g, f + \lambda g)_E = |\lambda|^2 \|g\|_E^2 + 2\operatorname{Re}(\bar{\lambda}(f, g)_E) + \|f\|_E^2 \geq 0.$$

Next, for $g \neq 0$ and $\lambda = -\frac{(f,g)}{\|g\|^2}$ (if $g = 0$, then (18.76) trivially holds), we have

$$\|f\|^2 - 2\operatorname{Re}\left(\frac{\overline{(f,g)}}{\|g\|^2}(f,g)\right) + \frac{|(f,g)|^2}{\|g\|^2} \geq 0 \quad \text{or} \quad \|f\|^2 - 2\frac{|(f,g)|^2}{\|g\|^2} + \frac{|(f,g)|^2}{\|g\|^2} \geq 0.$$

Thus, the inequality is proved. □

Problem 18.72. Prove that the equality in (18.76) holds if and only if $f = \mu g$ for some $\mu \in \mathbb{C}$.

Hint: Use the formula for inner product $(a, b) = |a||b| \cos\theta$, where θ is angle between a and b.

Definition 18.73. Let E be a Euclidean space and $f, g \in E$.
1. Then f is said to be *orthogonal* to g if $(f, g)_E = 0$.
2. A system $\{f_j\}_{j=1}^{\infty}, f_j \in E$ is said to be an *orthogonal system* (mutually orthogonal) if $(f_j, f_k)_E = 0, j \neq k$. If in this case $\|f_j\|_E = 1$ for all j then the system is called *orthonormal system.*

Legendre polynomials

Definition 18.74. *Legendre polynomials* are defined as

$$P_n(x) := \frac{1}{2^n n!}\left((x^2 - 1)^n\right)^{(n)}, \quad n = 0, 1, 2, \ldots, \tag{18.77}$$

where the derivative of order n with respect to x on the right-hand side is considered.

It is clear from (18.77) that $P_n(x)$ is really a polynomial of degree n. Moreover, since by Newton's binomial formula,

$$(x^2 - 1)^n = \sum_{k=0}^{n} \frac{(-1)^k n!}{k!(n-k)!} x^{2n-2k},$$

then we have

$$\begin{aligned} P_n(x) &= \frac{1}{2^n} \sum_{k=0}^{n} \frac{(-1)^k}{k!(n-k)!} (x^{2n-2k})^{(n)} \\ &= \frac{1}{2^n} \sum_{k=0}^{n} \frac{(-1)^k}{k!(n-k)!} (2n-2k)(2n-2k-1)\cdots(n-2k+1)x^{2n-2k-n}, \end{aligned}$$

where summation goes for $2n - 2k - n \geq 0$, i. e., $2k \leq n$ or $k \leq [\frac{n}{2}]$, since k is integer. Hence, we obtained the equivalent form for Legendre polynomial as

$$P_n(x) = \frac{1}{2^n} \sum_{k=0}^{[\frac{n}{2}]} \frac{(-1)^k (2n-2k)!}{k!(n-k)!(n-2k)!} x^{n-2k}. \tag{18.78}$$

Example 18.75. We have

$$\begin{aligned} P_0(x) &= 1, \\ P_1(x) &= x, \\ P_2(x) &= \frac{1}{2}(3x^2 - 1), \\ P_3(x) &= \frac{1}{2}(5x^3 - 3x), \\ P_4(x) &= \frac{1}{8}(35x^4 - 30x^2 + 3). \end{aligned}$$

It is easy to see also that

$$P_n(1) = 1, \quad P_n(-x) = (-1)^n P_n(x), \quad P_{2n-1}(0) = 0, \quad P_{2n}(0) = \frac{(-1)^n (2n)!}{2^{2n}(n!)^2}.$$

Theorem 18.76. *The generating function of* $\{P_n(x)\}_{n=0}^{\infty}$ *is*

$$G(t, x) = \frac{1}{\sqrt{t^2 - 2xt + 1}},$$

where $|x| \leq 1$, $|t| < \sqrt{2} - 1$.

Proof. Using the Taylor's expansion,

$$(1+y)^{-\frac{1}{2}} = \sum_{k=0}^{\infty} \frac{(-1)^k (2k)!}{2^{2k}(k!)^2} y^k,$$

which is valid for $|y| < 1$, we obtain for $y := t^2 - 2xt$ that

$$\begin{aligned} (1 - 2xt + t^2)^{-\frac{1}{2}} &= \sum_{k=0}^{\infty} \frac{(-1)^k (2k)!}{2^{2k}(k!)^2} (t^2 - 2xt)^k \\ &= \sum_{k=0}^{\infty} \frac{(-1)^k (2k)!}{2^{2k}(k!)^2} t^k \sum_{j=0}^{k} \frac{k!}{j!(k-j)!} t^j (-2x)^{k-j} \\ &= \sum_{n=0}^{\infty} t^n \sum_{j=0}^{[\frac{n}{2}]} \frac{(-1)^{2n-j}(2n-2j)! x^{n-2j} 2^{n-2j}}{2^{2n-2j} j!(n-2j)!(n-j)!} \end{aligned}$$

$$= \sum_{n=0}^{\infty} t^n P_n(x)$$

since $k = n - j$, $n = k + j \geq 2j$. This proves the theorem. □

Theorem 18.77 (Orthogonality of Legendre polynomials). *If $m \neq n$, then*

$$\int_{-1}^{1} P_n(x) P_m(x) \mathrm{d}x = 0, \quad m, n \in \mathbb{N}_0$$

but if $m = n$, then

$$\int_{-1}^{1} P_n^2(x) \mathrm{d}x = \frac{2}{2n+1}, \quad n \in \mathbb{N}_0. \tag{18.79}$$

Proof. Let us prove first (18.79). Indeed,

$$\int_{-1}^{1} P_n^2(x) \mathrm{d}x = \frac{1}{2^{2n}(n!)^2} \int_{-1}^{1} \big((x^2-1)^n\big)^{(n)} \big((x^2-1)^n\big)^{(n)} \mathrm{d}x.$$

Integration by parts n times (all substitutions at ± 1 are equal to 0) yield

$$\begin{aligned} \int_{-1}^{1} P_n^2(x) \mathrm{d}x &= \frac{(-1)^n}{2^{2n}(n!)^2} \int_{-1}^{1} \big((x^2-1)^n\big)^{(2n)} (x^2-1)^n \mathrm{d}x \\ &= \frac{(2n)!}{2^{2n}(n!)^2} \int_{-1}^{1} (1-x^2)^n \mathrm{d}x = \frac{(2n)!}{2^{2n}(n!)^2} B\left(\frac{1}{2}, n+1\right) \end{aligned}$$

since $((x^2-1)^n)^{(2n)} = (2n)!$. Here, B is Euler's beta function. Due to its properties (see Theorem 17.27 and Problem 17.32), we have finally that

$$\int_{-1}^{1} P_n^2(x) \mathrm{d}x = \frac{(2n)!}{2^{2n}(n!)^2} \frac{2^{n+1} n!}{(2n+1)!!} = \frac{2}{2n+1}.$$

This proves relation (18.79).

Let now $n \neq m$. Due to symmetricity, it is enough to consider $m < n$. Then integrating by parts n times as above, we obtain that

$$\int_{-1}^{1} P_n(x) P_m(x) \mathrm{d}x = \frac{(-1)^n}{2^n 2^m m! n!} \int_{-1}^{1} \big((x^2-1)^m\big)^{(m+n)} (x^2-1)^n \mathrm{d}x = 0$$

since $m+n > 2m$ and $(x^2-1)^m$ is a polynomial of degree $2m$. Thus, Theorem is completely proved. □

Theorem 18.78. *The Legendre polynomial $P_n(x)$ satisfies the following differential equation:*

$$(1-x^2)P_n''(x) - 2xP_n'(x) + n(n+1)P_n(x) = 0$$

or, equivalently,

$$((1-x^2)P_n'(x))' + n(n+1)P_n(x) = 0.$$

Proof. Let us consider a function $g(x) := ((1-x^2)P_n'(x))' + n(n+1)P_n(x)$, which is (at least formally) a polynomial of degree $\le n$. Next, we calculate the coefficient of $g(x)$, which is in front of x^n. Indeed, since

$$\begin{aligned}((1-x^2)P_n'(x))' &= -2xP_n(x) + (1-x^2)P_n''(x)\\ &= -\frac{2x}{2^n n!}(x^{2n}+\cdots)^{(n+1)} - P_n''(x) - \frac{x^2}{2^n n!}(x^{2n}+\cdots)^{(n+2)}\end{aligned}$$

then the needed coefficient (i. e., coefficient of $g(x)$ in front of x^n) is equal to

$$\begin{aligned}&\frac{1}{2^n n!}(-2(2n)(2n-1)\cdots(n+1)n - 2n(2n-1)\cdots n(n-1)\\ &\quad + (n+1)n(2n)(2n-1)\cdots(n+1))\\ &= \frac{2n(2n+1\cdots(n+1))}{2^n n!}(-2n - n(n-1) + n(n+1)) = 0.\end{aligned}$$

Hence, $g(x)$ is a polynomial of degree $\le (n-1)$, and thus we can represent it as

$$g(x) = \sum_{k=0}^{n-1} c_k P_k(x),$$

where the coefficients $c_k, k = 0,1,2,\dots,n-1$ are chosen appropriately. Indeed, since the coefficients of $g(x)$ are known, we can choose c_{n-1} such that $g(x) - c_{n-1}P_{n-1}(x)$ will be a polynomial of degree $(n-2)$. And this process can be continued to define all other coefficients c_k. This representation for $g(x)$ allows to conclude that for $j = 0,1,2,\dots,n-1$ (see Theorem 18.77),

$$\int_{-1}^{1} g(x)P_j(x)dx = \sum_{k=0}^{n-1} c_k \int_{-1}^{1} P_k(x)P_j(x)dx = \frac{2c_j}{2j+1}.$$

Hence, for these values of j the latter equality can be rewritten (see again Theorem 18.77)

$$\frac{2c_j}{2j+1} = \int_{-1}^{1} ((1-x^2)P_n'(x))' P_j(x)\mathrm{d}x, \quad j = 0, 1, 2, \ldots, n-1.$$

Integrating by parts twice on the right-hand side of the latter equality, we obtain

$$\frac{2c_j}{2j+1} = \int_{-1}^{1} P_n(x)((1-x^2)P_j'(x))'\mathrm{d}x, \quad j = 0, 1, 2, \ldots, n-1.$$

Since the function $((1-x^2)P_j'(x))'$ is a polynomial of degree $j \le n-1$, then (repeating the process with $g(x)$) we may represent this function as $\sum_{m=0}^{j} b_m P_m(x)$ with some coefficients. But then we will have for any $j = 0, 1, 2, \ldots, n-1$ that

$$\int_{-1}^{1} P_n(x)((1-x^2)P_j'(x))'\mathrm{d}x = \sum_{m=0}^{j} b_m \int_{-1}^{1} P_n(x)P_m(x)\mathrm{d}x = 0.$$

This implies that $g(x) = 0$. Hence, the theorem is proved. □

Corollary 18.79. *Let $P_n(\cdot)$ be a Legendre polynomial. Then the function $u(\theta) := P_n(\cos\theta)$, $\theta \in [-\pi, \pi]$ satisfies the differential equation:*

$$\frac{1}{\sin\theta}\left(\sin\theta \frac{\mathrm{d}u(\theta)}{\mathrm{d}\theta}\right)' + n(n+1)u(\theta) = 0.$$

Proof. It follows straightforwardly from Theorem 18.78. □

Now we prove the recursive properties of Legendre polynomials that are called the *Bonnet's recursion formulae.*

Theorem 18.80. *The following properties are valid* ($n = 1, 2, \ldots$)*:*
1. $(n+1)P_{n+1}(x) = (2n+1)xP_n(x) - nP_{n-1}(x)$,
2. $P_{n+1}'(x) - xP_n'(x) = (n+1)P_n(x)$,
3. $xP_n'(x) - P_{n-1}'(x) = nP_n(x)$,
4. $P_{n+1}'(x) - P_{n-1}'(x) = (2n+1)P_n(x)$.

Proof. First, (4) follows immediately from (2) and (3). We will prove only (1) since all others can be proved by the same manner. Using the generating function for Legendre polynomials (see Theorem 18.76) and differentiating it with respect to t, we obtain

$$\frac{x-t}{(1-2xt+t^2)^{\frac{3}{2}}} = \sum_{n=1}^{\infty} nP_n(x)t^{n-1}.$$

This equality can be rewritten (see again Theorem 18.76) as

$$x\sum_{n=0}^{\infty} P_n(x)t^n - \sum_{n=0}^{\infty} P_n(x)t^n = (1-2xt+t^2)\sum_{n=1}^{\infty} nP_n(x)t^{n-1}$$

or

$$\begin{aligned} x\sum_{n=0}^{\infty} P_n(x)t^n - \sum_{n=0}^{\infty} P_n(x)t^n \\ = \sum_{n=1}^{\infty} nP_n(x)t^{n-1} - 2x\sum_{n=0}^{\infty} nP_n(x)t^n + \sum_{n=0}^{\infty} nP_n(x)t^{n+1}. \end{aligned}$$

We may rewrite this equality as

$$\sum_{n=0}^{\infty} x(2n+1)P_n(x)t^n = \sum_{n=1}^{\infty} nP_n(x)t^{n-1} + \sum_{n=0}^{\infty}(n+1)P_n(x)t^{n+1},$$

which is equivalent to

$$\sum_{n=0}^{\infty} x(2n+1)P_n(x)t^n = \sum_{n=0}^{\infty}(n+1)P_{n+1}(x)t^n + \sum_{n=0}^{\infty} nP_{n-1}(x)t^n.$$

Equating now the coefficients in front of $t^n, n = 1, 2, \dots,$ we obtain that

$$x(2n+1)P_n(x) = (n+1)P_{n+1}(x) + nP_{n-1}(x).$$

This proves (1). The theorem is proved. □

The orthogonality of the family $\{P_n(x)\}_{n=0}^{\infty}$ (see Theorem 18.77) and formula (18.79) allow us to consider the system $\sqrt{\frac{2n+1}{2}}\{P_n(x)\}_{n=0}^{\infty}$ as an orthonormal basis in the Euclidean space $L^2(-1,1)$, and thus we may consider for any $f \in L^2(-1,1)$ the corresponding Fourier series (not the trigonometric Fourier series) with respect to Legendre polynomials as

$$f(x) \approx \sum_{n=0}^{\infty} c_n\sqrt{\frac{2n+1}{2}}P_n(x), \quad c_n = \int_{-1}^{1} f(x)\sqrt{\frac{2n+1}{2}}P_n(x)\mathrm{d}x.$$

It is clear that this series converges to f in $L^2(-1,1)$ (this property is called the *completeness* of the corresponding system in L^2) but the main result here is the pointwise (even uniform) convergence of this series. The following theorem holds.

Theorem 18.81. *If $f(x)$ is continuous, having a piecewise continuous derivative on the interval $[-1,1]$, then its Fourier series with respect to Legendre polynomials,*

$$\sum_{n=0}^{\infty} c_n\sqrt{\frac{2n+1}{2}}P_n(x) = f(x),$$

converges pointwise on the interval $(-1,1)$ *and uniformly on any closed subinterval* $[x_1, x_2] \subset (-1,1)$.

Proof. We first rewrite the Fourier series for $f(x)$ in the form

$$f(x) \approx \sum_{n=0}^{\infty} \tilde{c}_n P_n(x), \quad \tilde{c}_n = \frac{1}{d_n}\int_{-1}^{1} f(x)P_n(x)\mathrm{d}x, \quad d_n = \frac{2}{2n+1},$$

and second, prove the Bessel's inequality

$$\sum_{n=0}^{\infty} |\tilde{c}_n|^2 d_n \le \int_{-1}^{1} |f(x)|^2 \mathrm{d}x.$$

Indeed, since

$$\int_{-1}^{1} \left| f(x) - \sum_{n=0}^{N} \tilde{c}_n P_n(x) \right|^2 \mathrm{d}x \ge 0,$$

then, using the orthogonality of Legendre polynomials (see Theorem 18.77) and the fact that $P_n(x)$ are real, we have

$$\int_{-1}^{1} |f(x)|^2 \mathrm{d}x - 2\sum_{n=0}^{N} d_n |\tilde{c}_n|^2 + \sum_{n=0}^{N} d_n |\tilde{c}_n|^2 \ge 0.$$

Due to arbitrariness of N, the latter inequality yields the Bessel's inequality.

Consider now the first derivative of Legendre polynomials $P_n(x)$. It turns out that this system $\{P_n'(x)\}_{n=1}^{\infty}$ forms the system of the orthogonal polynomials with some weight. Prove the following result.

Problem 18.82. If $m \ne n$, then

$$\int_{-1}^{1} P_n'(x)P_m'(x)(1-x^2)\mathrm{d}x = 0, \quad m, n \in \mathbb{N}$$

but if $m = n$, then

$$\int_{-1}^{1} (P_n'(x))^2(1-x^2)\mathrm{d}x = d_n', \quad n \in \mathbb{N},$$

where d_n' are positive constants.

These constants d_n' can be calculated as follows (using integration by parts and Theorem 18.78):

$$d_n' = \int_{-1}^{1} P_n'(x)P_n'(x)(1-x^2)\mathrm{d}x = -\int_{-1}^{1} P_n(x)(P_n'(x)(1-x^2))'\mathrm{d}x$$
$$= n(n+1)\int_{-1}^{1} P_n^2(x)\mathrm{d}x = n(n+1)d_n.$$

Based on Problem 18.82, we will estimate now the Fourier coefficients c_n' of the function $f'(x)$ with respect to the system $\{P_n'(x)\}_{n=1}^{\infty}$ and the constants d_n. Applying Bessel's inequality for the system of $\{P_n'(x)\}$ (see above) and for $f'(x)$, we obtain

$$\sum_{n=1}^{\infty} |c_n'|^2 d_n' \le \int_{-1}^{1} |f'(x)|^2(1-x^2)\mathrm{d}x.$$

Moreover, under the conditions of the theorem, coefficients c_n' are equal to $\tilde{c}_n$ for $n = 1, 2, \ldots$. Indeed, by the definition (and due to integration by parts and Theorem 18.78),

$$c_n' = \frac{1}{d_n'}\int_{-1}^{1} f'(x)P_n'(x)(1-x^2)\mathrm{d}x = -\frac{1}{d_n'}\int_{-1}^{1} f(x)(P_n'(x)(1-x^2))'\mathrm{d}x$$
$$= \frac{n(n+1)}{d_n'}\int_{-1}^{1} f(x)P_n(x)\mathrm{d}x = \frac{n(n+1)}{d_n'}d_n\tilde{c}_n = \tilde{c}_n.$$

We can prove now that the series $\sum_{n=0}^{\infty} \tilde{c}_n P_n(x)$ converges uniformly on any closed interval $[x_1, x_2] \subset (-1, 1)$. Using the Cauchy–Schwarz–Bunjakovskii inequality yields ($M > N \ge 1$)

$$\left|\sum_{n=N}^{M} \tilde{c}_n P_n(x)\right| \le \left(\sum_{n=N}^{M} |c_n'|^2 d_n'\right)^{\frac{1}{2}} \left(\sum_{n=N}^{M} \frac{|P_n(x)|^2}{d_n'}\right)^{\frac{1}{2}}$$

since (as we proved above) $\tilde{c}_n = c_n'$, $n = 1, 2, \ldots$. In order to estimate the second sum on the right-hand side of the latter inequality, let us use the asymptotic behavior (18.117) (see below Problem 18.118) for large n of Legendre polynomials pointwise in $x \in (-1, 1)$ or uniformly in $x \in [x_1, x_2] \subset (-1, 1)$. Due to this estimate and taking into account Bessel's inequality for $f'(x)$ and the fact that $d_n' \approx n$, we obtain that

$$\left|\sum_{n=N}^{M} \tilde{c}_n P_n(x)\right| \le C\left(\sum_{n=N}^{M} |c_n'|^2 d_n'\right)^{\frac{1}{2}} \left(\sum_{n=N}^{M} \frac{1}{n^2}\right)^{\frac{1}{2}} \to 0$$

as $M > N \to \infty$. It means that the series $\sum_{n=0}^{\infty} \tilde{c}_n P_n(x)$ converges uniformly on any closed interval $[x_1, x_2] \subset (-1, 1)$ and, therefore, defines a continuous function on the interval $(-1, 1)$. It is not so difficult to show now (using the completeness of the orthonormal system of Legendre polynomials in $L^2(-1, 1)$) that the limiting function is just a given function $f(x)$. □

Remark. The problem of convergence of Fourier series with respect to Legendre polynomials naturally appears in solving the Dirichlet boundary value problem in $\mathbb{R}^3$ of Laplace equation by separation of variables in cylindrical coordinates.

Another application of the expansion by Legendre polynomials is the expansions of spherical and plane waves. Consider the function

$$v(R) := \frac{J_{\frac{1}{2}}(R)}{\sqrt{R}} = \sqrt{\frac{2}{\pi}} \frac{\sin R}{R},$$

where $J_{\frac{1}{2}}(\cdot)$ is the Bessel function of order $\frac{1}{2}$ and $R := \sqrt{r^2 + \rho^2 - 2r\rho\mu}$ with $|\mu| \le 1$. It is easy to check that with respect to R this function $v(R)$ satisfies the equation

$$Rv''(R) + 2v'(R) + Rv(R) = 0.$$

From this equation, it is easy to obtain a partial differential equation in the variables r and μ for fixed ρ. This equation has a form

$$r^2 \frac{\partial^2 v}{\partial r^2} + 2r \frac{\partial v}{\partial r} + r^2 v + (1 - \mu^2) \frac{\partial^2 v}{\partial \mu^2} - 2\mu \frac{\partial v}{\partial \mu} = 0. \tag{18.80}$$

Problem 18.83. Show the validity of equation (18.80).

We look for the bounded solution of the equation (18.80), using the separation of variables, in the form

$$v(R) \equiv v(r, \mu) = \phi(r)\psi(\mu).$$

Consequently, we have

$$\frac{r^2\phi''(r) + 2r\phi'(r) + r^2\phi(r)}{\phi(r)} = -\frac{(1 - \mu^2)\psi''(\mu) - 2\mu\psi'(\mu)}{\psi(\mu)} = \lambda,$$

where λ is some constant. This yields first that

$$\left((1 - \mu^2)\psi'(\mu)\right)' + \lambda\psi(\mu) = 0.$$

This equation can be considered as the eigenvalue problem (λ is a spectral parameter) with the homogeneous Dirichlet boundary conditions at the points $\mu = \pm 1$ (since $\psi(\mu)$ is bounded). Hence, the only solutions with respect to λ are $\lambda_n = n(n+1)$, $n = 0, 1, 2, \ldots$, and the corresponding eigenfunctions are the Legendre polynomials $P_n(\mu)$, $n = 0, 1, 2, \ldots$ (see Theorem 18.78). This implies that the corresponding equation for the function $\phi(r)$ is

$$r^2\phi''(r) + 2r\phi'(r) + (r^2 - n(n+1))\phi(r) = 0.$$

Replacing $\phi(r)$ by $r^a g(r)$ with some a to be determined, yields immediately to $a = -\frac{1}{2}$ and to the equation for the function $g(r)$,

$$r^2 g''(r) + rg'(r) + (r^2 - (n+1/2)^2)g(r) = 0.$$

But this equation has a unique bounded solution (up to the multiplicative constant), which is equal to the Bessel's function $J_{n+\frac{1}{2}}(r)$ (see (18.18)) of order $n + \frac{1}{2}$. Hence, the solutions for the function $\phi(r)$ are

$$\phi_n(r) = \frac{J_{n+\frac{1}{2}}(r)}{\sqrt{r}}, \quad n = 0, 1, 2, \ldots.$$

Therefore, the general solution of (18.80) is equal to

$$v(R) = \sqrt{\frac{2}{\pi}\frac{\sin R}{R}} = \sum_{n=0}^{\infty} f_n \frac{J_{n+\frac{1}{2}}(r)}{\sqrt{r}} P_n(\mu), \tag{18.81}$$

where the coefficients f_n are to be determined as the Fourier coefficients (together with $\frac{J_{n+\frac{1}{2}}(r)}{\sqrt{r}}$) of the expansion with respect to the Legendre polynomials $P_n(\mu)$, i. e.,

$$f_n \frac{J_{n+\frac{1}{2}}(r)}{\sqrt{r}} = \frac{1}{d_n}\int_{-1}^{1} \sqrt{\frac{2}{\pi}\frac{\sin R}{R}} P_n(\mu)\,d\mu,$$

where $d_n = \int_{-1}^{1} P_n^2(\mu)\,d\mu = \frac{2}{2n+1}$ (see Theorem 18.77). Since the Legendre polynomials $P_n(\mu) = \frac{(-1)^n}{2^n n!}((1-\mu^2)^n)^{(n)}$ (see (18.77)), then integrating n times with respect to μ in the latter integral we obtain (all substitutions at ± 1 will be equal to 0)

$$f_n \frac{J_{n+\frac{1}{2}}(r)}{\sqrt{r}} = \frac{1}{2^n n! d_n}\int_{-1}^{1} (1-\mu^2)^n \left(\sqrt{\frac{2}{\pi}\frac{\sin R}{R}}\right)_\mu^{(n)} d\mu.$$

Next, it can be proved by induction that

$$\left(\sqrt{\frac{2}{\pi}\frac{\sin R}{R}}\right)_\mu^{(n)} = (r\rho)^n \left(-\frac{1}{R}\frac{d}{dR}\right)^n \left(\sqrt{\frac{2}{\pi}\frac{\sin R}{R}}\right) = (r\rho)^n \frac{J_{n+\frac{1}{2}}(R)}{R^{n+1/2}}$$

if we also take into account that $\sqrt{\frac{2}{\pi}}\frac{\sin R}{R} = \frac{J_{\frac{1}{2}}(R)}{\sqrt{R}}$ and Example 18.12. This yields the equality

$$f_n \frac{J_{n+\frac{1}{2}}(r)}{r^{n+1/2}} = \frac{\rho^n}{2^n n! d_n} \int_{-1}^{1} (1-\mu^2)^n \frac{J_{n+\frac{1}{2}}(R)}{R^{n+1/2}} d\mu.$$

If we put $r \to 0$, then $R \to \rho$, and thus we obtain in the latter equality (see Proposition 18.4 for the asymptotic of Bessel's functions at zero)

$$\frac{f_n}{2^{n+1/2}\Gamma(n+3/2)} = \frac{J_{n+1/2}(\rho)}{\sqrt{\rho}2^n n! d_n} \int_{-1}^{1} (1-\mu^2)^n d\mu = \frac{J_{n+1/2}(\rho)}{\sqrt{\rho}2^n n! d_n} \frac{n!\sqrt{\pi}}{\Gamma(n+3/2)},$$

and finally we obtain that

$$f_n(\rho) = \sqrt{2\pi}(n+1/2)\frac{J_{n+1/2}(\rho)}{\sqrt{\rho}}.$$

Substituting this formula for $f_n(\rho)$ into (18.81) gives us

$$\frac{\sin R}{R} = \pi \sum_{n=0}^{\infty} (n+1/2) \frac{J_{n+1/2}(\rho)}{\sqrt{\rho}} \frac{J_{n+1/2}(r)}{\sqrt{r}} P_n(\mu).$$

Using a new parameter $k > 0$, the latter formula can be rewritten as

$$\frac{\sin(kR)}{R} = \pi \sum_{n=0}^{\infty} (n+1/2) \frac{J_{n+1/2}(k\rho)}{\sqrt{\rho}} \frac{J_{n+1/2}(kr)}{\sqrt{r}} P_n(\mu). \tag{18.82}$$

This formula (18.82) is said to be the *real representation of the spherical wave*. The complex version of this formula can be easily obtained from the real one and it has a form (with the same meaning and name)

$$\frac{e^{ikR}}{R} = i\pi \sum_{n=0}^{\infty} (n+1/2) \frac{J_{n+1/2}(kr)}{\sqrt{r}} \frac{H^{(1)}_{n+1/2}(k\rho)}{\sqrt{\rho}} P_n(\mu), \tag{18.83}$$

where $H^{(1)}_{n+1/2}(k\rho)$ is the Hankel function of order $n+\frac{1}{2}$ and of the first kind.

We are in the position now to obtain the representation for the plane wave. Let in the formula (18.83) $\rho \to +\infty$. In this case, we have

$$R = \rho - r\mu + O\left(\frac{1}{\rho}\right).$$

Taking the complex conjugate in the formula (18.83) and using the latter asymptotic for R, we obtain

$$\frac{e^{-ik\rho}e^{ikr\mu}}{\rho - r\mu} = -i\pi\sum_{n=0}^{\infty}(n+1/2)\frac{J_{n+1/2}(kr)}{\sqrt{r}}\frac{H^{(2)}_{n+1/2}(k\rho)}{\sqrt{\rho}}P_n(\mu) + o(1), \quad \rho \to +\infty.$$

Further, using the asymptotic for the Hankel function $H^{(2)}_{n+1/2}(k\rho)$ for a large argument (see (18.112) and its complex conjugate), we have

$$H^{(2)}_{n+1/2}(k\rho) = \sqrt{\frac{2}{\pi k\rho}}e^{-i(k\rho-\pi/2-\pi n/2)} + o\left(\frac{1}{\rho}\right) = i^{n+1}\sqrt{\frac{2}{\pi k\rho}}e^{-ik\rho} + o\left(\frac{1}{\rho}\right).$$

Substituting this to the previous asymptotic formula yields

$$\frac{e^{-ik\rho}e^{ikr\mu}}{\rho - r\mu} = \sqrt{2\pi}\sum_{n=0}^{\infty}i^n(n+1/2)\frac{J_{n+1/2}(kr)}{\sqrt{rk}}\frac{e^{-ik\rho}}{\rho}P_n(\mu) + o(1).$$

Hence, finally we have that

$$e^{ikr\mu} = \sqrt{2\pi}\sum_{n=0}^{\infty}i^n(n+1/2)\frac{J_{n+1/2}(kr)}{\sqrt{rk}}P_n(\mu), \quad k, r > 0, |\mu| \le 1. \tag{18.84}$$

This formula is the *representation of the plane wave* by the Legendre polynomials. It can be also rewritten in the vector form as

$$e^{i\vec{k}\cdot\vec{r}} = \sqrt{2\pi}\sum_{n=0}^{\infty}i^n(n+1/2)\frac{J_{n+1/2}(|\vec{k}||\vec{r}|)}{\sqrt{|\vec{r}||\vec{k}|}}P_n(\cos\theta), \tag{18.85}$$

where $\vec{k}\cdot\vec{r}$ is an inner product of two vectors $\vec{k}$ and $\vec{r}$ in some Euclidean space and θ is an angle between these two vectors.

Hermite polynomials

Definition 18.84. *Hermite polynomials* $H_n(x), n = 0, 1, 2, \ldots$ are defined as

$$H_n(x) := (-1)^n e^{x^2}\left(e^{-x^2}\right)^{(n)}, \quad x \in \mathbb{R},$$

where the derivative of order n with respect to x is considered.

This definition yields that $H_0(x) = 1$, $H_1(x) = 2x$, $H_2(x) = 4x^2 - 2$, $H_3(x) = 8x^3 - 12x, \ldots$. Moreover, $H_n(x), n = 0, 1, 2, \ldots$ are polynomials of degree n. Let us first prove the recursive formula

$$H_{n+1}(x) = 2xH_n(x) - 2nH_{n-1}(x), \quad n = 0, 1, 2, \ldots. \tag{18.86}$$

Indeed, Definition 18.84 and Leibnitz formula lead to

$$\begin{aligned} H_{n+1}(x) &= (-1)^{n+1}e^{x^2}(e^{-x^2})^{(n+1)} = (-1)^{n+1}e^{x^2}(-2xe^{-x^2})^{(n)} \\ &= (-1)^{n+1}e^{x^2}\sum_{k=0}^{n} C_n^k(-2x)^{(k)}(e^{-x^2})^{(n-k)} \\ &= (-1)^{n+1}e^{x^2}((-2x)(e^{-x^2})^{(n)} + n(-2)(e^{-x^2})^{(n-1)}), \end{aligned}$$

where symbol C_n^k denotes the binomial coefficients. This equality is precisely recursive formula (18.86). It also proves, by induction, that $H_n(x)$ are polynomials of degree n, since $H_0(x)$ is a polynomial of degree 0.

Problem 18.85. Show that for $n = 1, 2, \dots$ we have

$$H_n(x) = 2xH_{n-1}(x) - H'_{n-1}(x).$$

There are two more evident properties: $H_n(-x) = (-1)^n H_n(x)$ and $H_{2k-1}(0) = 0$. The first equality follows, e. g., from (18.86) by induction.

Theorem 18.86. *Hermite polynomials $H_n(x)$ satisfy the differential equation:*

$$y''(x) - 2xy'(x) + 2ny(x) = 0, \quad n = 1, 2, \dots. \tag{18.87}$$

Proof. Definition 18.84 leads to

$$\begin{aligned} H_n'(x) &= (-1)^n(e^{x^2}(e^{-x^2})^{(n)})' = (-1)^n(2xe^{x^2}(e^{-x^2})^{(n)} + e^{x^2}(e^{-x^2})^{(n+1)}) \\ &= 2xH_n(x) + (-1)^n e^{x^2}((-2x)e^{-x^2})^{(n)}. \end{aligned}$$

Using again the Leibnitz formula (as for proving formula (18.86)), we obtain

$$\begin{aligned} H_n'(x) &= 2xH_n(x) + (-1)^n e^{x^2}((-2x)(e^{-x^2})^{(n)} + n(-2)(e^{-x^2})^{(n-1)}) \\ &= 2xH_n(x) - 2xH_n(x) + 2nH_{n-1}(x) = 2nH_{n-1}(x). \end{aligned}$$

Differentiating the latter equality with respect to x and using Problem 18.85, we obtain

$$H_n''(x) = 2nH_n'(x) = 2n(2xH_{n-1}(x) - H_n(x)).$$

Hence, substituting these $H_n''(x)$ and $H_n'(x)$ into the corresponding equation (18.87), we obtain

$$H_n''(x) - 2xH_n'(x) + 2nH_n(x) = 4nxH_{n-1}(x) - 2nH_n(x) - 2x2nH_{n-1} + 2nH_n(x) = 0.$$

Thus, the theorem is proved. □

Corollary 18.87. *The following formula holds:*

$$H_n'(x) = 2nH_{n-1}(x).$$

Theorem 18.88. *The generating function for the sequence* $\{\frac{1}{n!}H_n(x)\}_{n=0}^{\infty}$ *is equal to* $G(z,x) = \mathrm{e}^{2xz-z^2}$, $x \in \mathbb{R}$, $z \in \mathbb{C}$.

Proof. Due to Taylor's expansion, we have

$$\begin{aligned}\mathrm{e}^{2xz-z^2} &= \mathrm{e}^{x^2}\mathrm{e}^{-(z-x)^2} = \mathrm{e}^{x^2}\sum_{n=0}^{\infty}\frac{(\mathrm{e}^{-(z-x)^2})^{(n)}_{|z=0}}{n!}z^n \\ &= \mathrm{e}^{x^2}\sum_{n=0}^{\infty}(-1)^n\frac{(\mathrm{e}^{-x^2})^{(n)}}{n!}z^n = \sum_{n=0}^{\infty}(-1)^n\frac{\mathrm{e}^{x^2}(\mathrm{e}^{-x^2})^{(n)}}{n!}z^n \\ &= \sum_{n=0}^{\infty}\frac{H_n(x)}{n!}z^n.\end{aligned}$$

This completes the proof of the theorem. □

Theorem 18.89. *As the polynomial of degree n, Hermite polynomials have the form*

$$H_n(x) = \sum_{k=0}^{[\frac{n}{2}]}\frac{(-1)^k n!}{k!(n-2k)!}(2x)^{n-2k}, \quad n = 0,1,2,\ldots. \tag{18.88}$$

Proof. We use first the following Taylor's expansion:

$$\mathrm{e}^{2xz-z^2} = \sum_{j=0}^{\infty}\frac{(2xz-z^2)^j}{j!}.$$

Next, using the binomial formula, we have

$$\begin{aligned}\mathrm{e}^{2xz-z^2} &= \sum_{j=0}^{\infty}\sum_{k=0}^{j}C_j^k\frac{(2xz)^{j-k}(-z^2)^k}{j!} = \sum_{j=0}^{\infty}\sum_{k=0}^{j}(-1)^k C_j^k\frac{(2x)^{j-k}z^{k+j}}{j!} \\ &= \sum_{n=0}^{\infty}\sum_{k=0}^{[\frac{n}{2}]}(-1)^k\frac{(n-k)!}{(n-k)!k!(n-2k)!}(2x)^{n-2k}z^n \\ &= \sum_{n=0}^{\infty}\frac{1}{n!}\left(\sum_{k=0}^{[\frac{n}{2}]}(-1)^k\frac{n!}{k!(n-2k)!}(2x)^{n-2k}\right)z^n = \sum_{n=0}^{\infty}\frac{H_n(x)}{n!}z^n\end{aligned}$$

due to Theorem 18.88. This proves (18.88). □

There is an orthogonality of Hermite polynomials $H_n(x)$, which is the following theorem.

Theorem 18.90. *If* $n \neq m$, *then*

$$\int_{-\infty}^{\infty}\mathrm{e}^{-x^2}H_n(x)H_m(x)\mathrm{d}x = 0, \quad m,n \in \mathbb{N}_0,$$

but if $m = n$, then

$$\int_{-\infty}^{\infty} e^{-x^2} H_n^2(x)dx = 2^n n! \sqrt{\pi}, \quad n \in \mathbb{N}_0.$$

Proof. Let $n \neq m$. Then we may assume (without loss of generality) that $m > n$, and hence we have

$$\int_{-\infty}^{\infty} e^{-x^2} H_n(x)H_m(x)dx = (-1)^m \int_{-\infty}^{\infty} H_n(x)(e^{-x^2})^{(m)}dx.$$

Integrating now the right-hand side by parts m times and using Theorem 18.88 (all substitutions at $\pm\infty$ are equal to 0), we obtain that

$$\int_{-\infty}^{\infty} e^{-x^2} H_n(x)H_m(x)dx = (-1)^{2m} \int_{-\infty}^{\infty} (H_n(x))^{(m)} e^{-x^2} dx = 0$$

since $H_n(x)$ is a polynomial of degree n and $m > n$. If $n = m$, we proceed similarly and obtain

$$\int_{-\infty}^{\infty} e^{-x^2} H_n^2(x)dx = (-1)^{2n} \int_{-\infty}^{\infty} (H_n(x))^{(n)} e^{-x^2} dx.$$

Next, formula (18.88) yields that

$$H_n(x) = (2x)^n + \text{terms of degree } j < n \text{ with respect to } x^j.$$

That is why $(H_n(x))^{(n)} = 2^n (x^n)^{(n)} = 2^n n!$. Hence,

$$\int_{-\infty}^{\infty} e^{-x^2} H_n^2(x)dx = 2^n n! \int_{-\infty}^{\infty} e^{-x^2} dx = 2^n n! \sqrt{\pi}$$

since the latter integral equals to $\Gamma(\frac{1}{2}) = \sqrt{\pi}$ (see Theorem 17.19). The theorem is proved. □

The orthogonality of the family of Hermite polynomials $\{H_n(x)\}_{n=0}^{\infty}$ (see Theorem 18.90) with the weight $w(x) = e^{-x^2}$ allow us to consider the system $\tilde{H}_n(x) := \sqrt{\frac{1}{2^n n! \sqrt{\pi}}} \{H_n(x)\}_{n=0}^{\infty}$ as an orthonormal basis in the Euclidean space $L_w^2(\mathbb{R})$ with this weight, and thus we may consider for any $f \in L_w^2(\mathbb{R})$ the corresponding Fourier series (not the trigonometric Fourier series) with respect to Hermite polynomials as

$$f(x) \approx \sum_{n=0}^{\infty} c_n \tilde{H}_n(x),$$

where the Fourier coefficients with respect to the system $\tilde{H}_n(x)$ are defined by

$$c_n = \int_{-\infty}^{\infty} f(x)\tilde{H}_n(x)e^{-x^2}\,dx.$$

It is clear that this series converges to f in $L_w^2(\mathbb{R})$ but the main result here is the pointwise convergence of this series. The following theorem holds.

Theorem 18.91. *If $f \in L_w^2(\mathbb{R})$ is continuous, having a piecewise continuous derivative on the line $\mathbb{R}$, then its Fourier series with respect to Hermite polynomials*

$$\sum_{n=0}^{\infty} c_n\tilde{H}_n(x) = f(x)$$

converges pointwise on the line $\mathbb{R}$ and uniformly on any closed interval $[x_1, x_2] \subset \mathbb{R}$.

Proof. We can rewrite the Fourier series for $f(x)$ in the form (see the proof of Theorem 18.81)

$$f(x) \approx \sum_{n=0}^{\infty} \tilde{c}_n H_n(x), \quad \tilde{c}_n = \frac{1}{d_n}\int_{-\infty}^{\infty} f(x)H_n(x)e^{-x^2}\,dx, \quad d_n = 2^n n!\sqrt{\pi}$$

and obtain then the Bessel's inequality (analogously to the proof of Theorem 18.81)

$$\sum_{n=0}^{\infty} |\tilde{c}_n|^2 d_n \leq \int_{-\infty}^{\infty} |f(x)|^2 e^{-x^2}\,dx.$$

Consider now the first derivative of Hermite polynomials $H_n(x)$. It turns out that this system $\{H_n'(x)\}_{n=1}^{\infty}$ forms the system of the orthogonal polynomials with the same weight (see Corollary 18.87). That is why we may consider the Fourier coefficients of the function $f'(x)$ with respect to the system $\{H_n'(x)\}$ such that (due to integration by parts and due to f' vanishing at the infinity faster than any polynomial)

$$c_n' = \frac{1}{d_n'}\int_{-\infty}^{\infty} f'(x)H_n'(x)e^{-x^2}\,dx = -\frac{1}{d_n'}\int_{-\infty}^{\infty} f(x)(H_n'(x)e^{-x^2})'\,dx,$$

where

$$d_n' := \int_{-\infty}^{\infty} (H_n'(x))^2 e^{-x^2}\,dx = -\int_{-\infty}^{\infty} H_n(x)(H_n'(x)e^{-x^2})'\,dx.$$

Using Theorem 18.86 (or, equivalently, Corollary 18.87) and the orthogonality condition (see Theorem 18.90), we obtain that

$$c'_n = \frac{2n\tilde{c}_n d_n}{d'_n}, \quad d'_n = 2nd_n, \quad \text{and} \quad c'_n = \tilde{c}_n.$$

Moreover, the Bessel's inequality for $f'(x)$ with respect to the system $\{H'_n\}$ has the form

$$\sum_{n=1}^{\infty} |c'_n|^2 d'_n \le \int_{-\infty}^{\infty} |f'(x)|^2 e^{-x^2} dx.$$

We can prove now that the series $\sum_{n=0}^{\infty} \tilde{c}_n H_n(x)$ converges uniformly on any closed interval $[x_1, x_2] \subset \mathbb{R}$. Using the Cauchy–Schwarz–Bunjakovskii inequality yields ($M > N \ge 1$)

$$\left|\sum_{n=N}^{M} \tilde{c}_n H_n(x)\right| \le \left(\sum_{n=N}^{M} |c'_n|^2 d'_n\right)^{\frac{1}{2}} \left(\sum_{n=N}^{M} \frac{|H_n(x)|^2}{2nd_n}\right)^{\frac{1}{2}}$$

since (as we proved above) $\tilde{c}_n = c'_n, d'_n = 2nd_n, n = 1, 2, \ldots$. In order to estimate the second sum on the right-hand side of the latter inequality let us use the following estimate for Hermite polynomials:

$$\frac{|H_n(x)|^2}{d_n} \le Ce^{x^2}\left(\frac{1}{n^{1/4}} + \frac{|x|^{5/2}}{n^{1/2}}\right)^2.$$

It can be obtained by some precise calculations and using induction with respect to n. Due to this estimate and taking into account the Bessel's inequality for $f'(x)$, we obtain that uniformly in $x \in [x_1, x_2] \subset \mathbb{R}$,

$$\left|\sum_{n=N}^{M} \tilde{c}_n P_n(x)\right| \le C\left(\sum_{n=N}^{M} |c'_n|^2 d'_n\right)^{\frac{1}{2}} \left(\sum_{n=N}^{M} \frac{1}{n^{3/2}}\right)^{\frac{1}{2}} \to 0$$

as $M > N \to \infty$. It means that the series $\sum_{n=0}^{\infty} \tilde{c}_n H_n(x)$ converges uniformly on any closed interval $[x_1, x_2] \subset \mathbb{R}$ and, therefore, defines a continuous function on the line $\mathbb{R}$. It is not so difficult to show now (using the completeness of the system of Hermite polynomials in $L^2_w(\mathbb{R})$) that the limiting function is just a given function $f(x)$. This proves the theorem. □

Chebyshev polynomials

Definition 18.92. *Chebyshev polynomials* are defined as

$$T_n(x) := \cos(n \arccos x), \quad x \in [-1, 1], n = 0, 1, 2, \ldots.$$

One can prove that $T_n(x)$ is a polynomial in x of degree n. Indeed, since for $x = \cos\theta$, $\theta \in [0, \pi]$, we have that

$$T_n(\cos\theta) = \cos(n\theta) = \mathrm{Re}(e^{in\theta}) = \mathrm{Re}(z^n), \quad |z| = 1,$$

then considering $z = x + iy = e^{i\theta}$ we obtain

$$\mathrm{Re}(z^n) = \mathrm{Re}((x+iy)^n) = \mathrm{Re}\left(\sum_{k=0}^{n} C_n^k x^{n-k}(iy)^k\right) = \sum_{m=0}^{[\frac{n}{2}]} C_n^{2m}(-1)^m x^{n-2m} y^{2m}.$$

Recalling (since $x + iy = e^{i\theta}$) that $x = \cos\theta, y = \sin\theta$, the latter equality can be rewritten as

$$\begin{aligned} T_n(\cos\theta) &= \sum_{m=0}^{[\frac{n}{2}]} C_n^{2m}(-1)^m(\cos\theta)^{n-2m}(1-\cos^2\theta)^m \\ &= \sum_{m=0}^{[\frac{n}{2}]} C_n^{2m}(-1)^m(\cos\theta)^{n-2m}\left(\sum_{j=0}^{m} C_m^j(-1)^j(\cos\theta)^{2j}\right). \end{aligned}$$

Returning to the previous variable x, the latter equality yields

$$T_n(x) = \sum_{m=0}^{[\frac{n}{2}]}\sum_{j=0}^{m} C_n^{2m} C_m^j (-1)^{m+j} x^{n-2m+2j}. \tag{18.89}$$

This formula (18.89) shows that $T_n(x)$ is really polynomial of degree n and this is an equivalent definition (compared with Definition 18.92) of Chebyshev polynomials. Furthermore, this formula (18.89) implies immediately that

$$T_n(-x) = (-1)^n T_n(x).$$

Problem 18.93. Prove that (18.89) is equivalent to

$$T_n(x) = \sum_{k=0}^{[\frac{n}{2}]} \frac{n!}{(2k)!(n-2k)!} x^{n-2k}(x^2-1)^k.$$

Example 18.94. These formulae imply that $T_0(x) = 1$, $T_1(x) = x$, $T_2(x) = 2x^2 - 1$, $T_3(x) = 4x^3 - 3x, \ldots$, and $T_{2k-1}(0) = 0$, $T_{2k}(0) = (-1)^k$, in particular, $|T_n(x)| \le 1$.

Proposition 18.95. *For any $n = 1, 2, \ldots$, we have*

$$T_{n+1}(x) = 2xT_n(x) - T_{n-1}(x), \quad x \in [-1, 1].$$

Proof. Using Definition 18.92, we have

$$\begin{aligned} T_{n+1}(x) &= \cos(n\arccos x + \arccos x) = \cos(n\arccos x)\cos(\arccos x) \\ &\quad - \sin(n\arccos x)\sin(\arccos x) = xT_n(x) - \sqrt{1 - T_n^2(x)}\sqrt{1-x^2}. \end{aligned}$$

This equality first implies that

$$xT_n(x) \geq T_{n+1}(x), \quad x \in [-1,1],$$

and, in addition,

$$x^2T_n^2(x) - 2xT_n(x)T_{n+1}(x) + T_{n+1}^2(x) = (1-x^2)(1-T_n^2(x)).$$

The latter equality can be rewritten (also after substituting $n-1$ instead of n) as two different equalities

$$T_{n+1}^2(x) - 2xT_n(x)T_{n+1}(x) = 1 - x^2 - T_n^2(x),$$
$$T_n^2(x) - 2xT_{n-1}(x)T_n(x) = 1 - x^2 - T_{n-1}^2(x).$$

Considering now their difference, we obtain

$$T_{n+1}^2(x) - T_{n-1}^2(x) = 2xT_n(x)(T_{n+1}(x) - T_{n-1}(x))$$

i. e.,

$$T_{n+1}(x) + T_{n-1}(x) = 2xT_n(x).$$

Hence, the proposition is proved. □

Theorem 18.96. *Chebyshev polynomials $T_n(x), n = 1, 2, \ldots$ satisfy the differential equation:*

$$(1-x^2)y''(x) - xy'(x) + n^2y(x) = 0, \quad x \in [-1,1]. \tag{18.90}$$

Proof. Using Definition 18.92, we obtain

$$T_n'(x) = -n\sin(n\arccos x)(\arccos x)' = n\sin(n\arccos x)\frac{1}{\sqrt{1-x^2}}.$$

Similarly,

$$T_n''(x) = -n^2\cos(n\arccos x)\frac{1}{1-x^2} + n\sin(n\arccos x)\frac{x}{(1-x^2)\sqrt{1-x^2}}.$$

The latter equality can be rewritten as

$$T_n''(x) = -\frac{n^2}{1-x^2}T_n(x) + \frac{x}{(1-x^2)}T_n'(x).$$

This is equivalent to (18.90). The theorem is proved. □

The orthogonality property of Chebyshev polynomials is the following theorem.

Theorem 18.97. *If $n \neq m, n, m \in \mathbb{N}_0$, then*

$$\int_{-1}^{1} \frac{1}{\sqrt{1-x^2}} T_n(x) T_m(x) \mathrm{d}x = 0,$$

but if $n = m$, $n, m \in \mathbb{N}_0$, then

$$\int_{-1}^{1} \frac{1}{\sqrt{1-x^2}} T_n^2(x) \mathrm{d}x = \frac{\pi}{2}, \quad n = m > 0, \quad \int_{-1}^{1} \frac{1}{\sqrt{1-x^2}} \mathrm{d}x = \pi. \tag{18.91}$$

Proof. Due to Definition 18.92 for any $n, m \in \mathbb{N}_0$, we have

$$\begin{aligned} \int_{-1}^{1} \frac{1}{\sqrt{1-x^2}} T_n(x) T_m(x) \mathrm{d}x &= \int_{-1}^{1} \frac{1}{\sqrt{1-x^2}} \cos(n \arccos x) \cos(m \arccos x) \mathrm{d}x \\ &= \int_{0}^{\pi} \cos(nt) \cos(mt) \mathrm{d}t \\ &= \frac{1}{2} \left(\int_{0}^{\pi} \cos((n-m)t) \mathrm{d}t + \int_{0}^{\pi} \cos((n+m)t) \mathrm{d}t \right), \end{aligned}$$

if we use a new variable $t := \arccos x$. Next, for $n \neq m$ both integrals on the right-hand side of the previous equality are equal to 0, while for $n = m > 0$, the first integral is equal to π, and the second one is equal to 0. For the case $n = m = 0$, the result (18.91) follows trivially. Hence, the theorem is proved. □

There is one more recursive property of Chebyshev polynomials (compare with Proposition 18.95).

Proposition 18.98. *For any $n = 2, 3, \ldots$,*

$$\frac{T'_{n+1}(x)}{n+1} - \frac{T'_{n-1}(x)}{n-1} = 2T_n(x).$$

Proof. Due to Definition 18.92, we have for $n = 2, 3, \ldots$ that

$$T'_{n+1}(x) = \sin((n+1) \arccos x) \frac{n+1}{\sqrt{1-x^2}},$$

and similarly,

$$T'_{n-1}(x) = \sin((n-1) \arccos x) \frac{n-1}{\sqrt{1-x^2}}.$$

Thus, we have

$$\frac{T'_{n+1}(x)}{n+1} - \frac{T'_{n-1}(x)}{n-1} = \frac{\sin((n+1)\arccos x) - \sin((n-1)\arccos x)}{\sqrt{1-x^2}}$$
$$= \frac{2T_n(x)\sin(\arccos x)}{\sqrt{1-x^2}} = \frac{2T_n(x)\sqrt{1-x^2}}{\sqrt{1-x^2}} = 2T_n(x).$$

The proposition is proved. □

Remark. If $n = 1$, then Proposition 18.98 reads as

$$T'_2(x) = 4T_1(x).$$

Theorem 18.97 (orthogonality property) allows us to construct an orthonormal system of Chebyshev polynomials in the Euclidean space $L^2_w(-1,1)$ with weight $w(x) = \frac{1}{\sqrt{1-x^2}}$ as

$$\left\{\sqrt{\frac{2}{\pi}}T_n(x)\right\}_{n=1}^{\infty} \cup \frac{1}{\sqrt{\pi}}T_0(x).$$

This yields that for any $f \in L^2_w(-1,1)$ we may consider the Fourier series with respect to this system as

$$f(x) \approx \frac{c_0}{\sqrt{\pi}} + \sum_{n=1}^{\infty} c_n \sqrt{\frac{2}{\pi}}T_n(x),$$

where the Fourier coefficients c_n, $n = 0, 1, 2, \dots$ are defined by

$$c_0 = \frac{1}{\sqrt{\pi}} \int_{-1}^{1} \frac{f(x)}{\sqrt{1-x^2}}\mathrm{d}x, \quad c_n = \sqrt{\frac{2}{\pi}} \int_{-1}^{1} \frac{f(x)T_n(x)}{\sqrt{1-x^2}}\mathrm{d}x.$$

It is clear that this series converges to f in the sense of the Euclidean space $L^2_w(-1,1)$, but the main result here is pointwise (even uniform) convergence.

Theorem 18.99. *If $f \in L^2_w(-1,1)$ is continuous, having a piecewise continuous derivative on the interval $[-1,1]$, then its Fourier series with respect to Chebyshev polynomials,*

$$\frac{c_0}{\sqrt{\pi}} + \sum_{n=1}^{\infty} c_n \sqrt{\frac{2}{\pi}}T_n(x) = f(x)$$

converges pointwise on the interval $[-1,1]$ and uniformly on any closed interval $[x_1, x_2] \subset [-1,1]$.

Proof. Denoting by $d_n = \frac{\pi}{2}$, $n = 1, 2, \dots$, and $d_0 = \pi$ (see (18.91)), we can rewrite the Fourier series with respect to Chebyshev polynomials as

$$f(x) \approx \sum_{n=0}^{\infty} \tilde{c}_n T_n(x), \quad \tilde{c}_n = \frac{1}{d_n} \int_{-1}^{1} \frac{f(x)T_n(x)}{\sqrt{1-x^2}} \mathrm{d}x, \quad n \in \mathbb{N}_0.$$

The next step is to obtain the Bessel's inequality with respect to this system as

$$\sum_{n=0}^{\infty} |\tilde{c}_n|^2 d_n \leq \int_{-1}^{1} \frac{|f(x)|^2}{\sqrt{1-x^2}} \mathrm{d}x.$$

Consider now the first derivative of Chebyshev polynomials $T_n(x)$. It turns out that this system $\{T_n'(x)\}_{n=1}^{\infty}$ forms the system of the orthogonal polynomials with the weight $\sqrt{1-x^2}$. This follows from the following differential equation (which is equivalent to (18.90)):

$$\left(\sqrt{1-x^2}T_n'(x)\right)' + n^2 \frac{T_n(x)}{\sqrt{1-x^2}} = 0.$$

That is why we may consider the Fourier coefficients of the function $f'(x)$ with respect to the system $\{T_n'(x)\}$ such that (due to integration by parts and due to corresponding differential equation for $T_n(x)$)

$$c_n' = \frac{1}{d_n'} \int_{-1}^{1} f'(x)T_n'(x)\sqrt{1-x^2}\mathrm{d}x = -\frac{1}{d_n'} \int_{-1}^{1} f(x)(T_n'(x)\sqrt{1-x^2})'\mathrm{d}x,$$

where

$$d_n' := \int_{-1}^{1} (T_n'(x))^2 \sqrt{1-x^2}\mathrm{d}x = n^2 \int_{-1}^{1} \frac{(T_n(x))^2}{\sqrt{1-x^2}} \mathrm{d}x = n^2 d_n.$$

Further, the definitions of $\tilde{c}_n$, c_n', and the latter equalities imply (as above) that $\tilde{c}_n = c_n'$. Moreover, the Bessel's inequality for $f'(x)$ with respect to the system $\{T_n'(x)\}$ has the form

$$\sum_{n=1}^{\infty} |c_n'|^2 d_n' \leq \int_{-1}^{1} |f'(x)|^2 \sqrt{1-x^2}\mathrm{d}x.$$

We can prove now that the series $\sum_{n=0}^{\infty} \tilde{c}_n T_n(x)$ converges uniformly on the closed interval $[-1,1]$. Using the Cauchy–Schwarz–Bunjakovskii inequality yields ($M > N \geq 1$)

$$\left|\sum_{n=N}^{M} \tilde{c}_n T_n(x)\right| \leq \left(\sum_{n=N}^{M} |c_n'|^2 d_n'\right)^{\frac{1}{2}} \left(\sum_{n=N}^{M} \frac{|T_n(x)|^2}{n^2 d_n}\right)^{\frac{1}{2}}$$

since (as we proved above) $\tilde{c}_n = c_n'$, $d_n' = n^2 d_n$, $n = 1, 2, \ldots$. In order to estimate the second sum on the right-hand side of the latter inequality, let us use the trivial estimate for

Chebyshev polynomials (see Definition 18.92), i. e., $|T_n(x)| \le 1$ uniformly in $x \in [-1,1]$. Due to this estimate, due to the Bessel's inequality for $f'(x)$ with respect to Chebyshev polynomials and due to the values for d_n, we may conclude that the series $\sum_{n=0}^{\infty} \tilde{c}_n T_n(x)$ converges uniformly on the interval $[-1,1]$ and, therefore, defines a continuous function on this interval. It is not so difficult to show now that (using the completeness of Chebyshev polynomials in $L^2_w(-1,1)$) that the limiting function on the interval $(-1,1)$ is just a given function $f(x)$. This proves the theorem. □

Problem 18.100. Show that the generating function of $\{T_n(x)\}_{n=0}^{\infty}$ is

$$G(t,x) = \frac{1-tx}{1-2tx+t^2},$$

where $|x| \le 1$, $|t| < \sqrt{2}-1$, and the generating function of $\{\frac{1}{n}T_n(x)\}_{n=1}^{\infty}$ is

$$G(t,x) = \log\left(\frac{1}{\sqrt{1-2tx+t^2}}\right),$$

where $|x| \le 1$, $|t| < \sqrt{2}-1$.

Problem 18.101. Find the Fourier series for function $|x|$ with respect to the orthonormal system of Chebyshev polynomials.

Problem 18.102. Find the Fourier series for function $\log(1+x)$ with respect to the orthonormal system of Chebyshev polynomials.

Problem 18.103. Show that for any $n = 1, 2, \dots$ among all polynomials of degree n with leading coefficient 1, the function $f(x) := \frac{1}{2^{n-1}}T_n(x)$ is the one whose maximal absolute value on the interval $[-1,1]$ is minimal. The maximal absolute value is $\frac{1}{2^{n-1}}$ and this maximum is attained exactly $n+1$ times at $x = \cos(\frac{k\pi}{n})$, $k = 0, 1, 2, \dots, n$.

Trigonometric polynomials

Definition 18.104. Consider the interval $[-L, L]$, $L > 0$. The system of the trigonometric polynomials of degree n is given by

$$\begin{aligned} &1, \quad \cos\frac{\pi x}{L}, \quad \cos\frac{2\pi x}{L}, \quad \dots, \quad \cos\frac{n\pi x}{L}, \quad \dots, \\ &\sin\frac{\pi x}{L}, \quad \sin\frac{2\pi x}{L}, \quad \dots, \quad \sin\frac{n\pi x}{L}, \quad \dots \quad = \left\{\cos\frac{n\pi x}{L}, \sin\frac{n\pi x}{L}\right\}_{n=0}^{\infty}. \end{aligned}$$

Remark. One can easily check by induction that $\cos\frac{n\pi x}{L}$ is a polynomial of degree n in $\cos\frac{\pi x}{L}$ and $\sin\frac{n\pi x}{L}$ is a polynomial of degree n in $\sin\frac{\pi x}{L}$.

The orthogonality of this system is the following proposition.

Proposition 18.105. *For any $n, m \in \mathbb{N}_0$, we have*

$$\int_{-L}^{L} \cos\frac{\pi x}{L}\,dx = \int_{-L}^{L} \sin\frac{\pi x}{L}\,dx = 0, \qquad \int_{-L}^{L} \cos\frac{n\pi x}{L}\sin\frac{m\pi x}{L}\,dx = 0,$$

and for $n = 1, 2, \dots,$

$$\int_{-L}^{L} dx = 2L, \qquad \int_{-L}^{L} \cos^2\frac{n\pi x}{L}\,dx = L, \qquad \int_{-L}^{L} \sin^2\frac{n\pi x}{L}\,dx = L.$$

Proof. It follows straightforwardly from the properties of odd functions on the symmetric intervals and the well-known properties of trigonometric functions. □

Proposition 18.105 allows us to construct orthonormal system of trigonometric polynomials in the Euclidean space $L^2(-L, L)$ as

$$\left\{\frac{1}{\sqrt{2L}}, \frac{\cos\frac{n\pi x}{L}}{\sqrt{L}}, \frac{\sin\frac{n\pi x}{L}}{\sqrt{L}}\right\}_{n=1}^{\infty}.$$

This yields that for any $f \in L^2(-L, L)$ we may consider the trigonometric Fourier series with respect to this system as

$$f(x) \approx \frac{\tilde{a}_0}{\sqrt{2L}} + \sum_{n=1}^{\infty}\left(\tilde{a}_n\frac{\cos\frac{n\pi x}{L}}{\sqrt{L}} + \tilde{b}_n\frac{\sin\frac{n\pi x}{L}}{\sqrt{L}}\right),$$

where the Fourier coefficients $\tilde{a}_n, \tilde{b}_n, n = 0, 1, 2, \dots$ are defined by

$$\tilde{a}_0 = \frac{1}{\sqrt{2L}}\int_{-L}^{L} f(x)\,dx, \quad \tilde{a}_n = \int_{-L}^{L} f(x)\frac{\cos\frac{n\pi x}{L}}{\sqrt{L}}\,dx, \quad \tilde{b}_n = \int_{-L}^{L} f(x)\frac{\sin\frac{n\pi x}{L}}{\sqrt{L}}\,dx.$$

This trigonometric Fourier series can be rewritten in simpler form as follows:

$$f(x) \approx \frac{a_0}{2} + \sum_{n=1}^{\infty}\left(a_n\cos\frac{n\pi x}{L} + b_n\sin\frac{n\pi x}{L}\right), \tag{18.92}$$

where the coefficients $a_n, b_n, n = 0, 1, 2, \dots$ are equal to

$$a_0 = \frac{1}{L}\int_{-L}^{L} f(x)\,dx, \quad a_n = \frac{1}{L}\int_{-L}^{L} f(x)\cos\frac{n\pi x}{L}\,dx, \quad b_n = \frac{1}{L}\int_{-L}^{L} f(x)\sin\frac{n\pi x}{L}\,dx.$$

Since the system of normalized trigonometric polynomials is an orthonormal basis in $L^2(-L, L)$, then for any $f \in L^2(-L, L)$ the trigonometric Fourier series (18.92) converges in $L^2(-L, L)$ to f, i. e.,

$$\lim_{N\to\infty}\int_{-L}^{L}\left|\frac{a_0}{2}+\sum_{n=1}^{N}\left(a_n\cos\frac{n\pi x}{L}+b_n\sin\frac{n\pi x}{L}\right)-f(x)\right|^2 dx=0.$$

This fact is equivalent to the identity

$$\frac{1}{L}\int_{-L}^{L}|f(x)|^2 dx=\frac{|a_0|^2}{2}+\sum_{n=1}^{\infty}(|a_n|^2+|b_n|^2), \tag{18.93}$$

which is called the *Parseval equality*. We prove these two facts in the following theorem.

Theorem 18.106. *For every* $f \in L^2(-L,L)$, *its Fourier series* (18.92) *converges to* f *in* $L^2(-L,L)$ *and the Parseval equality* (18.93) *holds.*

Proof. If $f \in L^2(-L,L)$ and $g(x) := \frac{c_0}{2}+\sum_{n=1}^{N}(c_n\cos\frac{n\pi x}{L}+d_n\sin\frac{n\pi x}{L})$, $N=1,2,\dots$ with an arbitrary coefficient $c_n, d_n, n=0,1,2,\dots$, then using Proposition 18.105 (the orthogonality condition), we obtain

$$\begin{aligned}\frac{1}{L}\int_{-L}^{L}|f(x)-g(x)|^2 dx &= \frac{1}{L}\int_{-L}^{L}|f(x)|^2 dx+\frac{1}{L}\int_{-L}^{L}|g(x)|^2 dx\\ &\quad-\frac{2}{L}\operatorname{Re}\int_{-L}^{L}f(x)\overline{g(x)}dx\\ &=\frac{1}{L}\int_{-L}^{L}|f(x)|^2 dx-\left(\frac{|a_0|^2}{2}+\sum_{n=1}^{N}(|a_n|^2+|b_n|^2)\right)\\ &\quad+\frac{|a_0-c_0|^2}{2}+\sum_{n=1}^{N}(|a_n-c_n|^2+|b_n-d_n|^2).\end{aligned}$$

The latter equality has the following two important consequences:

1. The minimum error of

$$\min_{g(x)=\frac{c_0}{2}+\sum_{n=1}^{N}(c_n\cos\frac{n\pi x}{L}+d_n\sin\frac{n\pi x}{L})}\frac{1}{L}\int_{-L}^{L}|f(x)-g(x)|^2 dx$$

is equal to

$$\frac{1}{L}\int_{-L}^{L}|f(x)|^2 dx-\left(\frac{|a_0|^2}{2}+\sum_{n=1}^{N}(|a_n|^2+|b_n|^2)\right) \tag{18.94}$$

and it is attained when $c_n=a_n$, $d_n=b_n$, $n=0,1,2,\dots$.

2. The following inequality holds (since N is arbitrary in (1) from above):

$$\frac{|a_0|^2}{2} + \sum_{n=1}^{\infty}(|a_n|^2 + |b_n|^2) \le \frac{1}{L}\int_{-L}^{L} |f(x)|^2 dx.$$

This inequality is called the *Bessel's inequality.*

Consider now instead of $f(x)$ the polynomial $\frac{a_0}{2} + \sum_{n=1}^{N}(a_n \cos\frac{n\pi x}{L} + b_n \sin\frac{n\pi x}{L})$ and instead of $g(x)$ the polynomial $\frac{a_0}{2} + \sum_{n=1}^{M}(a_n \cos\frac{n\pi x}{L} + b_n \sin\frac{n\pi x}{L})$, where $a_n, b_n, n = 0, 1, 2, \ldots$ are the Fourier coefficients of $f \in L^2(-L, L)$, we obtain, using the Bessel's inequality, that

$$\frac{1}{L}\int_{-L}^{L}\left|\sum_{n=1}^{N}\left(a_n \cos\frac{n\pi x}{L} + b_n \sin\frac{n\pi x}{L}\right) - \sum_{n=1}^{M}\left(a_n \cos\frac{n\pi x}{L} + b_n \sin\frac{n\pi x}{L}\right)\right|^2 dx$$

tends to zero as $N > M \to \infty$. It means that

$$\left\{\frac{a_0}{2} + \sum_{n=1}^{N}\left(a_n \cos\frac{n\pi x}{L} + b_n \sin\frac{n\pi x}{L}\right)\right\}$$

is a Cauchy sequence in $L^2(-L, L)$, and thus there is a function $F \in L^2(-L, L)$ such that

$$\frac{1}{L}\int_{-L}^{L}\left|F(x) - \sum_{n=1}^{N}\left(a_n \cos\frac{n\pi x}{L} + b_n \sin\frac{n\pi x}{L}\right)\right|^2 dx \to 0, \quad N \to \infty.$$

Problem 18.107. Prove that function $F(x)$ (from above) is equal to a given function $f \in L^2(-L, L)$.

Hence, the first part of this theorem is proved. The Parseval equality (18.93) follows from the first part of the theorem and (18.94) if we let $N \to \infty$. Thus, the theorem is completely proved. □

The main result here is uniform and the absolute convergence of the trigonometric Fourier series for continuous functions having piecewise continuous derivative.

Theorem 18.108. *If $f(x)$ is periodic (i. e., $f(-L) = f(L)$) with period $2L$ and continuous, having piecewise continuous derivative on the interval $[-L, L]$, then uniformly in $x \in \mathbb{R}$,*

$$\frac{a_0}{2} + \sum_{n=1}^{\infty}\left(a_n \cos\frac{n\pi x}{L} + b_n \sin\frac{n\pi x}{L}\right) = f(x).$$

Moreover, this convergence is absolute, i. e.,

$$\frac{|a_0|}{2} + \sum_{n=1}^{\infty}(|a_n| + |b_n|) < \infty.$$

Proof. It suffices to show that the number series

$$\frac{|a_0|}{2} + \sum_{n=1}^{\infty}(|a_n| + |b_n|)$$

converges because it follows, first, that the series (18.92) converges uniformly on $\mathbb{R}$ (due to periodicity), and, second, it converges to a given function $f(x)$ (see Problem 18.107).

Denoting by $\alpha_n, \beta_n, n = 0, 1, 2, \dots$, the Fourier coefficients of the piecewise continuous function $f'(x)$, which is redefined arbitrarily at a finite number of points where the derivative does not exist. Further integrating by parts (we have to divide the interval $[-L, L]$ to subintervals if the derivative of $f(x)$ does not exist), we obtain (taking into account the equality $f(-L) = f(L)$) that

$$\alpha_0 = 0, \quad \alpha_n = \frac{1}{L}\int_{-L}^{L} f'(x)\cos\frac{n\pi x}{L}dx = \frac{n\pi}{L^2}\int_{-L}^{L} f(x)\sin\frac{n\pi x}{L}dx = \frac{n\pi}{L}b_n,$$

and

$$\beta_n = \frac{1}{L}\int_{-L}^{L} f'(x)\sin\frac{n\pi x}{L}dx = -\frac{n\pi}{L^2}\int_{-L}^{L} f(x)\cos\frac{n\pi x}{L}dx = -\frac{n\pi}{L}a_n,$$

where $n = 1, 2, \dots$. These equalities allow us to conclude that

$$\frac{\pi}{L}\sum_{n=1}^{\infty}(|a_n| + |b_n|) = \sum_{n=1}^{\infty}\left(\frac{|\alpha_n|}{n} + \frac{|\beta_n|}{n}\right) \le \frac{1}{2}\sum_{n=1}^{\infty}(\alpha_n^2 + \beta_n^2) + \sum_{n=1}^{\infty}\frac{1}{n^2} < \infty.$$

The first series on the right-hand side converges due to Proposition 18.105 (see Parseval equality (18.93)), and the second is a well-known convergent number series. Here, we have used also the elementary inequality $2|ab| \le a^2+b^2$. Thus, the theorem is proved. □

Summarizing the orthogonal polynomials (including Bessel's functions and trigonometric polynomials), we can interpret (correspondingly) them as solutions of some boundary value problems for a differential equation of the second order. Namely, let us consider the following operator of order two in the form

$$Ly(x) := (p(x)y'(x))' + q(x)y(x), \quad x \in (a, b), \tag{18.95}$$

with boundary conditions (quite general)

$$\alpha_1 y(a) + \alpha_2 y'(a) = 0, \quad \beta_1 y(b) + \beta_2 y'(b) = 0, \tag{18.96}$$

where $p(x), q(x)$ are real-valued functions, and $\alpha_j, \beta_j, j = 1, 2$ are real constants with the conditions

$$\alpha_1^2 + \alpha_2^2 > 0, \quad \beta_1^2 + \beta_2^2 > 0.$$

The equation

$$Ly(x) + \lambda r(x)y(x) = 0 \tag{18.97}$$

(function $r(x)$ is real-valued, λ is *spectral parameter*) with operator L from (18.95) and with boundary conditions (18.96) is said to be a *Sturm–Liouville boundary value problem* in the self-adjoint form.

Definition 18.109. If there is $\lambda \in \mathbb{R}$, such that (18.97) with (18.96) has a non-trivial solution $y(x)$, then this λ is called an *eigenvalue* and this $y(x)$ is called an *eigenfunction* of the Sturm–Liouville boundary value problem, which is said to be in that case an *eigenvalue problem.*

We consider some examples.

1. Let (18.97) have the form

$$((1-x^2)y'(x))' + \lambda y(x) = 0, \quad x \in (-1,1), \quad y(-1) = y(1) = 0.$$

Then for eigenvalues $\lambda_n = n(n+1)$, $n = 0,1,2,\dots$, there exist eigenfunctions $y_n(x) = P_n(x)$, which are Legendre polynomials (see Theorem 18.78 and (18.77)).

2. Let (18.97) have the form

$$(e^{-x^2}y'(x))' + \lambda e^{-x^2}y(x) = 0, \quad x \in \mathbb{R}, \quad e^{-x^2}y(x) \in L^2(\mathbb{R}).$$

Then for eigenvalues $\lambda_n = 2n$ there exist eigenfunctions $y_n(x) = H_n(x)$, which are Hermite polynomials (see Theorem 18.86 and Definition 18.84).

3. Let (18.97) have the form ($\nu > 0$ is given)

$$(xy'(x))' + \left(\lambda x - \frac{\nu^2}{x}\right)y(x) = 0, \quad x \in (0,1),$$
$$y(x) \quad \text{is bounded at } 0, \quad y(1) = 0.$$

Then there exist eigenfunctions $y_n(x) = J_\nu(\mu_n x)$, where J_ν is Bessel function of order ν, $J_\nu(\mu_n) = 0$ and $\lambda_n = \mu_n^2$ are eigenvalues (see (18.18)). The Fourier–Bessel expansion is valid as in the remark after Problem 18.28.

4. Let (18.97) have the form

$$y''(x) + \lambda y(x) = 0, \quad x \in (0,L), \quad y'(0) = y'(L) = 0.$$

Then there exist eigenvalues $\lambda_n = (\frac{n\pi}{L})^2$, $n = 0,1,2,\dots$ and corresponding eigenfunctions $y_n = \cos\frac{n\pi x}{L}$, $n = 0,1,2,\dots$. If we have different boundary conditions, namely

$$y''(x) + \lambda y(x) = 0, \quad x \in (0,L), \quad y(0) = y(L) = 0$$

then there exist eigenvalues $\lambda_n = (\frac{n\pi}{L})^2$, $n = 1, 2, \ldots$ and corresponding eigenfunctions $y_n = \sin\frac{n\pi x}{L}$, $n = 1, 2, \ldots$.

Remark. Based on the orthogonality condition of Bessel's functions (see Problem 18.28), we may consider the Fourier expansions with respect to the orthogonal system of Bessel's functions as follows (see for comparison example (4) from above). Assume that a function $f(x)$ satisfies the conditions

$$\int_0^1 |f(x)|x^{\nu-1}\mathrm{d}x < \infty, \quad \int_1^\infty |f(x)|x^{-3/2}\mathrm{d}x < \infty.$$

Then under the conditions for $f(x)$ from above, and for any fixed $\nu > 0$, we may consider the Fourier expansion as

$$f(x) \approx \sum_{n=0}^{\infty} a_{\nu,n} J_{\nu+2n}(x),$$

where the Fourier coefficients are given by

$$a_{\nu,n} = 2(\nu + 2n)\int_0^\infty \frac{f(x)}{x} J_{\nu+2n}(x)\mathrm{d}x.$$

Problem 18.110. Let function $f(x)$ be defined as

$$f(x) = 0, \quad x \in [-\pi, 0), \quad f(x) = x, \quad x \in [0, \pi].$$

Show that pointwise in $x \in (-\pi, \pi)$,

$$f(x) = \frac{\pi}{4} - \frac{2}{\pi}\sum_{n=1}^{\infty} \frac{\cos(2n-1)x}{(2n-1)^2} - \sum_{n=1}^{\infty} \frac{(-1)^n \sin(nx)}{n}.$$

Show, in particular, that

$$\sum_{n=1}^{\infty} \frac{1}{(2n-1)^2} = \frac{\pi^2}{8}, \quad \sum_{n=1}^{\infty} \frac{(-1)^{n-1}}{2n-1} = \frac{\pi}{4}.$$

Problem 18.111. Let function $f(x) = |x|$, $x \in [-\pi, \pi]$. Show that

$$|x| = \frac{\pi}{2} - \frac{4}{\pi}\sum_{n=1}^{\infty} \frac{\cos(2n-1)x}{(2n-1)^2}$$

uniformly in $x \in [-\pi, \pi]$.

Problem 18.112. Show that uniformly in $x \in [-\pi, \pi]$,

$$\sum_{n=1}^{\infty} \frac{(-1)^n \cos nx}{n^2} = \frac{x^2}{4} - \frac{\pi^2}{12}.$$

Show, in particular, that

$$\frac{\pi^2}{12} = \sum_{n=1}^{\infty} \frac{(-1)^{n+1}}{n^2}.$$

Problem 18.113. Show that uniformly in $x \in [-\pi, \pi]$,

$$\sum_{n=1}^{\infty} \frac{\sin^2 nx}{n^2} = \frac{\pi|x| - x^2}{2}.$$

18.5 Laplace's method

The Laplace's method or the *saddle-point method* is widely used for constructing an asymptotic of certain curve integrals of functions of a complex (and also real) variable. We examine first the integrals of a real variable of the form

$$\Psi(\lambda) = \int_a^b \phi(t) e^{\lambda f(t)} dt, \tag{18.98}$$

where $\lambda > 0$ large enough, a function $f(t)$ is real and $\phi(t)$ might be real or complex (in the latter case, we consider the asymptotic of the integral for $\operatorname{Re}\phi$ and $\operatorname{Im}\phi$ separately). Here, one or both of the limits of integration might be infinite. The following result holds for the integral (18.98).

Theorem 18.114. *Suppose a function $f(t)$ attains its absolute maximum at some interior point $t_0 \in (a, b)$. Let $f''(t_0) < 0$, and assume that there exists $\delta_0 > 0$ such that for $|t - t_0| \le \delta_0$, we have*

$$\begin{aligned} &f(t) = f(t_0) + \frac{f''(t_0)}{2}(t - t_0)^2 + \mu(t), \quad |\mu(t)| < -\frac{f''(t_0)}{4}(t - t_0)^2, \\ &\phi(t) = c_0 + c_1(t - t_0) + O((t - t_0)^2), \quad \mu(t) = c_2(t - t_0)^3 + O((t - t_0)^4). \end{aligned} \tag{18.99}$$

Assume in addition that for $|t - t_0| > \delta_0$, $f(t_0) - f(t) \ge h > 0$, with some constant h, and for some $\lambda_0 > 0$, that

$$\int_a^b |\phi(t)| e^{\lambda_0 f(t)} dt \le M, \quad M > 0. \tag{18.100}$$

Then the following asymptotic formula, as $\lambda \to +\infty$, holds:

$$\int_a^b \phi(t) e^{\lambda f(t)} dt = e^{\lambda f(t_0)} \left(\phi(t_0) \sqrt{-\frac{2\pi}{\lambda f''(t_0)}} + O(\lambda^{-\frac{3}{2}}) \right). \tag{18.101}$$

Proof. Let us split the integral $\Psi(\lambda)$ in (18.98) into five terms:

$$\begin{aligned}\Psi(\lambda) = &\int_a^{t_0-\delta_0} \phi(t)e^{\lambda f(t)} dt + \int_{t_0-\delta_0}^{t_0-\delta(\lambda)} \phi(t)e^{\lambda f(t)} dt + \int_{t_0-\delta(\lambda)}^{t_0+\delta(\lambda)} \phi(t)e^{\lambda f(t)} dt \\ &+ \int_{t_0+\delta(\lambda)}^{t_0+\delta_0} \phi(t)e^{\lambda f(t)} dt + \int_{t_0+\delta_0}^{b} \phi(t)e^{\lambda f(t)} dt =: I_1 + I_2 + I_3 + I_4 + I_5,\end{aligned}$$

where the function $\delta(\lambda)$ is chosen as

$$\lambda\delta^2(\lambda) \to \infty, \quad \lambda\delta^3(\lambda) \to 0, \quad \lambda \to +\infty. \tag{18.102}$$

Since for $a < t \le t_0 - \delta_0$ and $t_0 + \delta_0 \le t < b$, we have for $\lambda > 0$ that

$$\lambda(f(t_0) - f(t)) \ge h\lambda,$$

then the integrals I_1 and I_5 can be estimated for $\lambda > \lambda_0$ as (see (18.100))

$$\begin{aligned}|I_1| &\le \left| \int_a^{t_0-\delta_0} \phi(t)e^{\lambda f(t)} dt \right| \le e^{\lambda f(t_0)} \int_a^{t_0-\delta_0} |\phi(t)| e^{-\lambda(f(t_0)-f(t))} dt \\ &= e^{\lambda f(t_0)} \int_a^{t_0-\delta_0} |\phi(t)| e^{-(\lambda-\lambda_0)(f(t_0)-f(t))} e^{\lambda_0(f(t)-f(t_0))} dt \\ &\le M e^{\lambda f(t_0)} e^{-h(\lambda-\lambda_0)} e^{-\lambda_0 f(t_0)} = e^{\lambda f(t_0)} O(e^{-\lambda h}), \quad \lambda \to \infty,\end{aligned}$$

and

$$\begin{aligned}|I_5| &\le \left| \int_{t_0+\delta_0}^{b} \phi(t)e^{\lambda f(t)} dt \right| \le e^{\lambda f(t_0)} \int_{t_0+\delta_0}^{b} |\phi(t)| e^{-\lambda(f(t_0)-f(t))} dt \\ &= e^{\lambda f(t_0)} \int_{t_0+\delta_0}^{b} |\phi(t)| e^{-(\lambda-\lambda_0)(f(t_0)-f(t))} e^{\lambda_0(f(t)-f(t_0))} dt \\ &\le M e^{\lambda f(t_0)} e^{-h(\lambda-\lambda_0)} e^{-\lambda_0 f(t_0)} = e^{\lambda f(t_0)} O(e^{-\lambda h}), \quad \lambda \to \infty,\end{aligned}$$

respectively. Next, due to the conditions for $f(t)$ and $\mu(t)$ for $|t - t_0| \le \delta_0$, we have that (see (18.99))

$$f(t_0) - f(t) = -\mu(t) - \frac{f''(t_0)}{2}(t - t_0)^2 \geq -|\mu(t)| - \frac{f''(t_0)}{2}(t - t_0)^2$$
$$\geq \frac{f''(t_0)}{4}(t - t_0)^2 - \frac{f''(t_0)}{2}(t - t_0)^2 = -\frac{f''(t_0)}{4}(t - t_0)^2.$$

Hence, repeating the same procedure as for the estimation of the integrals I_1, I_5, and taking into account that for the integrals I_2, I_4, we really have that $(t - t_0)^2 \geq \delta^2(\lambda)$, and we obtain (using the fact that $f''(t_0) < 0$)

$$|I_2|, |I_4| \leq e^{\lambda f(t_0)} O(e^{-\frac{f''(t_0)}{4}\lambda\delta^2(\lambda)}), \quad \lambda \to \infty.$$

It remains now to examine the principal integral I_3 of $\Psi(\lambda)$. By virtue of the conditions (18.99), this integral can be rewritten, using a new variable $\tau := t - t_0$, and second condition of (18.102), as

$$I_3 = e^{\lambda f(t_0)} \int_{-\delta(\lambda)}^{\delta(\lambda)} \phi(t_0 + \tau) e^{\lambda(\frac{1}{2}f''(t_0)\tau^2 + c_2\tau^3 + O(\tau^4))} d\tau$$
$$= e^{\lambda f(t_0)} \int_{-\delta(\lambda)}^{\delta(\lambda)} (c_0 + c_1\tau + O(\tau^2)) e^{\lambda(\frac{1}{2}f''(t_0)\tau^2 + c_2\tau^3 + O(\tau^4))} d\tau$$
$$= e^{\lambda f(t_0)} \int_{-\delta(\lambda)}^{\delta(\lambda)} (c_0 + c_1\tau + c_0c_2\lambda\tau^3 + O(\tau^2) + O(\lambda\tau^4) + O(\lambda^2\tau^6)) e^{\lambda f''(t_0)\tau^2/2} d\tau$$
$$= e^{\lambda f(t_0)} \int_{-\delta(\lambda)}^{\delta(\lambda)} (c_0 + O(\tau^2) + O(\lambda\tau^4) + O(\lambda^2\tau^6)) e^{\lambda f''(t_0)\tau^2/2} d\tau.$$

Here we have used the fact that the terms with $c_1\tau$ and $c_2\tau^3$ of the resulting expression vanish since the integrands are odd. The first part of the latter integral yields ($f''(t_0) < 0$),

$$c_0 e^{\lambda f(t_0)} \int_{-\delta(\lambda)}^{\delta(\lambda)} e^{\lambda f''(t_0)\tau^2/2} d\tau$$
$$= 2c_0 e^{\lambda f(t_0)} \int_0^{\delta(\lambda)} e^{\lambda f''(t_0)\tau^2/2} d\tau$$
$$= c_0 \sqrt{-\frac{2}{\lambda f''(t_0)}} e^{\lambda f(t_0)} \int_0^{-\lambda\delta^2(\lambda) f''(t_0)/2} e^{-\xi} \xi^{-\frac{1}{2}} d\xi$$
$$= c_0 \sqrt{-\frac{2}{\lambda f''(t_0)}} e^{\lambda f(t_0)} \left(\int_0^\infty e^{-\xi} \xi^{-\frac{1}{2}} d\xi - \int_{-\lambda\delta^2(\lambda) f''(t_0)/2}^\infty e^{-\xi} \xi^{-\frac{1}{2}} d\xi \right)$$

$$= c_0\sqrt{-\frac{2}{\lambda f''(t_0)}}\mathrm{e}^{\lambda f(t_0)}\left(\Gamma\left(\frac{1}{2}\right) + O(\mathrm{e}^{-\lambda\delta^2(\lambda)f''(t_0)/4})\right)$$

$$= \phi(t_0)\sqrt{-\frac{2}{\lambda f''(t_0)}}\mathrm{e}^{\lambda f(t_0)}(\sqrt{\pi} + O(\mathrm{e}^{-\lambda\delta^2(\lambda)f''(t_0)/4})), \quad \phi(t_0) = c_0. \tag{18.103}$$

To obtain (18.103), we have used the first condition of (18.102) and (18.99). The other terms in the previous integral can be estimated similarly. Consider, e. g., the term with $O(\tau^2)$. Then we will have for this term, using a new variable $\xi := \lambda\tau^2$, that

$$O(1)\mathrm{e}^{\lambda f(t_0)}\int\limits_0^{\delta(\lambda)} \tau^2\mathrm{e}^{\lambda f''(t_0)\tau^2/2}\mathrm{d}\tau = O(1)\mathrm{e}^{\lambda f(t_0)}\frac{1}{\lambda^{\frac{3}{2}}}\int\limits_0^{\lambda\delta^2(\lambda)} \xi^{\frac{1}{2}}\mathrm{e}^{f''(t_0)\xi/2}\mathrm{d}\xi$$

$$= O\left(\frac{1}{\lambda^{\frac{3}{2}}}\right)\mathrm{e}^{\lambda f(t_0)}, \quad \lambda \to +\infty, \tag{18.104}$$

since $f''(t_0) < 0$ and $\lambda\delta^2(\lambda) \to \infty$ (see (18.102)). The terms with $O(\lambda\tau^4)$ and $O(\lambda^2\tau^6)$ can be estimated by the same manner as the previous one and we can obtain for them the same estimate (18.104). Choosing $\delta(\lambda) = \frac{1}{\lambda^{\frac{2}{5}}}$ (in that case (18.102) is satisfied) and combining the estimates of I_1, I_2, I_4, I_5, and (18.103), (18.104), we obtain the final result (18.101). Thus, the theorem is proved. □

Example 18.115. As an example of the application of Theorem 18.114, we consider the asymptotic for Euler's gamma function $\Gamma(p+1)$ when $p \to +\infty$. Since (here, $t = px$ is a new variable)

$$\Gamma(p+1) = \int\limits_0^{\infty} x^p\mathrm{e}^{-x}\mathrm{d}x = p^{p+1}\int\limits_0^{\infty} \mathrm{e}^{p(\log t - t)}\mathrm{d}t,$$

then direct application of Theorem 18.114 with $\phi(t) \equiv 1, f(t) = \log t - t, t_0 = 1, f''(1) = -1$, $f(1) = -1$, and with $\lambda = p$ gives us

$$\Gamma(p+1) = p^{p+1}\mathrm{e}^{-p}\left(\sqrt{\frac{2\pi}{p}} + O\left(\frac{1}{p\sqrt{p}}\right)\right) = \sqrt{2\pi p}\left(\frac{p}{\mathrm{e}}\right)^p\left(1 + O\left(\frac{1}{p}\right)\right). \tag{18.105}$$

This formula (18.105) is called *Stirling's formula*. Moreover, it can be extended not only for real $p \to +\infty$ but also for complex $p \to \infty$ such that $|\arg z| \le \pi - \epsilon, \epsilon > 0$. In addition, since $\Gamma(n+1) = n!$ for $n \in \mathbb{N}$, we have the following asymptotic:

$$n! = \sqrt{2\pi n}\left(\frac{n}{\mathrm{e}}\right)^n\left(1 + O\left(\frac{1}{n}\right)\right), \quad n \to +\infty.$$

Remark. The essential addition to the Laplace's method is the so-called method of *stationary phase*, which concerns the asymptotic behavior of the integral

$$\Phi(\lambda) = \int_a^b \phi(t) e^{i\lambda f(t)} dt,$$

where $f(t)$ is a real-valued function with one stationary point $t_0 \in (a,b)$, i. e., $f'(t_0) = 0, f''(t_0) \neq 0$, under the assumption (at the moment) that the interval (a,b) is finite. Assuming that $\phi(t)$ is continuous on the interval $[a,b]$, the following asymptotic holds (in the neighborhood of t_0):

$$\Phi(\lambda) = e^{i\lambda f(t_0) + i \operatorname{sgn} f''(t_0)\frac{\pi}{4}} \phi(t_0) \sqrt{\frac{2\pi}{\lambda |f''(t_0)|}} + O\left(\frac{1}{\lambda\sqrt{\lambda}}\right), \quad \lambda \to +\infty. \tag{18.106}$$

As an application of the method of the stationary phase, we consider the behavior of the Bessel functions for large argument. Using the integral representation for $J_n(x), n \in \mathbb{N}_0$ (see, e. g., Problem 18.27(2)), we have (using the properties of odd and even functions)

$$\begin{aligned} J_n(x) &= \frac{1}{2\pi} \int_{-\pi}^{\pi} \cos(n\theta - x\sin\theta) d\theta = \frac{1}{2\pi} \int_{-\pi}^{\pi} e^{i(n\theta - x\sin\theta)} d\theta \\ &= \frac{1}{2\pi} \int_{-\pi}^{\pi} e^{in\theta} e^{-ix\sin\theta} d\theta. \end{aligned}$$

In this case, $f(\theta) = -\sin\theta$, $\phi(\theta) = e^{in\theta}$, $t = \theta$, and $\lambda = x$. That is why we have two stationary points $\theta = \pm\frac{\pi}{2}$ and, therefore (see (18.106)) we have that

$$\begin{aligned} J_n(x) &= \frac{1}{2\pi} \left(e^{-ix + i\frac{\pi}{4} + i\frac{n\pi}{2}} + e^{ix - i\frac{\pi}{4} - i\frac{n\pi}{2}}\right) \sqrt{\frac{2\pi}{x}} + O\left(\frac{1}{x\sqrt{x}}\right) \\ &= \cos\left(x - \frac{\pi}{4} - \frac{n\pi}{2}\right) \sqrt{\frac{2}{\pi x}} + O\left(\frac{1}{x\sqrt{x}}\right), \quad x \to +\infty. \end{aligned} \tag{18.107}$$

Furthermore, the asymptotic (18.107) is valid not only for $n \in \mathbb{N}_0$, but for any $\nu \in \mathbb{R}$, if we substitute ν instead of n.

We now examine the saddle point method for obtaining asymptotic expansions of integrals in the form

$$F(\lambda) = \int_\gamma \phi(z) e^{\lambda f(z)} dz, \tag{18.108}$$

where the functions $\phi(z), f(z)$ are analytic in some domain $D \subset \mathbb{C}$ and $\gamma \subset D$ is a piecewise smooth Jordan curve. Due to the Cauchy theorem, integral (18.108) is independent

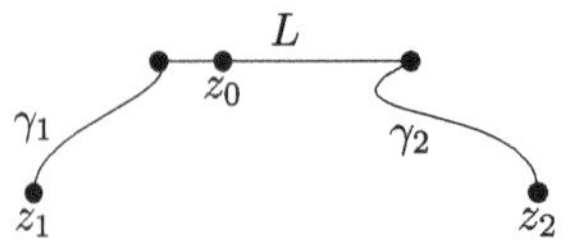

Figure 18.2: Curve of integration in Laplace's method.

on the curves with the same end points (initial and terminal). So, we may assume that this curve γ is deformed into the curve $\Gamma = L + \gamma_1 + \gamma_2$ such that L is a line on the complex plane (it might be small enough) and γ_1 and γ_2 (with finite length) are joined with L starting from the points z_1 and z_2, respectively; see Figure 18.2. The range of application of the saddle point method is so that L is chosen such that z_0 is not an end point of L and

$$f'(z_0) = 0, \quad f''(z_0) \neq 0, \quad \operatorname{Im} f(z) = \text{constant}, \quad z \in L, \tag{18.109}$$

i. e., z_0 is a saddle point for $f(z)$, and we assume that it is unique.

The following theorem holds (it is called Laplace's method in the complex case).

Theorem 18.116. *Suppose that all conditions for ϕ, f, γ, L, z_0 from above are satisfied. Assume in addition that there exists $\delta > 0$ (small enough) such that for all $|z - z_0| \geq \delta$,*

$$\operatorname{Re} f(z_0) - \operatorname{Re} f(z) \geq h > 0, \quad h = \text{constant}, \tag{18.110}$$

and also for some $\lambda_0 > 0$, the following integral converges:

$$\int_\gamma |\phi(z)| \mathrm{e}^{\lambda_0 \operatorname{Re} f(z)} |\mathrm{d}z| \leq M. \tag{18.111}$$

Then for all $\lambda \geq \lambda_0$, the following asymptotic formula holds:

$$\int_\gamma \phi(z) \mathrm{e}^{\lambda f(z)} \mathrm{d}z = \mathrm{e}^{\lambda f(z_0)} \left(\sqrt{\frac{2\pi}{\lambda |f''(z_0)|}} \phi(z_0) \mathrm{e}^{\mathrm{i}\theta_m} + O(\lambda^{-\frac{3}{2}}) \right), \tag{18.112}$$

where $\theta_m = \frac{\pi - \arg f''(z_0)}{2} + \pi m$, $m = 0, 1$. Here, the choice of m determines the sign in (18.112) *and actually depends only on the direction of integration along curve γ.*

Proof. The conditions of the theorem (and considerations from above) imply that

$$\begin{aligned} F(\lambda) &= \int_L \phi(z) \mathrm{e}^{\lambda f(z)} \mathrm{d}z + \int_{\gamma_1} \phi(z) \mathrm{e}^{\lambda f(z)} \mathrm{d}z + \int_{\gamma_2} \phi(z) \mathrm{e}^{\lambda f(z)} \mathrm{d}z \\ &= \mathrm{e}^{\mathrm{i}\lambda \operatorname{Im} f(z_0)} \int_L \phi(z) \mathrm{e}^{\lambda \operatorname{Re} f(z)} \mathrm{d}z + \int_{\gamma_1} \phi(z) \mathrm{e}^{\lambda f(z)} \mathrm{d}z + \int_{\gamma_2} \phi(z) \mathrm{e}^{\lambda f(z)} \mathrm{d}z. \end{aligned}$$

By conditions (18.110) and (18.111), the integrals along the curves γ_1 and γ_2 can be estimated as

$$\left|\int_{\gamma_1,\gamma_2} \phi(z)e^{\lambda f(z)}dz\right| \le \int_{\gamma_1,\gamma_2} |\phi(z)|e^{\lambda \operatorname{Re} f(z)}|dz|$$
$$\le \int_{\gamma_1,\gamma_2} |\phi(z)|e^{\lambda_0 \operatorname{Re} f(z)}e^{(\lambda-\lambda_0)\operatorname{Re} f(z)}|dz| \le Me^{(\lambda-\lambda_0)\operatorname{Re} f(z_0)}e^{-(\lambda-\lambda_0)h}$$
$$= e^{\lambda \operatorname{Re} f(z_0)}O(e^{-\lambda h}), \quad \lambda \to +\infty.$$

On the line L, we introduce the parametrization $z := z(s)$ so that $z(0) = z_0$, $z(-a) = z_1$, and $z(b) = z_2$ (see the description of Γ above). Using this parametrization, the integral along L can be written as (note that $\operatorname{Im} f(z) = \operatorname{Im} f(z_0)$ on the line L)

$$F_1(\lambda) := \int_L \phi(z)e^{\lambda f(z)}dz = e^{i\lambda \operatorname{Im} f(z_0)} \int_{-a}^{b} \phi(z(s))e^{\lambda \operatorname{Re} f(z(s))}z'(s)ds.$$

One can see that the latter integral satisfies all conditions of Theorem 18.114: the function $\operatorname{Re} f(z(s))$ attains its maximum at the point $s = 0$ and we have $(\operatorname{Re} f(z(s)))''(0) < 0$. Then according to (18.101), we have

$$F_1(\lambda) = e^{i \operatorname{Im} f(z_0)}e^{\operatorname{Re} f(z_0)}\left(\operatorname{Re} f(z(0))z'(0)\sqrt{-\frac{2\pi}{\lambda(\operatorname{Re} f(z(s)))''(0)}} + O(\lambda^{-\frac{3}{2}})\right). \tag{18.113}$$

It remains to express the quantities appearing in (18.113) in terms of the values of the functions $\phi(z)$ and $f(z)$ at the point z_0. It is clear (due to parametrization of the line L) that $z - z_0 = se^{i\pi m}$, $m = 0, 1$ depending on the direction of integration along L. Moreover, ($\operatorname{Im} f(z) = \operatorname{Im} f(z_0)$ on the line L), for $z \in L$, we have

$$z'(0) = ie^{i\pi m}, (\operatorname{Re} f(z))_s''|_{s=0} = f''(z(0))(z'(0))^2 = -|f''(z_0)|e^{i \arg f''(z_0)}.$$

Taking into account the considerations from above and the fact that $i = e^{i\frac{\pi}{2}}$, and collecting the estimates for the integrals along γ_1, γ_2, and the integral F_1, we can easily obtain (18.112). Thus, the theorem is completely proved. □

Remark. The parameter λ might be complex. If λ is complex, i. e., $\lambda = |\lambda|e^{i \arg \lambda}$ we can absorb the exponential factor into $f(z)$ and proceed with a new function $\tilde{f}(z) := e^{i \arg \lambda} f(z)$.

Example 18.117. We will obtain the asymptotic of the Hankel function for a large positive argument. Using Problem 18.27(9) (see (18.26)), we have that

$$H_\nu^{(1)}(x) = \frac{1}{i\pi}\int_{-\infty}^{0} e^{-\nu t + x \sinh t}dt + \frac{1}{i\pi}\int_0^{\pi} e^{-i\nu t + x \sinh(it)}idt$$
$$+ \frac{1}{i\pi}\int_0^{\infty} e^{-\nu(t+i\pi)+x\sinh(t+i\pi)}dt =: I_1 + I_2 + I_3.$$

Since $\sinh t \approx \frac{1}{2}e^t$ as $t \to +\infty$ and $\sinh(t + i\pi) = -\sinh t$, then both integrals I_1 and I_3 vanish exponentially as $x \to +\infty$, i. e.,

$$I_1, I_3 = O(e^{-c_0 x}), \quad x \to +\infty, \text{ with some } c_0 > 0.$$

For the integral I_2, we use the method of stationary phase (see (18.106)). Then we have that (the unique stationary point is $t_0 = \frac{\pi}{2}, f''(t_0) = -1, \phi(t) = e^{-i\nu t}$)

$$I_2 = \frac{1}{\pi} e^{ix - i\frac{\pi}{4}} e^{-i\frac{\nu\pi}{2}} \sqrt{\frac{2\pi}{x}} + O(x^{-\frac{3}{2}}), \quad x \to +\infty.$$

Combining asymptotic for I_1, I_3, and I_2, we obtain finally

$$H_\nu^{(1)}(x) = e^{i(x - \frac{\pi}{4} - \frac{\nu\pi}{2})} \sqrt{\frac{2}{\pi x}} + O(x^{-\frac{3}{2}}), \quad x \to +\infty. \tag{18.114}$$

In addition, since $H_\nu^{(2)}(x) = \overline{H_\nu^{(1)}(x)}$, and $J_\nu(x) = \frac{1}{2}(H_\nu^{(1)}(x) + H_\nu^{(2)}(x))$, $Y_\nu = \frac{1}{2i}(H_\nu^{(1)}(x) - H_\nu^{(2)}(x))$ (see Definition 18.14), we can obtain easily the asymptotic for all these functions also. For example,

$$J_\nu(x) = \cos\left(x - \frac{\pi}{4} - \frac{\nu\pi}{2}\right)\sqrt{\frac{2}{\pi x}} + O(x^{-\frac{3}{2}}), \quad x \to +\infty \tag{18.115}$$

and

$$Y_\nu(x) = \sin\left(x - \frac{\pi}{4} - \frac{\nu\pi}{2}\right)\sqrt{\frac{2}{\pi x}} + O(x^{-\frac{3}{2}}), \quad x \to +\infty. \tag{18.116}$$

Problem 18.118. Using Laplace's method (see Theorem 18.114), show that for any fixed $\theta \in (0, \pi)$ the following asymptotic formula with respect to n for Legendre polynomials holds:

$$P_n(\cos\theta) = \sqrt{\frac{2}{\pi n \sin\theta}}(\cos((n + 1/2)\theta - \pi/4) + O(n^{-1})), \quad n \to +\infty, \tag{18.117}$$

i. e., for any fixed $x \in (-1, 1)$ and even uniformly in $x \in [x_1, x_2] \subset (-1, 1)$,

$$P_n(x) = O(n^{-1/2}), \quad n \to +\infty.$$

Hint: First, use the following integral representation of Legendre polynomials:

$$P_n(\cos\theta) = \frac{1}{\pi\sqrt{2}} \int_{-\theta}^{\theta} \frac{e^{i(n+1/2)\phi}}{\sqrt{\cos\phi - \cos\theta}} d\phi, \quad \theta \in (0, \pi),$$

then, second, choose the closed curve γ (Figure 18.3) consisting of the interval ($y = 0, -\theta < x < \theta$) of the real axis, the vertical intervals ($x = -\theta, 0 < y < L$), ($x = \theta, 0 < y < L$)

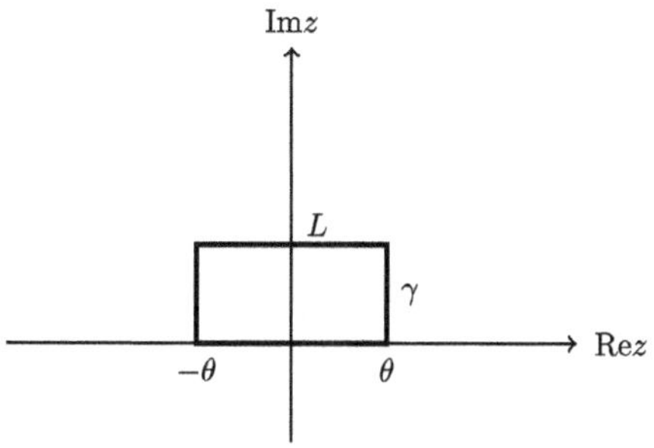

Figure 18.3: Curve of integration in Problem 18.118.

parallel to the imaginary axis, and the horizontal interval ($y = L, -\theta < x < \theta$), and finally, consider on the complex plane the function

$$g(z) := \frac{e^{i(n+1/2)z}}{\sqrt{\cos z - \cos\theta}}, \quad z \in \mathbb{C},$$

inside of this curve γ. The final step is to put $L \to +\infty$.

Exercises

1. Let $w(z) = \frac{az+b}{cz+d}$ be a nondegenerate ($ad \neq bc$, $c \neq 0$) linear-fractional transformation. A point z is said to be a fixed point if $w(z) = z$.
   a) Prove that this transformation has at most two fixed points. If $(a-d)^2 + 4bc \neq 0$, $c \neq 0$, then there are two different fixed points, otherwise they coincide.
   b) Show that if there are two different fixed points, p and q, then

$$\frac{w-p}{w-q} = k_1 \frac{z-p}{z-q} \quad \text{with } k_1 = \frac{cq+d}{cp+d}.$$

   But if there is only one fixed point p, then

$$\frac{1}{w-p} = \frac{1}{z-p} + k_2 \quad \text{with } k_2 = \frac{c(a+d)}{2(ad+bc)}.$$

2. Show that any equation of the form

$$\left|\frac{z-p}{z-q}\right| = k, \quad k \neq 1,$$

defines the circle with such conjugate p and q that p and q lie on the same line together with z_0 and

$$|p - z_0||q - z_0| = r^2,$$

where z_0 is a center of this circle with radius r.

3. Show that the general nondegenerate linear-fractional transformation of the domain $\{z : \operatorname{Re} z \geq 0\}$ onto the unit disk $\{w : |w| \leq 1\}$ has the form

$$w(z) = \mathrm{e}^{\mathrm{i}\lambda} \frac{z-a}{z+\bar{a}}, \qquad \operatorname{Re} a > 0, \lambda \in \mathbb{R}.$$

4. Show that the general nondegenerate linear-fractional transformation of the disk $\{z : |z| \leq r_1\}$ onto the disk $\{w : |w| \leq r_2\}$ has the form

$$w(z) = r_1 r_2 \mathrm{e}^{\mathrm{i}\lambda} \frac{z-a}{\bar{a}z - r_1^2}, \qquad 0 < |a| < r_1, \lambda \in \mathbb{R}.$$

5. Let function $f(z)$ be analytic in the open unit disk $\{z : |z| < 1\}$. Show that if $\operatorname{Re} f(z) > 0$ and $f(0) = a > 0$, then $|f'(0)| \leq 2a$.
6. Show that

$$\mathcal{L}(\sin(\omega t + \phi))(p) = \frac{p \sin\phi + \omega \cos\phi}{p^2 + \omega^2}, \quad \mathcal{L}(\cos(\omega t + \phi))(p) = \frac{p \cos\phi - \omega \sin\phi}{p^2 + \omega^2}.$$

7. Using Mellin's formula, show that:

https://doi.org/10.1515/9783111632278-021

a)

$$\mathcal{L}^{-1}\left(\frac{\lambda}{p^2-\lambda^2}\right) = \sinh \lambda t, \quad \mathcal{L}^{-1}\left(\frac{p}{p^2-\lambda^2}\right) = \cosh \lambda t, \quad \operatorname{Re} p > \operatorname{Re} \lambda.$$

b)

$$\mathcal{L}^{-1}\left(\frac{\omega}{(p+a)^2+\omega^2}\right) = \mathrm{e}^{-at}\sin(\omega t), \quad \operatorname{Re} p > |\operatorname{Im} \omega| - \operatorname{Re} a.$$

c)

$$\mathcal{L}^{-1}\left(\frac{1}{p}\tanh(p/2)\right) = \begin{cases} 1, & 2k \le t < 2k+1, \\ -1, & 2k+1 \le t < 2k+2, \end{cases} \qquad k = 0, 1, 2, \ldots,$$

where $\operatorname{Re} p > 0$.

d)

$$\mathcal{L}^{-1}\left(\frac{1}{\sqrt{p+a}}\right) = \frac{\mathrm{e}^{-ia}}{\sqrt{\pi t}}, \quad \operatorname{Re} p > 0.$$

e)

$$\mathcal{L}^{-1}\left(\frac{1}{p}\mathrm{e}^{-a\sqrt{p}}\right) = 1 - \int_0^a \mathrm{e}^{-y^2/(4t)}\mathrm{d}y, \quad \operatorname{Re} p > 0.$$

8. Let $F(p)$ be the Laplace transform of some function $f(t)$ so that $f(t) = \mathcal{L}^{-1}(F(p))$. Assume that $F(p)$ is considered for $p \in \mathbb{R}_+$, and is integrable there. Prove that

$$\int_0^\infty \frac{f(t)}{t}\mathrm{d}t = \int_0^\infty F(p)\mathrm{d}p.$$

9. Using the previous exercise, show that for $a, \beta \neq 0$,

$$\int_0^\infty \frac{\cos(at) - \cos(\beta t)}{t}\mathrm{d}t = \log\left|\frac{a}{\beta}\right|,$$

and

$$\int_0^\infty \frac{\sin(at) - \sin(\beta t)}{t}\mathrm{d}t = \begin{cases} 0, & a\beta > 0, \\ \pi, & a > 0, \beta < 0, \\ -\pi, & a < 0, \beta > 0. \end{cases}$$

10. Let $F(p)$ be the Laplace transform of some function $f(t)$. Show that

$$\lim_{p\to\infty, \operatorname{Re} p > 0} pF(p) = f(0), \quad \lim_{p\to 0} pF(p) = \lim_{t\to+\infty} f(t).$$

11. Using the Laplace transform, show that

$$u(x,t) := \frac{x}{2a\sqrt{\pi}} \int_0^t e^{-x^2/4a^2(t-\tau)} \frac{q(\tau)}{(t-\tau)^{3/2}} d\tau$$

and solves the following equation with additional conditions:

$$\begin{cases} u_t(x,t) = a^2 u_{xx}(x,t), & x > 0, t > 0, \\ u(x,0) = 0, \quad u(0,t) = q(t), \end{cases}$$

where function $q(t), t > 0$ satisfies the needed properties.

12. Prove the following representation of the Euler's gamma function $\Gamma(z)$:

$$\Gamma(z) = \frac{e^{-\gamma z}}{z} \prod_{n=1}^{\infty} \frac{e^{z/n}}{1 + z/n}, \quad z \in \mathbb{C} \setminus \{-\mathbb{N}_0\},$$

where γ is Euler's constant. Hint: Use Example 17.28.

13. Show that:

a)

$$|\Gamma(\mathrm{i}y)|^2 = \frac{\pi}{y \sinh(\pi y)}.$$

b)

$$|\Gamma(1/2 + \mathrm{i}y)|^2 = \frac{\pi}{\cosh(\pi y)}.$$

c)

$$|\Gamma(x + \mathrm{i}y)|^2 = 2\pi e^{-\pi|y|} |y|^{2x-1}(1 + o(1)), \quad y \to \infty, |x| \le \text{constant}.$$

Hint: Use the property $\Gamma(\overline{z}) = \overline{\Gamma(z)}$ and the formula $\Gamma(z)\Gamma(1-z) = \frac{\pi}{\sin(\pi z)}$.

14. Show that for any complex α and β,

$$\frac{\Gamma(z+\alpha)}{\Gamma(z+\beta)} \approx z^{\alpha-\beta}, \quad z \to \infty,$$

and even more precisely

$$\frac{\Gamma(z+\alpha)}{\Gamma(z+\beta)} = z^{\alpha-\beta}\left(1 + \frac{(\alpha-\beta)(\alpha+\beta-1)}{2z} + O(z^{-2})\right), \quad z \to \infty.$$

Hint: Use the Stirling's formula (18.105) for complex p.

15. Show that for $\operatorname{Re} z > 0$,

$$(\Gamma(z))^2 = 2^{2-2z} \int_0^{\infty} t^{2z-1} K_0(t) dt,$$

where $K_0(\cdot)$ is the MacDonald function of order 0. In particular,

$$\int_0^\infty tK_0(t)\mathrm{d}t = 1.$$

Hint: Use (17.17) and (17.20).

16. Prove that

$$\int_0^{\pi/2} \cos^\mu\theta \sin^\nu\theta \mathrm{d}\theta = \frac{1}{2}\frac{\Gamma(\frac{\mu+1}{2})\Gamma(\frac{\nu+1}{2})}{\Gamma(\frac{\mu+\nu}{2}+1)}, \quad \operatorname{Re}\mu, \operatorname{Re}\nu > -1.$$

In particular,

$$\int_0^{\pi/2} \cos^\nu\theta \mathrm{d}\theta = \int_0^{\pi/2} \sin^\nu\theta \mathrm{d}\theta = \frac{\sqrt{\pi}}{2}\frac{\Gamma(\frac{\nu+1}{2})}{\Gamma(\frac{\nu}{2}+1)}, \quad \operatorname{Re}\nu > -1.$$

17. Prove that:

a)

$$\int_0^x t^\mu J_\nu(t)\mathrm{d}t = x^\mu J_{\nu+1}(x) - (\mu-\nu-1)\int_0^x t^{\mu-1}J_\nu(t)\mathrm{d}t, \quad \operatorname{Re}(\mu+\nu) > 0.$$

b)

$$\nu\int_0^x \frac{J_\nu(t)}{t}\mathrm{d}t = \int_0^x J_{\nu-1}(t)\mathrm{d}t - J_\nu(x) - 1, \quad \operatorname{Re}\nu > 0.$$

Hint: Use the integration by parts and the recursion formulas for Bessel's functions.

18. Show that for $n \in \mathbb{N}_0$ we have:

a)

$$J_n^2(x) = \frac{2}{\pi}\int_0^{\pi/2} J_{2n}(2x\cos\theta)\mathrm{d}\theta \equiv (-1)^n\frac{2}{\pi}\int_0^{\pi/2} J_0(2x\cos\theta)\cos(2n\theta)\mathrm{d}\theta.$$

b)

$$J_n(x)J_n(y) = \frac{1}{\pi}\int_0^\pi J_0(\sqrt{x^2+y^2-2xy\cos\theta})\cos(n\theta)\mathrm{d}\theta.$$

19. Prove that

$$\int_0^x J_\nu(t)\mathrm{d}t = 2\sum_{k=0}^\infty J_{\nu+2k+1}(x), \quad \operatorname{Re}\nu > -1.$$

Hint: Use differentiation with respect to x and the recursion formulas for Bessel's functions.

20. Let $x > 0$. Show that:
    a)
$$J_\nu(-x + \mathrm{i}0) - J_\nu(-x - \mathrm{i}0) = 2\mathrm{i}\sin(\nu\pi)J_\nu(x).$$
    b)
$$Y_\nu(-x + \mathrm{i}0) - Y_\nu(-x - \mathrm{i}0) = 2\mathrm{i}\big(J_\nu(x)\cos(\nu\pi) + J_{-\nu}(x)\big).$$
    In particular,
$$J_n(-x + \mathrm{i}0) = J_n(-x - \mathrm{i}0), \quad Y_n(-x + \mathrm{i}0) = Y_n(-x - \mathrm{i}0) + 4\mathrm{i}(-1)^n J_n(x).$$

21. Show that for $k \geq 0$,
$$\frac{\mathrm{e}^{-k\sqrt{x^2+r^2}}}{\sqrt{x^2+r^2}} = \int_0^\infty \mathrm{e}^{-|x|\sqrt{\lambda^2+k^2}} \frac{\lambda J_0(\lambda|r|)}{\sqrt{\lambda^2+k^2}} \mathrm{d}\lambda.$$

22. Show that the constants K and K' from (18.68) and (18.69) can be written as
$$K = \frac{1}{2}\int_{1/k^2}^\infty \frac{1}{\sqrt{(t^2-t)(k^2t-1)}} dt, \quad K' = \frac{1}{2}\int_1^{1/k^2} \frac{1}{\sqrt{(t^2-t)(1-k^2t)}} dt.$$

23. Show that the constants K and K' from (18.68) and (18.69) are the solutions of the following ordinary differential equation:
$$\frac{\mathrm{d}}{\mathrm{d}k}\left(k(1-k^2)\frac{\mathrm{d}u}{\mathrm{d}k}\right) = ku.$$

24. Let function $E(z;k)$ be defined as
$$E(z;k) := \int_0^z \mathrm{dn}^2(t;k)\mathrm{d}t.$$
    Show that:
    a)
$$E(z + 2nK; k) = E(z;k) + 2nE(K;k), \quad n \in \mathbb{Z}.$$
    b)
$$E(K;k)K' + E(K;\sqrt{1-k^2})K - KK' = \frac{\pi}{2}.$$

25. Show that the Legendre polynomials admit the following representation:

$$P_n(x) = \frac{1}{\pi}\int_0^{\pi}\left(x + \mathrm{i}\sqrt{1-x^2}\cos\phi\right)^n \mathrm{d}\phi, \quad |x| \le 1, n = 0, 1, 2, \ldots.$$

Hint: Use the fact that

$$\int_0^{\pi}(\cos\phi)^k \mathrm{d}\phi = \begin{cases} 0, & k \text{ is odd}, \\ 2\int_0^{\pi/2}(\cos\phi)^k \mathrm{d}\phi, & k \text{ is even}. \end{cases}$$

26. Show that:
   a)

$$|P_n(x)| \le 1, \quad |x| \le 1.$$

   b)

$$(1-x^2)^{1/4}|P_n(x)| \le \sqrt{\frac{2}{\pi n}}, \quad |x| \le 1.$$

   Hint: Use Exercise 25.
27. Prove that the expansion with respect to the Legendre polynomials for the function $\sqrt{\frac{1-x}{2}}$ is equal to

$$\sqrt{\frac{1-x}{2}} = \frac{2}{3}P_0(x) - 2\sum_{n=1}^{\infty}\frac{P_n(x)}{(2n-1)(2n+3)}, \quad |x| < 1.$$

28. Show that all roots of the equation $P_n(x) = 0$, $n \in \mathbb{N}$ are real. Hint: Use Rolle's theorem.
29. Prove that for the Hermite polynomials the following is true:

$$H_{2n}(0) = (-1)^n\frac{(2n)!}{n!}, \quad H_{2n+1}(0) = 0,$$
$$H'_{2n}(0) = 0, \quad H'_{2n=1}(0) = 2(-1)^n\frac{(2n+1)!}{n!}.$$

   Hint: Use Theorem 18.86 and its corollary.
30. Show that all roots of the equation $H_n(x) = 0$, $n \in \mathbb{N}$ are real. Hint: Use Rolle's theorem.
31. Show the following integral representation for the Hermite polynomials:

$$H_{2n}(x) = \frac{(-1)^n 2^{2n+1} e^{x^2}}{\sqrt{\pi}}\int_0^{\infty} e^{-t^2} t^{2n}\cos(2xt)\mathrm{d}t,$$

$$H_{2n+1}(x) = \frac{(-1)^n 2^{2n+2} e^{x^2}}{\sqrt{\pi}} \int_0^\infty e^{-t^2} t^{2n+1} \sin(2xt) dt,$$

where $n = 0, 1, 2, \ldots$. Hint: Use the well-known integral

$$e^{-x^2} = \frac{2}{\sqrt{\pi}} \int_0^\infty e^{-t^2} \cos(2xt) dt.$$

32. Prove that the expansion with respect to the Hermite polynomials for the function e^{-ax^2}, $a > 0$ is equal to

$$e^{-ax^2} = \sum_{n=0}^\infty \frac{(-1)^n a^n}{2^{2n}(1+a)^{n+1/2}} H_{2n}(x), \quad x \in \mathbb{R}.$$

33. Prove the following expansions:

$$e^{t^2} \cos(2xt) = \sum_{n=0}^\infty \frac{(-1)^n H_{2n}(x)}{(2n)!} t^{2n}, \quad t \in \mathbb{R},$$

$$e^{t^2} \sin(2xt) = \sum_{n=0}^\infty \frac{(-1)^n H_{2n+1}(x)}{(2n+1)!} t^{2n+1}, \quad t \in \mathbb{R}.$$

Hint: Use Theorem 18.88.

34. Prove that

$$H_n(x) H_m(x) = n!m! \sum_{k=0}^{\min(n,m)} \frac{2^k H_{n+m-2k}(x)}{k!(n-k)!(m-k)!}, \quad n, m \in \mathbb{N}_0.$$

In particular,

$$H_n^2(x) = (n!)^2 \sum_{k=0}^n \frac{2^k H_{2n-2k}(x)}{k!((n-k)!)^2}.$$

35. Show that the function

$$f(x) := \sum_{n=1}^\infty \frac{\sin(nx) \sin^2 n}{n}, \quad 0 < x < 2\pi$$

is equal to

$$f(x) = \begin{cases} \frac{\pi-2}{4}, & 0 < x < 2, \\ 0, & 2 < x < \pi. \end{cases}$$

Explain the behavior of this series at the points 0, 2, and π. Show also that

$$\sum_{n=1}^{\infty} \frac{\sin(2n)\sin^2 n}{n} = \frac{\pi - 2}{8}.$$

36. Show that

$$\sum_{n=1}^{\infty} \frac{\sin(\pi(2n-1)x)}{2n-1} = \frac{\pi}{4} \operatorname{sgn} x$$

pointwise in $x \in (-1, 1)$.

37. Find the trigonometric Fourier expansion with respect to the system $\{\cos(nx)\}_{n=0}^{\infty}$ for the function

$$f(x) = \begin{cases} \sin x + \cos x, & 0 \le x \le \frac{\pi}{2}, \\ \sin x - \cos x, & \frac{\pi}{2} \le x \le \pi. \end{cases}$$

Hint: Extend this function as an even function to the interval $[-\pi, \pi]$.

38. Prove that:

a)

$$\log\left|2\cos\frac{x}{2}\right| = \sum_{n=1}^{\infty} (-1)^{n-1} \frac{\cos(nx)}{n}, \qquad -\pi < x < \pi.$$

b)

$$\log\left|2\sin\frac{x}{2}\right| = -\sum_{n=1}^{\infty} \frac{\cos(nx)}{n}, \qquad -\pi < x < \pi.$$

In particular,

$$\log 2 = \sum_{n=1}^{\infty} \frac{(-1)^{n-1}}{n}.$$

39. Show that

$$\frac{\pi - x}{2} = \frac{\pi}{2} + \sum_{n=1}^{\infty} \frac{(-1)^n \sin(nx)}{n}, \qquad -\pi < x < \pi,$$

and consequently,

$$x = 2\sum_{n=1}^{\infty} (-1)^{n+1} \frac{\sin(nx)}{n}, \qquad -\pi < x < \pi.$$

40. Show that for any $a \in \mathbb{R}$ it is true that

$$e^{ax} = \frac{e^{2\pi a} - 1}{\pi} \left(\frac{1}{2a} + \sum_{n=1}^{\infty} \frac{a\cos(nx) - n\sin(nx)}{a^2 + n^2} \right), \qquad 0 < x < 2\pi.$$

Compare with Exercise 39 when we let $a \to 0$.

41. Consider Exercise 40 and change a to $-a$ to obtain that

$$\sum_{n=1}^{\infty} \frac{n\sin(nx)}{n^2+a^2} = -\frac{\pi}{2}\frac{\sinh(a(x-\pi))}{\sinh(\pi a)}, \quad 0 < x < 2\pi,$$

$$\sum_{n=1}^{\infty} \frac{a\cos(nx)}{n^2+a^2} = \frac{\pi}{2}\frac{\cosh(a(x-\pi))}{\sinh(\pi a)} - \frac{1}{2a}, \quad 0 < x < 2\pi.$$

In particular,

$$\sum_{n=1}^{\infty} \frac{(-1)^n}{n^2+1} = \frac{\pi}{2\sinh\pi} - \frac{1}{2}, \quad \sum_{n=1}^{\infty} \frac{1}{n^2+1} = \frac{\cosh\pi}{2\sinh\pi} - \frac{1}{2}.$$

42. Using the first equality of Exercise 41 and putting there ix instead of a and $a+\pi$ instead of x, prove the following equality:

$$\sum_{n=1}^{\infty} (-1)^{n+1} \frac{n\sin(an)}{n^2-x^2} = \frac{\pi}{2}\frac{\sin(ax)}{\sin(\pi x)}, \quad -\pi < a < \pi,$$

where x is not integer but it might be equal to 0. In particular, when $x = 0$, we will have that

$$\sum_{n=1}^{\infty} (-1)^{n+1} \frac{\sin(an)}{n} = \frac{a}{2}.$$

Bibliography

[1] Titchmarsh E C, *The theory of functions* (in Russian), Nauka, Moscow, 1960.
[2] Whittaker E T and Watson G N, *A course of modern analysis*, Vols. 1 and 2 (in Russian), Fizmatgiz, Moscow, 1963.
[3] Sveshnikov A and Tikhonov A, *The theory of functions of a complex variable*, Mir Publisher, 1973.
[4] Nikiforov A F and Uvarov V B, *Special functions of mathematical physics* (in Russian), Nauka, Moscow, 1978.

https://doi.org/10.1515/9783111632278-022

Index

https://doi.org/10.1515/9783111632278-023